AF864053

Data Capital

Chunlei Tang

Data Capital

How Data is Reinventing Capital for Globalization

Springer

Chunlei Tang
Harvard University
Cambridge, MA, USA

ISBN 978-3-030-60191-1 ISBN 978-3-030-60192-8 (eBook)
https://doi.org/10.1007/978-3-030-60192-8

© Springer Nature Switzerland AG 2021
This work is subject to copyright. All rights are reserved by the Publisher, whether the whole or part of the material is concerned, specifically the rights of translation, reprinting, reuse of illustrations, recitation, broadcasting, reproduction on microfilms or in any other physical way, and transmission or information storage and retrieval, electronic adaptation, computer software, or by similar or dissimilar methodology now known or hereafter developed.
The use of general descriptive names, registered names, trademarks, service marks, etc. in this publication does not imply, even in the absence of a specific statement, that such names are exempt from the relevant protective laws and regulations and therefore free for general use.
The publisher, the authors, and the editors are safe to assume that the advice and information in this book are believed to be true and accurate at the date of publication. Neither the publisher nor the authors or the editors give a warranty, expressed or implied, with respect to the material contained herein or for any errors or omissions that may have been made. The publisher remains neutral with regard to jurisdictional claims in published maps and institutional affiliations.

This Springer imprint is published by the registered company Springer Nature Switzerland AG
The registered company address is: Gewerbestrasse 11, 6330 Cham, Switzerland

Highlights

Data Capital, *noun.*

(1) A human-created resource that is naturally one capital.
(2) A digital, intangible capital form that claims to cover almost the digital part of all existing capital, from tangibles' digital twin and intangibles' measurable aspect, to financials.
(3) The strategic economic resources for the data economy.
(4) A parasitic economic logic to develop new forms of business that serve the industries within the first three categories of Fisher-Clark's classification.
(5) An intangible wealth marked by concentrations of information, knowledge, and wisdom unprecedented in human history.
(6) A possible sovereign power that is subordinated to modern global architecture but has no physical boundaries.
(7) The origin of a decentralized instrumentation power that asserts dominance over society and brings the opportunities for market democracy.

Abstract

What is the next grand impetus that injects into economic globalization? Big societal problems and other transformations ushered in the digital era lie at the heart of economics. But satisfactory answers have been hard to find for lack of research across disciplinary divides. In *Data Capital*, Chunlei Tang, Ph.D., lets data speak for itself ranging in data nature, to uncover how the features make a data-rich economy fundamentally different from the previous ones. Her findings will transform debate and set the agenda for the next strategic economic development plan about globalization. Tang shows that data capital as a human-created resource is naturally a capital that will reshape the world economy throughout the next decades. Understanding the importance of data as natural capital is vital in order to counteract and balance dangerous inequities in the future as decentralized instrumentations of power assert dominance over society. Because a decentralized instrumentation power asserting dominance over society will bring the opportunities for curbing dangerous inequalities in the future. As a work for originality and rigor, *Data Capital: How Data is Reinventing Capital for Globalization* presents possible scenarios for what a data product-rich world might be like and reorients our understanding of economic future.

Acknowledgments

Several years ago, I wrote a book about what to expect from an increasingly data-driven economy. Although signs of the data economy are everywhere now, there remain two challenges with regard to data profits that require attention. The first challenge is figuring out how to derive profits from data. While some argue that bitcoin futures trading represents a highly profitable business, it just provides an arbitrage opportunity characterized by price volatility because their value is based solely on supply and demand. One proper way to obtain profit from data includes not merely the transfer of raw data but add value through the creation of raw data. To put it another way, data profit is achievable from the valuation of both raw data and its derivative data products or services generated through data processing on raw data. In this sense, a digital currency tends not to be the data; meanwhile, cryptocurrency trading via an exchange is treated similarly to an investment that pursues a possible risk premium only.

The second challenge is that data-rich profits can come from various sources—personal, private, or third party—but each stakeholder may not share equitably in the profit return.

Meeting the above challenges regarding data profits will allow us to bridge the gap between a data asset and its capital value. Capital in economics is regarded as one of the factors of production (alongside the other factors: land and labor). Today when capital moves from tangible (e.g., physical capital) to be intangible (e.g., intellectual capital), the data is presented as an amount of wealth that is used in making respective profits to affect the sum of wealth. A clear understanding of this gap in wealth creation motivates me to define and develop the concept of "data capital." My point of view regarding data capital had the extremely good fortune to win the enthusiastic support of Springer's Lorraine Klimowich and of Yike Guo from Imperial College London. Guo and I collaborated with on this book proposal, and Klimowich accepted it to make a book contract with me (in 2016).

This book would be impossible without the years of keenly insightful work by Yangyong Zhu and Yajun Huang who glimpsed the world economics under "data nature (that is an artificial ecosystem in cyberspace)." Zhu and Huang taught me to appreciate letting data speak for itself regarding the economic notion of such a new form of capital. Specifically, a discourse on data capital is a data-driven process. Historically, economic research was the prophetic product of Thomas R. Malthus' overpopulation, David Ricardo's scarcity of land, and Karl Marx's unlimited accumulation of capital. One of the obvious drawbacks of these works is the lack of empirical data to support their studies. Inspired by Kuznets curve, modern economists began to employ representative data and abstract mathematical models. With the availability of big data, abstract mathematical models have to be verified and refined with data, for which I offer strategies and solutions in this book.

It's my pleasure to thank my friends and close colleagues, specifically at Harvard, over the years. Beyond their direct contribution of the thinking to the research on which this book draws, their enthusiasm and energy have fostered a climate of intellectual excitement in which the work matured. I have considered the criticisms of this study which have come to my notice, and have, consequently, modified in this book substantially. I would like to thank in particular David W. Bates, Lynn A. Volk, Joseph M. Plasek, and Eugene W. Mont. More specifically, without the full support from Bates and Volk, I couldn't systematically present healthcare industrial examples in this book to demonstrate the possibilities of a transformation. Without the efficiency, rigor, wordsmithing ability, and talents of Plasek, our originality on "data sovereigns" might not progress smoothly. Without the careful attention to detail and an impressive capacity for work of Mont, I would never have considered the core issue of how to drive data profit return on equity, which plays a vital role in this book.

Above all, I am grateful to my parents Yiming Tang and Hongjun Lu for their boundless support and encouragement throughout the writing process. I dedicate this book, with love, to them.

Contents

Chapter 1
Introduction

The difficulty lies not so much in developing new ideas as in escaping from old ones.
John M. Keynes in The General Theory of Employment, Interest and Money (1936)

One fundamental reality of our current economic system is globalization. This round of globalization is from the 1970s marked the dawn of the modern era in digitalization, creating a "disruptive innovation" to reinvent the developments of earlier beliefs and theories about business. Compared with the prior round of globalization (that was between the late nineteenth century and World War I), the endogenous dynamics of and endogenous mechanisms of this round is led by information technology (IT) that impacts resource allocation with a given international division of labor, instead of electricity and communication techniques. We survived the use of digital technologies to transform our world but failed to see any change in the rate of return on capital exceeds the rate of growth of output and income. "As it did in the nineteenth century and seems quite likely to do again in the twenty-first,"[1] this phenomenon is the tendency to exacerbate the global contradictions brought by economic efficiency, social justice, and individual liberty. The distribution of wealth has been further widely controversial.

Everything has two sides: the wrong side and the right side. The wave of the digital revolution has swept the world in recent decades. This causes our physical reality able to store as "data" in cyberspace. Data is becoming the world's

[1] I reference directly here from the original sentence published in the book of *Capital in the Twenty-First Century*, whose author is a French economist Thomas Piketty (born in 1970). ***See:*** PiKetty, T. translated by Goldhammer A. Capital in the Twenty-First Century. Cambridge, MA: the Belknap Press of Harvard University Press; Reprint edition (August 14, 2017), 2017: 816 pages.

© Springer Nature Switzerland AG 2021
C. Tang, *Data Capital*, https://doi.org/10.1007/978-3-030-60192-8_1

most valuable resource[2] and promises to be the oil of the digital era. Data-driven innovation and data entrepreneurship have the potential to alter the current landscapes dramatically, the economy itself included. But do the data have different economic implications from factor augmenting technological progress? Should we see the data from a perspective on the cost of capital or elements of production, and what is the critical issue? Or, do the balancing forces of data and data-driven technological progress offer an unexpected way to reconcile the risks and contradictions brought by deglobalization? What lessons can we derive from data science to establish a raising channel to narrow the rich-poor gap?

These are the questions I attempt to answer in this book. I must admit that the answers contained herein are imperfect and incomplete since I am an expert in data science instead of in economics. Disciplinary disputes and turf wars are both pointless. Although being a data optimist benefitted of data on markets and innovation, I won't deny any potential drawbacks to the data economy such as surveillance. I envision that the answers should "let data speak for itself" on finding a suitable solution, which may be much more constructive than the one based on an abundance of prejudice and a scarcity of facts. In any case, I hope that economists, data scientists, and amateurs of all fields will find something of interest in those answers, including those who claim to "know nothing" and who nevertheless have honest opinions about this topic.

Globalization and Digital Success in Global Flows

We are now surrounded by unprecedented globalization. "Globalization" can mean many things but isn't assimilation (that may result in a low-dimensional or single thematic framework). The road to global prosperity is somewhat through the complementarity that depends upon difference.[3] Of which, the process of heterogenization and homogenization occurs at the same time.[4] I agree

---

[2]There is of course a metaphor, and in fact the reference doesn't not the earliest one. In 2006, Clive R. Humby (born in 1955) declared "data is the new oil!" at the AMA (American Marketing Association) Senior marketer's summit. He is a British mathematician and entrepreneur in the field of data science and customer-centric business strategies. Hereafter he also pointed out that, "… data needs processing, just as oil needs refining before its true value can be unlocked." ***See:*** The Economist. The world's most valuable resource is no longer oil, but data. May 2017. Retrieved from https://www.economist.com/news/leaders/21721656-data-economy-demands-new-approach-antitrust-rules-worlds-most-valuable-resource.

[3]There exist indepth study and discussion of globalization. I choose John R. Short's dialectic opinions published during the new millennium. ***See:*** Short, J.R. Global Dimensions: Space, Place and the Contemporary World. London: Reaktion Books, November 1, 2004, 192 pages.

[4]The Indian-American anthropologist Arjun Appadurai (born 1949) who is recognized as a major theorist in globalization studies, submitted that with globalization, the processes of "homogenization and heterogenization" occur at the same time. ***See:*** Appadurai, A. The capacity to aspire: Culture and the terms of recognition, in Rao, V. and Walton, M., (eds.) Culture and Public Action. Stanford, CA: Stanford University Press, 2004: 59-84.

with common views among those who take a sociological stance as globalization "is best thought of as a multidimensional set of social processes."[5] Multiple dimensions, such as economic, technological, political, social, and cultural, shape people's understanding of globalization for variety. Due to my academic background, I'm not entirely familiar with political, social, and cultural globalization, so that for this book, I will limit my discussion to economic and technological aspects.

The integration and interaction of globalization on the economic field is much more realistic than the others due to the motivation that is to identify new market opportunities; more explicitly, it is to "discover wealth with plunder."[6] Globalization as a term began emerging in the 1960s and came to prominence in the 1980s. Nevertheless, most historians attach globalization "big bang" significance to the great geographical discoveries in the mid-to-late fifteenth century. In 1492, Christopher Columbus led his three ships—the Nina, the Pinta, and the Santa Maria—out of the Spanish port of Palos and discovered America. His "Discovery of the New World" led people to have a sense of existence in "global capital markets" from more land, labor and other natural resources. Human attacks our own kind through colonialism, plunder, and enslavement to achieve globalization in a later 500-year rampage. Despite historical sins, the global trading network with primary economic transmission channels was created objectively which offered an essential part of the industrial revolution in Europe. To some extent, this also stimulates some Asian or African countries to learn from advanced Western countries, resulting in significant change and transformation from traditional to modern societies globally, but conflicts and frictions.

Globalization, hence, does not stem from technological developments rather than policy choices. Since recorded history (around 10,000 B.C.E., i.e., before common era), as shown in Table 1.1,[7] there exists a total of 31 general purpose technologies (GPT) that can affect an entire economy at a national or global level. Of them, the six related to computing are accounted for the half of a total 12 from the twentieth century, in which big data shows the most unique corresponding to all three GPT types as process, product, and organization. Table 1.1 does not include quantum computing. Although quantum computing

---

[5] Manfred B. Steger said in his 2015 paper after having summarized 8 literature related to globalization. The original word is "Indeed, most scholars of globalization have defined their key concept along those lines as a multidimensional set of social processes that create, multiply, stretch, and intensify worldwide social interdependencies and exchanges while at the same time fostering in people a growing awareness of deepening connections between the local and the distant," ***See:*** Steger, M. B. Ideologies of globalization. Journal of Political Ideologies, 2005;10(1): 11–30.

[6] ***See:*** Xiang, S. The New Capitalism: Rising, Crisis and Future of the Global Financial Capitalism. Beijing: China Citic Press; 1 edition, December 1, 2014, 616 pages.

[7] I redrew and modified Figure 6 that is in Pater Marsh's book (pp. 221) as *The new industrial revolution: Consumer, globalization, and the end of mass production*. ***See:*** Marsh, P. The new industrial revolution: Consumer, globalization, and the end of mass production. New Haven, CT: Yale University Press; Reprint edition (October 8, 2013), 2013: 320 pages.

Table 1.1 Types of general purpose technologies

General purpose technology (GPT)	Computing related GPT	Time period	Type of GPT
1. The domestication of plants		9000–8000 B.C.E.	Process
2. The domestication of animals		8500–7500 B.C.E.	
3. Smelting of ore		8000–7000 B.C.E.	
4. The wheel		4000–3000 B.C.E.	Product
5. Writing		3400–3200 B.C.E.	Process
6. Bronze		2800 B.C.E.	Process
7. Iron		1200 B.C.E.	
8. Mechanical pulleys/gears		500 B.C.E.	
9. Heavy plough		Eighth century	
10. The water wheel		Early medieval period	
11. The three-masted sailing ship		Fifteenth century	
12. Printing		Sixteenth century	Product
13. The steam engine		Late eighteenth–early nineteenth century	Process
14. The factory system		Late eighteenth–early nineteenth century	Organization
15. The railway		Mid-nineteenth century	Process
16. The iron steam ship			
17. The internal combustion engine		Late nineteenth century	
18. Mass production steelmaking			Product
19. Electricity			Process
20. The motor vehicle		Twentieth century	
21. Chemical synthesis			Product
22. Aircraft			Process
23. Factory production			Organization
24. The computer			Process
25. Semiconductors	√		
26. Laser	√		
27. Lean production			Organization
28. The Internet			Process
29. Biotechnology	√		Product
30. Nanotechnology	√		
31. Big data	√	Twenty-first century	Process, product, organization

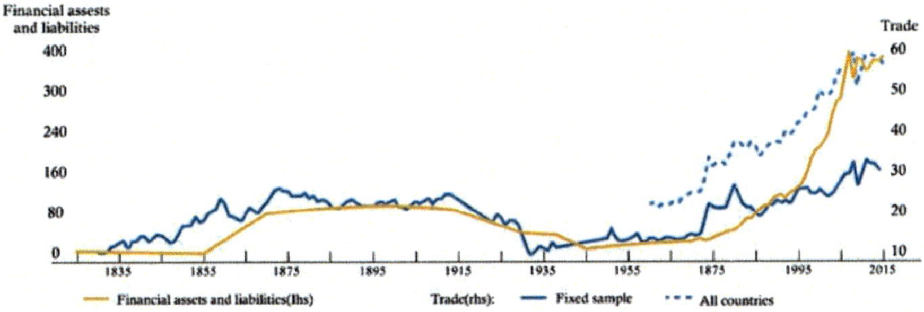

Figure 1.1 The recent wave of economic globalization has outstripped the prior one. Note: Prior to 1970, calculated as external financial asses multiplied by 2. (Data source: Author redrew this figure according to the BIS' 87th annual report, p. 100)

is impressive, this field is still in its infancy and the quantum economy remains far on the horizon. This round of economic globalization (that is from the 1970s) has outstripped the prior one (that was between the late nineteenth century and World War I) according to the 87th Annual Report[8] made by the Bank for International Settlements (BIS) (Figure 1.1). As a speech that the English economist John Williamson gave while he was the Chief Economist for the South Asia Region at the World Bank, "trying to reverse globalization would be rather like playing at King Canute."[9] Williamson also gave an example that, the experience of widespread growth alongside rapidly growing trade in the postwar period serves to substantiate this. The costs of transport, of travel, and above all the costs of communicating information have fallen dramatically in the postwar period, almost entirely because of the progress of technology. A 1-min telephone call from the New York (USA, i.e., United States of American) to London (UK, i.e., United Kingdom) cost US$4 in 1946, whereas today it can cost as little as 2.4 cents, despite the fact that consumer prices have multiplied by close to 12 times (11.9 times, consumer price index as CPI in 1946 as 19.5%, and in 2018 as 251.1%)[10] in the intervening period. It should be noted

[8] The data I use here is based on BIS' 87th Annual Report, both Chinese and English versions. ***See:*** Bank for International Settlements. 87th Annual Report: from April 2016 to March 2017. June 25, 2017. Retrieved from (1) https://www.bis.org/publ/arpdf/ar2017e.pdf; (2) https://www.bis.org/publ/arpdf/ar2017_zh.pdf.

[9] John Williamson addressed at the Congress of the Sri Lankan Association for the Advancement of Science Colombo, Sri Lanka, and the original word is "If I am right in arguing that globalization stems from technological developments rather than policy choices, trying to reverse it would be rather like playing at King Canute." ***See:*** Williamson, J. Globalization: The concept, causes, and consequences. Peterson Institute for International Economics. December 15, 1998. Retrieved from https://piie.com/commentary/speeches-papers/globalization-concept-causes-and-consequences.

[10] The data I use here is based on the US Inflation Calculator. ***See:*** US Inflation Calculator Consumer Price Index Data from 1913 to 2019. Retrieved from https://www.usinflationcalculator.com/inflation/consumer-price-index-and-annual-percent-changes-from-1913-to-2008.

that, a major contribution to this round of economic globalization lies on digital success in global flows.

International exchanges and flows of goods, services, capitals, technology transfer, and human movements,[11] we call global flows involving global economic and trade activities, create an alternative understanding of globalization. Global flows are a common thread in economic growth for multiple rounds of globalization including the commercial, industrial, and information revolution. The role of global flows lies not only in creating or rebuilding degrees of connectedness among economies (both emerging and advanced economies), but in determining the fate of nations, companies, and individuals. Both inflows and outflows matter for growth since they expose economies to ideas, research, technologies, talent, and best practices globally, and to be unconnected is to fall behind.

The majority think that global financial flows present essential characteristics for this round of global economic system, which began from the collapse of the Bretton Woods system in 1971 getting started with the emergence of "anchorless" currencies. And even, Britain leading financial digital media *Financial Times* chose "We are living in Financial Times"[12] as its advertising slogan in 2007. But the 2008 financial crisis accidentally revealed the truth. Since the 2008 financial crisis, the global goods trade has flattened, and cross-border capital flows have declined sharply (see Figure 1.2).[13] Conventional wisdom says that globalization is heading into reverse, but maybe it is not correct. A McKinsey's report[14] showed that soaring flows of services represented by data (e.g., free content and instant information represented by data) driven by SMEs (i.e., small and medium-sized enterprises) or individuals now generate more

---

[11] Referenced definition from the International Monetary Fund. ***See:*** Global Staff, Globalization: A brief overview. International Monetary Fund. 2008. Retrieved from https://www.imf.org/external/np/exr/ib/2008/053008.htm.

[12] Referenced definition from the International Monetary Fund. ***See:*** Ronbinson, G., We live in Financial Times. The Financial Times, April 2007. Retrieved from https://ftalphaville.ft.com/2007/04/23/4014/we-live-in-financial-times.

[13] With the deceleration of financial assets in global liquidity and the emergence of accelerating flows of data and information since the 2008 financial crisis, many people began to realize digital globalization. Recognizing digital globalization instead of financial globalization is the process of cognitive development. Take McKinsey as an example; there may exist a watershed moment in 2016. ***See:*** Lund, S., Windhagen, E., Manyika, J., Härle, P., Woetzel, J., Goldshtein, D. McKinsey Global Institute. The new dynamics of financial globalization. August 2017. Retrieved from https://www.mckinsey.com/~/media/McKinsey/Industries/Financial%20Services/Our%20Insights/The%20new%20dynamics%20of%20financial%20globalization/MGI-Financial-globalization-Executive-summary-Aug-2017.ashx.

[14] Global flows have a process from digital successful to data. ***See:*** Manyika, J., Bughin, J., Lund, S., Nottebohm, O., Poulter, D., Jauch, S., Ramaswamy, S. McKinsey Global Institute. Global flows in a digital age: How trade, finance, people, and data connect the world economy. April 2014. Retrieved from https://www.mckinsey.com/~/media/McKinsey/Business%20Functions/McKinsey%20Digital/Our%20Insights/Digital%20globalization%20The%20new%20era%20of%20global%20flows/MGI-Digital-globalization-Full-report.ashx.

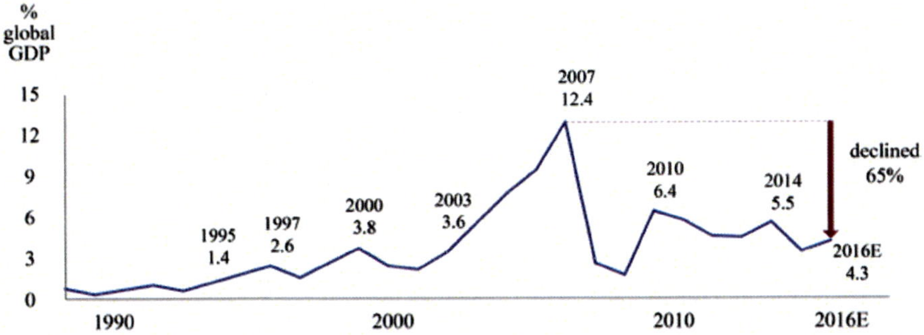

Figure 1.2 Global cross-border capital flows, 1990–2016E (E as estimate). Note: From the 2008 financial crisis, gross capital flows (including FDI, debt securities, equity, and other investment) have declined 65% since the 2007 peak. (Data source: Author redrew this figure according to data from the IMF Balance of Payments)

Table 1.2 Globalization: Then (twentieth century) vs. now (twenty-first century)

Then	Now
Tangible flows of physical goods	Intangible flows of data and information
Flows mainly between advanced economies	Greater participation by emerging economies
Capital- and labor-intensive flows	More knowledge-intensive flows
Transportation infrastructure is critical for flows	Digital infrastructure becomes equally important
Multinational companies drive flows	Growing role of small enterprises and individuals
Flows mainly of monetized transactions	More exchanges of free content and services
Ideas diffuse slowly across borders	Instant global access to information
Innovation flows from advanced to emerging economies	Innovation flows in both directions

economic value than traditional flows of physical goods, capital, and monetized transaction, as shown in Table 1.2. McKinsey (see Footnote 14) also estimated that global flows could reach US$54 trillion to US$85 trillion by 2025, more than double or triple their current scale, of which the most proportion has been entirely digitized.

In fact, the prior so-called "financial globalization" didn't exist and only represents a manifestation of "digital globalization." Because finance is (and always has been) the most thorough field of digitalization. Cross-border financial flow—the inflows and outflows of global liquid financial assets and gross cross-border capital—link together national financial markets, connecting borrowers and savers from different countries, which include lending, foreign direct investment (FDI), and purchases of equities and bonds. Financial flows are essentially represented by the form of the data via digital infrastructure. They only accounted for about 20.7% of global gross domestic product (GDP) even though in 2007 (that was proved to be the height of a global credit bubble), and fell from US$11.9 trillion of 2007 to less than half their previous value (about US$5.2 trillion) in 2014, which is only one-third as high relative to 2017 global GDP.[15] Also, financial flows are not a separate phenomenon from digital success in global flows; all trade and goods-producing value chains have been transforming into the cross-border data flows as well (see Table 1.2) (see Footnote 15).

A Dispute on Cross-Border Data Flows

Digital success in global flows causes arising cross-border data disputes—a realistic problem. Figure 1.3 shows that there exists a global interest in the words related to cross-border data flows following the same rough trend. This is because; no longer only play a central role in the data economy distinct, cross-border data flows are inextricable aspects of the general economic growth and development. Every industry—whether agriculture, manufacturing, retail, or services—relies on cross-border data flows that are fundamental enablers of cloud-based data technology. Industrial giants not only Siemens and General Electric but also dairy producer Land O'Lake now sell themselves as data firms.[16] Studies have shown that the free flow of cross-border data has contributed

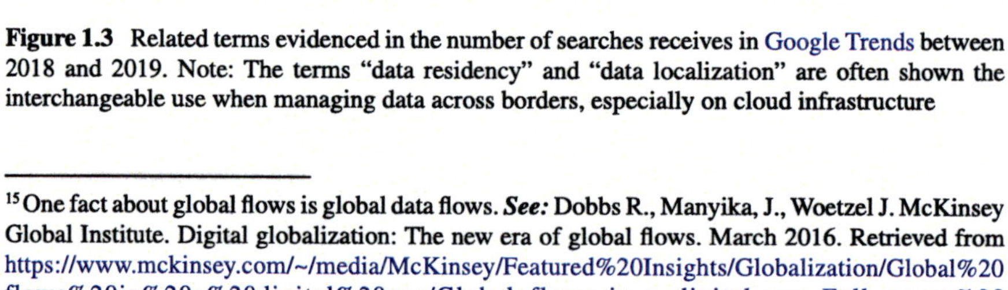

Figure 1.3 Related terms evidenced in the number of searches receives in Google Trends between 2018 and 2019. Note: The terms "data residency" and "data localization" are often shown the interchangeable use when managing data across borders, especially on cloud infrastructure

[15] One fact about global flows is global data flows. ***See:*** Dobbs R., Manyika, J., Woetzel J. McKinsey Global Institute. Digital globalization: The new era of global flows. March 2016. Retrieved from https://www.mckinsey.com/~/media/McKinsey/Featured%20Insights/Globalization/Global%20flows%20in%20a%20digital%20age/Global_flows_in_a_digital_age_Full_report%20March_2015.ashx.

[16] This is the first article on the data economy published in a popular magazine related to economics. Of which, many opinions are coincidentally in line with my first book published by Wiley in 2016 – *The Data Industry: The Business and Economics of Information and Big Data*. ***See:*** The Economist. The world's most valuable resource is no longer oil, but data. May 2017. Retrieved fromhttps://www.economist.com/news/leaders/21721656-data-economy-demands-new-approach-antitrust-rules-worlds-most-valuable-resource.

an estimated US$2.8 trillion to the global economy in 2014 (see Footnote 15), a figure could reach US$11 trillion by 2025.[17] Such disputes initially emerged in SMEs' cross-border trade in services, and then extended to big data giants' daily business that cross national boundaries (e.g., Google's nondomestic search results and Amazon's overseas trading records). According to the business community, trade in online information-oriented services is now facing obstacles with greater frequency than in the past. Spurred, in part, by concerns raised by industry, each single agreement between the importer and exporter might not contain a provision aimed at helping to alleviate the problem. Current dispute settlement has partially been transferred to numerous governments and major multilateral rule-making bodies (e.g., World Trade Organization, WTO). For example, the "summary of specific negotiating objective"[18] between the Trump administration and the European Union (EU) released by the US Trade Representative in January 2019 highlighted the clauses related to cross-border data flows in the first chapter. These clauses include that the US hopes the EU "does not impose measures (e.g., customs duties) that restrict cross-border data flows" and "does not require the use or installation of local computing facilities." However, almost all attempts to resolve these disputes stop in data privacy ethics.

Indeed, data privacy ethics are too important an issue to be left to data scientists and economists, and additionally require sociologists, philosophers, and so forth. But only are naturally sharp but contradictory political judgments on corporate social responsibility made, even if it is of interest to everyone. Many countries including the Western world and emerging economies have been regulating personal data protection in law. For instance,[19] the EU, China, and Brazil successively promulgated similar regulations to become active in 2018 alone (the former two came into effect both in May, only 25 days apart), respectively. These regulations can lead to extremely high legal risks (e.g., reaching a criminal responsibility) in transferring personal data among enterprises, which combate the choice of producers in the product market to a great extent. Focusing on

[17] ***See:*** Meltzer J., Lovelock P. Regulating for a digital economy: Understanding the importance of cross-border data flows in Asia. Brookings Institution. March 20, 2018. Retrieved from https://www.brookings.edu/wp-content/uploads/2018/03/digital-economy_meltzer_lovelock_working-paper.pdf.

[18] ***See:*** The United States Trade Representative. Summary of specific negotiating objectives on United States-European Union negotiations. January 2019. Retrieved from https://ustr.gov/sites/default/files/01.11.2019_Summary_of_U.S.-EU_Negotiating_Objectives.pdf.

[19] I listed the three references together to reduce space. ***See:*** (1) European Union. General Data Protection Regulation. 2016, April 27. Retrieved from https://gdpr-info.eu. (2) China. Information security technology – Personal information security specification. December 29, 2017. Retrieved from https://www.tc260.org.cn/upload/2018-01-24/1516799764389090333.pdf. (3) Monteiro, R. International Association of Privacy Professionals. The new Brazilian General Data Protection Law – A detailed analysis. August 15, 2018. Retrieved from https://iapp.org/news/a/the-new-brazilian-general-data-protection-law-a-detailed-analysis.

data privacy ethics to resolve cross-border data disputes, therefore, is not conducive to unlocking well productivity drivers.

Defining data rights (e.g., ownership rights in data) may be a suitable solution for resolving disputes in a systematic and methodical fashion. Put it in other words; there exist two issues. One is to determine the data who possess; the other is on how to enjoy data industrial empowerment moderately. There is no escaping the fact that, however, studying data rights often stuck in a variety of terminologies. These terms may be ambiguous, complicated, and misleading with rapid advancements in information technology. The terms "data residency" and "data localization" (see Figure 1.3) are often a source of confusion for business managing data across borders, especially on cloud infrastructure. There are in fact used so interchangeably that their individual meaning has become lost. Data residency refers to where a corporate, the business community, or industry bodies that their data is stored in a geographical location of their own choice, usually for taking advantage regulatory or policy reasons (e.g., a better tax regime). Data localization is a version of data residency predicated on legal obligations. It is also the notion that is growing the fastest internationally. For example, Russia's On Personal Data Law (OPD-Law)[20] requires the storage, update and retrieval of data on its citizens to be limited to data center resources within the Russian Federation. Typical criticism of these laws is that they use the mask of enhanced cybersecurity or citizen's privacy concerns to conceal the real motivation of national protectionism. We believe that, such border-based data silos obstruct businesses and governments from realizing the full potential that data stands to offer and contribute to digital factionalism and the "splinternet."[21]

When it comes to data, "accessing data trumps owning it,"[22] thus data rights are often the main commodity sold or traded to external entities. Before turning in greater detail to the source I tried to assemble in preparation for writing this book. I proposed the term "data sovereign" as a person or entity with the ability

[20] ***See:*** Bogaerts B., Segers K. The "localisation" of Russian citizens' personal data: Compliance with the Russian law on personal data. KPMG Global Sustainability Institute. 2018. Retrieved from https://home.kpmg/be/en/home/insights/2018/09/the-localisation-of-russian-citizens-personal-data.html.

[21] ***See:*** Box J. Data sovereignty vs data residency vs data localization. Insights For Professionals. March 12, 2019. Retrieved from https://www.insightsforprofessionals.com/it/storage/data-sovereignty-data-residency-data-localization.

[22] This viewpoint is based on a consensus in the data science community. I here quoted the two sources: one is from my doctoral advisor Yangyong Zhu's a popular science book, and the other comes from Kevin Kelly has made a career in predicting the future of technological advances. ***See:*** (1) Zhu Y. *Beautiful Data: Understanding Big Data in 100 Minutes* (《旖旎数据: 100分钟读懂大数据》). Shanghai: Shanghai Scientific & Technical Publishers. (October 1, 2018). 2018: 184 pages; (2) Ogilvy J. The future according to "Wired" editor Kevin Kelly. Forbes. 2017, March 30. Retrieved from https://www.forbes.com/sites/stratfor/2017/03/30/the-future-according-to-kevin-kelly/#46b07cbf76bf.

to possess and protect the data. Here, the term inherits the of the nation's characteristics on protecting the resource following Weber's formulations.[23] A perspective article about "data sovereignty" was accepted by *XX* in early 2020. I hope this original work for this book can help clarify misunderstanding on data rights, provide a feasible solution to resolve cross-border data disputes, and lead data economy pathing to prosperity.

Global History of Capitalism

Capitalism started to emerge during the seventeenth century in Northwestern Europe, especially in Great Britain and the Netherlands. The historiography of capitalism has many debated roots (and depend partly on how capitalism is defined), but is generally thought by scholars to be divided into two broad schools. One is with economic liberalism, of which the Adam Smith is the foundational figure. The other is with Marxism, drawing particular inspiration from Karl Marx. Liberals tend to see capitalism as originating in trade and commerce, and freeing people to exercise their entrepreneurial natures (as the most beneficial way of promoting human wellbeing). Marxists tend to view capitalism as a historically unusual system of relationships between classes—"more powerful people take control of the means of production and compel others to sell their labor as a commodity,"[24] which could be replaced by other economic systems that would serve human wellbeing better.

Adam Smith contained all his idea in his *An Inquiry into the Nature and Causes of the Wealth of Nations* (generally referred to by its shortened title *The Wealth of Nations*) first published in 1776—the early industrial revolution. That time capitalism and the market economy just ran a budding. Smith believed in the science of natural laws to propose the three major ones in economic affairs involving market mechanism, economic growth, and social order. Smith advocated the principle of "laissez-faire" on market mechanism, which requires that state should not impose any restriction on freedom of an individual. He opined that it is safe to leave the economy to be propelled, regulated, and controlled by

[23] Following Max Weber's formulations, here, I believe that data as an economic resource should be preserved through "the sole grantor of the right to physical force" with the legitimacy of domination. Apart from Waters' *Weber's Rationalism and Modern Society*, the reference also included the entry of "Politics as Vocation" on Wikipedia. Also, I envision that the notion of the geography-based country will be weaker than a data organization in Cyberspace soon. ***See:*** (1) Waters T. *Weber's rationalism and modern society: New translations on politics, bureaucracy, and social stratification*. London: Palgrave Macmillan. (April 8, 2015). 2015; 233 pages; (2) Retrieved from https://en.wikipedia.org/wiki/Politics_as_a_Vocation.

[24] The primary reference here is the entry of "History of Capitalism" on Wikipedia. ***See:*** Retrieved from https://en.wikipedia.org/wiki/History_of_capitalism.

"invisible hand." The "invisible hand" is used to describe the existence, stability and optimality of general equilibrium on markets (see Footnote 6). Smith emphasized labor as an important factor of production along with other factors (i.e., capital and land) and asserted, in essence, technological progress keeps us depending in division of labor. The division of labor becomes the true dynamic force in Adam Smith's theory of growth—the ultimate source of economic growth. He also recognized that division of labor greatly depends on the extent of the market as well and the size of market (i.e., social order that is determined by the available stock and the institutional restrictions placed upon both domestic and international trade).

Joseph A. Schumpeter pioneered innovation, entrepreneurship, and "creative destruction" in *Business Cycle: A Theoretical, Historical and Statistical Analysis of the Capitalist Process* which was first published in 1939. At least ten fundamental propositions are discussed in his book, such as the cycles of economic life that is nonlinear dynamics of non-equilibrium complex systems, innovation as the critical dimension of economic change also meaning destruction, and the entrepreneur that plays the role of protagonist of capitalism. In Schumpeter's interpretation of capitalism, the entrepreneur "innovates and applies new combinations of factors of production."[25] For Schumpeter, the entrepreneur both organizes innovative activities and represents "spirit of the contract" because he seeks his profits "outside the convention-bound circular flow by introducing new products and production techniques" (see Footnote 25). In addition, Schumpeter sought to prove that invention is not an external factor and has a fundamentally difference from innovation. Invention is admittedly important; but innovation is the key to increase and transform productivity for capitalism.

Karl Marx at his unfinished six-volume series of *Capital*, first published in 1867 that critically analyzes capitalism, is undoubtedly ambitious. Marx's fundamental discovery is on the characteristics of capital, such as the particularity (that is a product during the transition period in human societies), contradiction (that inherits from the commodity that contains a contradiction), and its nature (that is self-valorizing value and eternal movement). As a result, Marx's analysis achieves high effectiveness in the inherent contradictions of capitalism, to form his principal conclusion was what one might call the "Principle of Infinite Accumulation." That means "the inexorable tendency for capital is to accumulate and become concentrated in ever fewer hands, with no natural limit to the process" (see Footnote 1). With either the rate of return on capital would steadily diminish (thereby killing the engine of accumulation and causing violent conflict among capitalists), or capital's share of national income would increase indefinitely (which would unite the workers in revolt), "no stable socioeco-

[25] ***See:*** Mee, S. Joseph Schumpeter and the business cycle: An historical synthesis. 2015. Retrieved from https://www.tcd.ie/Economics/assets/pdf/SER/2009/simon_mee.pdf.

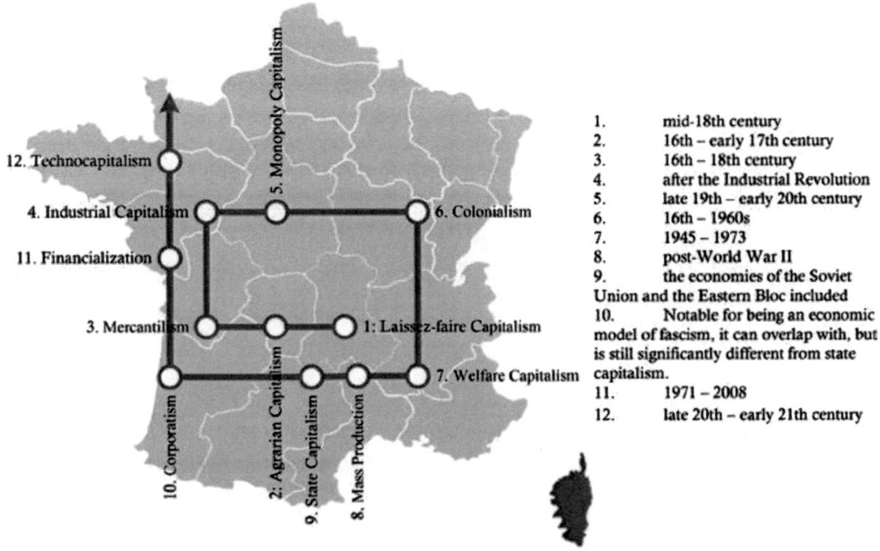

Figure 1.4 Various forms of capitalism globally. Note: These definitions are easily found on Wikipedia. (Source: Author summarized this figure via a visualization method designed by herself)

nomic or political equilibrium was possible" (see Footnote 1). Besides, Marx accounted for the trend of durable technological progress as capitalism that has a natural tendency towards maximizing profits.

Over the following centuries, capitalism has gradually predominated over much of the globe. People are keen on an ongoing debate within the fields of economics and sociology, which is concerned with the development of capitalism in its various forms. These forms (see Figure 1.4) might not be mutually exclusive and do not represent a fixed order of historical change, but represent a broadly chronological trend. Debates sometimes circle around the processes by which capitalism emerged, evolved, and spread. Key parameters of debate also include: whether or not capitalism is the most beneficial way to organize human societies and how far is a natural human behaviors; the role of class conflict; the role of the state; the extent to which capitalism is a distinctively European innovation, and its relationship with European imperialism. Capitalism has been much debated on tracing its origin to the middle ages and on expanding its range worldwide. However, the subject of extensive research remains stuck at how to bring substantive historical data to bear the key questions.

Truly interdisciplinary in my perspective, the renewed interest since the twenty-first century in the history of capitalism invigorates narrative to depth for capitalism and enrich topics to broaden among more traditional disciplines, involving business history, labor history, economic history, political economy, and institutional sociology. Some topics focus on environmental destruction, enormous waste, overpowered corporations, exploitation of works, or outrageous inequality. Others are more positive, telling a story about unparallelly prosperity, longer life expectancies, connectivity among peoples, and poverty alleviation. By looking at capitalism through a global lens, researchers also examined a range of different topics that impact on local, national, regional, and global levels. The topics include but not limited to commodities, labor relations, migration, consumption, finance, development, energy, and the environment.

Sidney W. Mintz, an anthropologist, was best known for his bestselling book *Sweetness and Power: The Place of Sugar in Modern History*. In "The Sweet Life of Sidney Mintz"—an article originally posted in the Boston Review shortly after Mentz's death in 2015, Sarah Hill summarized:[26]

> *Published in 1985, Sweetness and Power accounted for New and Old-World histories, the rise of Atlantic slavery and industrialism, and more than five hundred years of elite and plebeian tastes, folding them into one easily digestible confection. Sweetness and Power explained how we live—how world market systems shape taste and vice versa—in ways that no previous book had managed. With it, Mintz's accomplishments went far beyond launching a new mode of history writing; the book also contributed to a sea change that placed the human production of inequality at the center of anthropology and offered a profound correction to how we understand modern history.*

In short, Minta analyzed anthropology and history in the context of the political economy of global capitalism to break a path of culture studying. For the subject of commodities and consumption in capitalism, Mintz also explained an economic principle. That is, consumption across social groups, regardless of their chosen type or associate degrees, has plasticity and is not static. Mintz's using devastatingly clear language to clarify that culture became both a historical product (i.e., sugar), and a producer of future history that can interact with others across regions and over time.

One of similar studies with an intertwining of history and economics includes Seven Becket's important new book published in 2015, *Empire of Cotton: A Global History*. Becket, a Harvard historian, uses an optimistic story of cotton to distinguish two forms of capitalism: "war capitalism" and "industrial capitalism." What he terms war capitalism was often discussed in other literatures as "mercantilism." The availability of cheap and washable cotton textiles contributes significantly to improved health and thus greatly lengthened life expectancy. The product is an innovation that we easily take for granted, but that brings immense gains. After the US Civil War, however, cotton production became globalized, with a replacement of American slave-produced cotton with cotton grown in a Global South. Becket believe:[27]

[26] ***See:*** Hill, S. The Sweet Life of Sidney Mintz. Boston Review. 2015:31.

[27] ***See:*** Beckert S. Empire of Cotton: A New History of Global Capitalism. Penguin: 01 edition (October 29, 2015). 2015:640 pages.

Constantly changing character of economic reality as the feature make capitalism so difficult to define: The geographic rearrangement of economic relations is not just a noteworthy element of capitalism or an interesting aspect of its history; rather the constant shifting recombination of various systems of labor, and various compositions of capital and politics is the very essence of capitalism.

Becket started from the charge of historicizing concepts as his stories and cut into core debates of capitalism (depend mostly on how capitalism is defined). Obviously, it is to blaze a trail.

To sum up what has been said thus far: turning from the eighth- and ninth-century classics of Smith and Marx, to the twentieth-century analysis of Schumpeter, and to the twenty-first-century investigation of renewed interest in history, we may say that scholars' efforts are stumbling over an unexpected problem as "there's no agreement on what capitalism is":[28]

Capitalism started in the mid-19th century as a term applied by revolutionaries and social critics to describe the transformation of work practices and the inhuman subjection of mankind to violence and abuse. By the end of the century it had become a simple description of a new, rationalized form of conduct in which taking a wage was the most normal thing in the world. In the 20th century, capitalism again became a term of opprobrium, until a few businessmen thought it would be iconoclastic to present themselves as "capitalist tools." Then, after the 2008 financial crisis, critics diagnosed (yet again) a "crisis of capitalism." As a concept, capitalism hovers uneasily between the normative (in practice condemnatory) and the descriptive.

The newly definition of capitalism is given by *The Cambridge History of Capitalism*, a recent volume edited by economists Larry Neal and Jeffrey G. Williamson, as constituted by four elements: (1) private property rights, (2) contracts enforceable by third parties, (3) markets with responsive prices, and (4) supportive governments. The first three aspects are easily found in many human societies (that means they are near universal). While being very much open to debate, the fourth implies the current research trend of capitalism that is only to consider the relationship between the government and markets without a state's political system. Of course, it is the correct way to treat almost all economies worldwide as capitalist counties.

From Technology to Data Science, or Primary Productivity to Ecosystem Productive Forces

Since the age of discovery, there have appeared two irreversible trends of world-historical development. One is the trend pertaining to globalization, and the other is the transformation trend from a traditional, rural, agrarian society to a secular, urban, industrial society. The core of the latter trend is the modernization

[28] ***See:*** James H. Capitalism Da Capo. The American Interest. April 9, 2015. Retrieved from https://www.the-american-interest.com/2015/04/09/capitalism-da-capo.

model caused by the industrial revolution. This model has been dominating the historical development of recent hundreds of years and becomes the only one. Roughly, the first three industrial revolutions of humanity are with each of these three technological advances—the steam engine from 1750 to 1840, the age of science and mass production from 1840 to 1950, and the rise of digital technology from 1950 to 2000.

We now all know that science and technology constitute a primary productive force. The historical relationship between science technology and capitalism has long stood as a scholarly interest, at least since the 1930s. Numerous researchers participated in solving the classic quandary, foregrounding the entanglements between these two and tracing the diverse ways they mutually sharped each. But before the industrial revolution, this viewpoint lay at the level of uncertainty more than of certainty. Douglass C. North noted that, for example, not only David Ricardo but also even the latter economists within the European industrial transition had no way of anticipating the importance of industrial growth and technological progress in the years ahead. Why? This is because this transition from an agricultural to an industrial economy is too long. Karl Marx and Frederick Engels did not believe that human history featured a continuous growth of the "productive forces" (i.e., German: *produktivkräfte* as a central idea in Marxism), although the idea they proposed was probably derived the concept from Adam Smith's reference to the "productive power of labor." Rather, the development of the productive forces was characterized by social conflicts. Some productive forces destroyed the others, sometimes productive techniques were lost, and sometimes productive forces could be turned into destructive powers. In fact, technical productivity is not purely an economic phenomenon but rather a multidimensional process involving reorganization and reorientation of our complicated society. It is one of the most common and toughest challenges for a traditional single discipline to give a satisfactory answer.

The evasiveness of the present generation of analysis of capitalism is "in stark contrast to the certainties of prior interpreters" (see Footnote 28) in mainstream economics. Despite this, later scholars' new discussions on the science, technology, and capitalism nexus are via across disciplines, which represent major turning points and processes of studying in the global history of capitalism. Put in other words, interconnectedness as the fundamental difference between the pre-modern and modern world breaks discourse on European construction to drive a shift from the nation-state to globalization, or from within a single discipline to interdisciplinary subjects. The new version of studies across uses a full factual description of objects and the result of comparative research to put forward many unique insights rather than the apocalypse. However, interdisciplinary does little to "overcome the need for specialization."[29]

[29] ***See:*** Jacobs J. Why the disciplines still matter. The Chronicle of Higher Education. May 27, 2014. Retrieved from https://www.chronicle.com/article/Why-the-Disciplines-Still/146777.

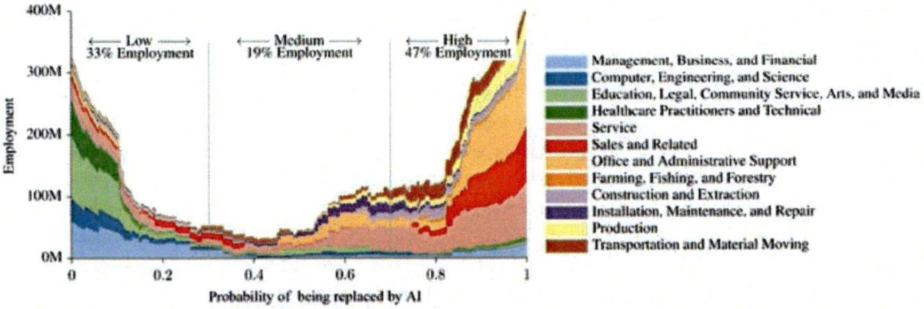

Figure 1.5 The distribution of the US Bureau of Labor Statistics 2010 occupational employment (a total of 702) over the probability of being replaced by AI, along with the share in low, medium and high probability categories. Note: The total area under all curves is equal to total US employment in 2010. (Data source: Author redrew this figure according to data from a study from Oxford University at Footnote 31)

Interdisciplinary research sites are often discrete niches on the boundary between two fields. I expect data science to hold for such a research focus because "data science is not a single discipline by itself,"[30] as Harvard statistician Xiao-Li Meng also wrote in the inaugural issue of Harvard Data Science Review:

> *Data science is best understood as a collection of disciplines with complementary foundations, perspective, approaches, and aims, but with a share grand mission. That is, to use data technology of any kind to advance human society as a harmonious, responsible, and vital ecosystem.*

With the article about the "sexiest job of the twenty-first century" released in the *Harvard Business Review* in 2012, most of us have some knowledge about data scientists (granted that such knowledge can be quite flawed); that is, artificial intelligence (AI) is the must-have skills that qualify an applicant for the data scientist. According to a high-citation study[31] performed by Oxford University (see Figure 1.5, with over 5000 citations), many existing positions (about 47% of 702 detailed occupations listed by the US Bureau of Labor Statistics (BLS) in 2010) are most likely to be AI replaced "repetitive" jobs. Interestingly, this study confirmed that computer programmers are high-risk (48%) because human programmers will be "eventually proved inferior to automatically optimized compilers" (see Footnote 31). So, artificial intelligence cannot operate outside of the programming but is not meant to be the programming solely. Similarly, data science is not just artificial intelligence or just statistics.

---

[30] I here quoted the viewpoint or some examples from Xiao-Li Meng's commentary. ***See:*** Meng X. Data science: An artificial ecosystem. Harvard Data Science Review, 2019;Issue 1.1. Retrieved from https://hdsr.mitpress.mit.edu/pub/jhy4g6eg.

[31] ***See:*** Frey C. B., Osborne M. A. The future of employment: How susceptible are jobs to computerisation? Technological Forecasting and Social Change, 2017;114:254-280.

Philosopher Sabina Leonelli reminds us in her perspective on "data governance" that,[32]

> *The concept of data is a subject of philosophical research because there is no such thing as objective raw data. Questions of what to collect, how to collect and measure, and how to process data all involve human motivation, judgments, and preferences; each of these questions directly impact data collection, analysis, and the interpretation of results.*

In response to the issue of the "data life cycle," another example, information scholar Christine L. Borgman points out the "after lives" of data:[33]

> *Data, especially those that are complex and large, often get reused to address new problems or to make deeper probing. It is therefore crucial to have appropriate mechanisms to store and curate data, with proper provenance to inform future users and investigators.*

These issues go beyond what is learned traditionally in STEM (Science, Technology, Engineering, and Mathematics) fields; but they are vital for the validity and applicability of data science, especially in terms of long-term ethical ramifications. While highlighting the diverse nature, data science education has a concrete aim: to help gain data acumen that is the leading factor driving data productive powers.

I would argue data productive powers belong to ecosystem productive forces. There are two reasons for this. First is that the artificial ecosystem in cyberspace has been evolving into the data "nature."[34] We should always be mindful of the volatile and complex nature of "data nature" as an ecosystem, which is consistent with the principle of "laissez-faire" market advocated by Adam Smith. According to Max Weber's statement "capitalism rested on the elimination of differences,"[35] I envision data nature would provide greater "cross-border"[36] convenience for engaging in transactions and exchanges with very different cultures and societies. Only one difference from Weber's prediction is "the local and the familiar" may bore special privilege.

---

[32] ***See:*** Leonelli S. Data governance is key to interpretation: Reconceptualizing data in data science. Harvard Data Science Review, 2019;Issue 1.1. Retrieved from https://hdsr.mitpress.mit.edu/pub/4ovhpe3v.

[33] ***See:*** Borgman C. L. The lives and after lives of data. Harvard Data Science Review, 2019;Issue 1.1. Retrieved from https://hdsr.mitpress.mit.edu/pub/4giycvvj.

[34] The concept of "data nature" was coined by my doctoral advisor Yongyong Zhu in his book *Dataology* published in 2009. ***See:*** Zhu Y., Xiong Y. Dataology. Fudan University Press; 1 edition (December 1, 2019, in Chinese). 2009, 131 pages.

[35] This viewpoint is referenced from the new biography of Max Weber wrote by Jürgen Kaube. ***See:*** Kaube J. Max Weber: ein Leben zwischen den Epochen. Berlin: Rowohlt Berline. January 17, 2014. 2014.

[36] The meaning of "border" not only is national or continental, but also refers to any restriction on various industrial sections or domains. In addition, there is no geographical distance in cyberspace.

Secondly, data productive powers could transform the economic intrinsic division of production from the traditional "H" hierarchy into a network-like structure; and a more adaptive network is possibly much closer to the "perfect competition." The hierarchical structure of the division of production that we understand today was invented in the last century, and it is an incredible invention. It can direct and coordinate the actions of thousands of enterprises and entrepreneurs making and selling thousands of products or services across thousands of miles, and do so effectively, efficiently, and profitably, week after week after week. The industrial value chain "H" hierarchy partitions the production process within the industry into a series of stages, each further refining the specializing the product of the previous stage, beginning with raw materials, and ending with the completed product delivered to the customer. Still, it can only extend vertically or horizontally. Enterprises and entrepreneurs at the high-end of a value chain have the discourse of power and play a role of determinants on the pricing, the allocation of resources, and income distribution. Being at the top of the value chain means often a monopoly or at least an oligopoly. The further down the value chain, the more intense the competition could be, and the lowest end of the value chain should be very close to the "pure or perfect competition" studied by economists. Horizontal and vertical integration as the two ways businesses can expand depends on technological advancements and innovations, reflected in the deepening and refinement of the division of production. Alfred C. Chandler Jr. stated in his masterpiece *The Visible Hand: The Managerial Revolution in American Business* that,

> *The growth and complexity of managerial hierarchies ("visible hands") in large business firms involve innovative ways of integrating existing or developing technologies into high-value solution. These innovative ways span across the entire value chain and require integration across internal and external business boundaries.*

Data innovation may break the traditional "H" hierarchy. Because data science tends to understand the complex world of interconnected entities we live in from comprehensive (i.e., different and complementary) viewpoints, with entities being modeled as nodes and their connections as edge, comprising large networks. These networked choices are a new and rich source mostly hidden with the value chain. Just as one principle of economics, "the more options that mean the lower cost of actions, the higher value could be created." Also, it is difficult to distinguish between the top and bottom of the value chain. The key factors that shape benefits is relevant on data technology.

Putting the Data at the Heart of Economic Analysis

The 2014 Thomas Piketty phenomenon caused no small commotion in the Western mainstream economic community. It also made economists pay increasing attention to the "data" at their core discussion. Piketty's *Capital in the Twenty-First Century* is direct to the rapid increase of wealth inequality during the neoliberal period, relying on the two datasets he participated in establishing. One is the World Top Incomes Database (WTID, see: https://wid.world) built in 2011 coving about 30 countries since the twentieth century, and the other is to collect individual estate tax returns from the US and a handful of other countries (France included) since the nineteenth century. Piketty began by criticizing classical political economies, such as Thomas R. Malthus' overpopulation, David Ricardo's scarcity of land, and Karl Marx's unlimited accumulation of capital. He believes all the conclusions belong to apocalypse or fairy tale with various nationalist prejudice and misleading comparison due to a complete lack of data support. He also scoffed at the "mathematical tendency"[37] of current mainstream economics.

Economics has always characterized multiple schools of economic thought, with different schools having different prominence across countries and over time. The "mainstream economics," according to 2001 edition of the seminal textbook *Economics* by Paul Samuelson and William D. Nordhaus, depicts arrows into neoclassical economics and neoclassical synthesis, and since the mid-twentieth century has included the Keynesian approach to macroeconomics. Part of the core of mainstream economics is the rational utility maximization hypothesis and law of diminishing marginal utility, in which essential tools are equilibrium with transaction costs. Methodologically, mainstream economics requires conforming to the language of mathematical models, featuring calculus, optimization, and comparative statics. Under this framework, areas of though which share very little common ground with the initial axioms of neoclassical economics, such as the game theory, behavioral economics, industrial organization, and evolutionary economics, can be considered mainstream, because of using mainstream methods. There exist some starting points of orthodox economic analysis: originally, it was the individual. Individuals and firms were generally defined as units with a common goal: maximizing (utility for individuals or profits for firms) through rational behavior. Of which, the process of maximization may have a set of constraints (i.e., limited income or commodity prices for individuals or technology or availability of inputs for firms). To deliver an intervention to the constraints, mainstream economics now includes descriptive theories of market and government failure, and private and public goods as well.

Mainstream economists expressed a lot of fierce opposition when Piketty slammed the discipline of economics for mathematics and purely theoretical speculation; Jean M. Tirole included who awarded the 2014 Nobel Prize in

---

[37]The references are the entry of "Mainstream Economics" and the opinion from Cambridge professor Tony Lawson. ***See:*** (1) Retrieved from https://en.wikipedia.org/wiki/Mainstream_economics; (2) Lawson, T. Mathematical Modelling and Ideology in the Economics Academy: competing explanations of the failings of the modern discipline?. Economic Thought, 2012, 1(1,2012).

Economics. They particularly blamed Piketty's book as "a puzzle" (see Footnote 28) that doesn't require a great deal of mathematical sophistication. For example, *Capital in the Twenty-First Century* distills capital accumulates into a simple equation by no considering inflation and the others: the rate of return on capital is persistently greater than growth, and explains it by retelling Honoré de Balzac's short novel Le Père Goriot:

> *Master criminal Vautrin tells the ambitious but impoverished young nobleman Eugène de Rastignac that it is pointless to try to achieve riches by study and hard work; that would only produce a modest marginal income. Better to marry an heiress, and then connive in a crime that would increase her inheritance.*

This model obviously cannot be projected backwards and forwards indefinite. Because:

> *An accumulation growing at geometric or exponential rate would swallow up all production. Indeed, there is a substantial literature devoted to the demonstration that, in the long term, the interest rate moves to set the growth rate of capital as equal to the growth rate of incomes.*

Besides, they criticized Piketty's sloppiness or omission. One robust criticism, by MIT's (Massachusetts Institute of Technology) Matthew Rognlie, reworks the figures to show diminishing net returns on capital. Another accusation is that Piketty makes no effort to place a sociological filter on the dataset.

> *There is little consideration of reasons why some significant cohorts in a population might choose not to earn more money, thus skewing the data – reasons like living alone because of divorce or extended age, extended educational periods for odyssey-years youth, and so forth.*

In fact, we must pay heed to the experienced analyst Andrew Gelman's dictum: "the most important aspect of an analysis is not what you do with the data, it's what data you use." We will explain this via an example starting from a classic case of "beauty and coin" in the game theory.

Mathematics is beautiful, in fact; and it matters. Scientists through the ages have noted, often with some astonishment, the aesthetic in mathematics: balancing form and content, syntax and semantics, utility and autonomy. Mathematical principles may help us think rationally beyond intuition. We first create a payoff matrix with two rows and two columns: $[\{3, -2\}, \{-2, 1\}]$ according to the rules. Note that you are the row player, and the beauty is the column player. Suppose the probability of heads is x, we can get the possibility of tails as $1 - x$. To maximize the payoff, your returns should be equal whatever the beauty chooses heads or tails; otherwise, the beauty may change the probability of heads or tails to reduce your total revenue. When there are only two rows, the only variable for your (i.e., the row player) linear program is x so that the equation of x is $3x + (-2) \times (1 - x) = (-2) \times x + 1 \times (1 - x)$. Solving the equation yields $x = 3/8$ which means your optimal strategy is three heads and five trails on every eight times. Substituting $x = 3/8$ into your payoff expression

$3 \times x + (-2) \times (1 - x)$ can obtain the expected revenue for each time as \$−1/8. A similar approach works for the column player (i.e., the beauty). Put in other words: if the beauty adopts the best strategy, you won't win no matter which plans you use; furthermore, you only can guarantee to lose the least, if you use the best as well. In general, game theory cannot be refuted by any empirical observations; it is only to understand the nature of competitive behaviors using mathematics. Game theorists could not think of the players who involved games as some flesh-and-blood actors, rather than as part of the body of mathematics to model the entities who consistently select mutually exclusive actions. Of course, observation and experience could lead someone favoring the conclusion is that "game theory is of little help in describing actual human behavior."

We change the rules of Case 1.1 to be a game involving bluffing (and asymmetric information). Two players can only toss one coin to make heads and tails appear randomly. The challenge in here is on generating random values of heads' or tails' appearances. The bet is set to \$100. You win \$100 if it comes up heads; otherwise, you lose \$100. When a coin is tossed, you are permitted to see the outcome. Your opponent does not see the outcome. You may double the bet from \$100 to \$200. If you double the bet, your opponent may accept the doubled bet or turn it down. If your opponent turns it down, you win \$100. If your opponent accepts the double, then you win \$200 if it is heads; otherwise you lose \$200. Table 1.3 indicates the six possible outcomes if considering a theoretical probability of heads' or tails' appearances if having no random values. Obviously, it guarantees an improved performance regardless of whether the bluffs are accepted or not if bluffing is done optimally as part of a mix strategy.

However, bluffing in practice only works if your opponent believed in you when you are pretending. In this case, mining all available data to eliminate asymmetric information may help obtain the optimal proportion of time to bluff resulting in success. The data includes but not limited to the opponent

Case 1.1: The Game Theory of "Beauty and Coin"

Say you're in a public library, and a beautiful stranger strikes up a conversation with you.

Beauty says: Let's show pennies to each other, either heads or tails. If we both show heads, I pay you \$3. If we both show tails, I pay you \$1. If they don't match, you pay me \$2.

3	-2
-2	1

At this point, the beauty is shushed.

You think: With both heads 1/4 of the time, I get \$3. And with both tails 1/4 of the time, I get \$1. So 1/2 of the time, I get \$4. And with no matches 1/2 of the time, she gets \$4. So it's a fair game.

As the game is quiet, you decide to play in the library.

Table 1.3 The six possible outcomes in the game involving bluffing

Coin tossed	Double		No double
	Accepted	Not accepted	
Heads	200	100	100
Tails	−200	100	−100

preferences, the total number of heads' and tails' appearances after the coin was tossed multiple times (e.g., 1000 times).

We, in the three scenarios of two above games, adopt different approaches due to the data. Specifically, the game theory of Case 1.1 to choose mathematical reasoning is because there is no data available to use. The purely logical approach to mathematical theorem lets us understand the nature of strategic behaviors with premises and hypotheses. Secondly, the modified game involving bluffing allows us to collect random samples by spending minutes to generate random values. That means we can follow Douglass C. North to use the traditional statistical approach based on a hypothesis-and-test paradigm, apart from pure mathematics. In other words, a hypothesis is proposed, a carefully designed experiment is designed to gather the data, and then the data is analyzed with respect to the hypothesis. The third has the amount of data representing opportunistic samples. We therefore can implement data analysis tasks with the generation and evaluation of thousands of hypotheses.

Let's envision the three possible scenarios applied in economics. The first two are entirely the current mainstream economic methods. That is to conform to the language of mathematical models showed no consideration for random disturbances (e.g., market and government failure). So a gap exists in real situations. Data science corresponds to the third one. It considers fundamental uncertainty ignored by mainstream economics through describing (or predicting) products (e.g., the total number of heads' and tails' appearances after having tossed one coin), customer habits or preferences (e.g., information of the opponent), the markets (e.g., someone might stop the players due to the fact that they are inside a public library), and so forth. The data is important, and more data means less inference. I believe that it is best to judge whether a method is more applicable or less applicable according to the data, instead of judging the way as good or bad.

Apart from widely affecting the methods, data itself would be the leading object of economic research (and capitalism). HBS's (Harvard Business School) Shoshana Zuboff claimed in her new book *The Age of Surveillance Capitalism: The Fight for a Human Future at the New Frontier of Power* that,[38] capitalism

[38] There exists some commentaries on Zuboff's book. ***See:*** (1) Zuboff S. The Age of Surveillance Capitalism: The Fight for a Human Future at the New Frontier of Power. New York, NY: PublicAffairs; 1 edition (January 15, 2019). 2019; 704 pages; (2) Naughton J. "The goal is to automate us": Welcome to the age of surveillance capitalism. The Guardian.Com. January 20, 2019. Retrieved from https://www.theguardian.com/technology/2019/jan/20/shoshana-zuboff-age-of-surveillance-capitalism-google-facebook.

transforms "human experience" so that they become productive to capital one again; and the "free raw material" can be used for nearly all business activities involving acquisition, prediction, and transaction. We understand that the processing start point of "human experience" as a new type of capital is the data. Personal data she worries to be misused merely accounts for a very little portion of big data. A real data scientist, in general, might cull or filter this part before dealing with the data. If the data is to become central, of course, we may easily respond to her appeal and would begin by improving "algorithm accountability"[39] as responsibility, explainability, accuracy, auditability, and fairness.

The Scope of This Book

Before proceeding, it may be useful to talk a little bit about the scope of this research as well as the intellectual itinerary that guided my thought on writing this book.

Since the global financial crisis of 2008, there has been an explosion of interest in capitalism. As far as I know, at least three books related to digital techno-capitalism touched a doorstop that seems to have achieved bestseller status after 2018. They are *Capitalism without Capital: The Rise of the Intangible* by Jonathan Haskel and Stian Westlake, *Reinventing Capitalism in the Age of Big Data* by Viktor Mayer-Schönberger and Thomas Ramge, and Zuboff's *Surveillance Capitalism* I mentioned above. I failed to finish drafting this book before the early of 2019 despite having a publishing contract. I would like to thank these authors and the argument they put forward before me, which brought more space to consider and then to perfect my book. Also, thanks to Springer's Lorraine Klimowich who gave me a grace period, so I have enough time to think and craft a manuscript that satisfied me. But I have no interest in exploring the capitalist system with its inequality, just like Piketty's opinion—"since social inequalities are not in themselves a problem as long as they are justified" (see Footnote 1). A partial reason lies in my background. The pragmatic data scientist generally takes the first place in data. My discussion is limited to data (not information and knowledge) because data deserves consideration in its own right. I would like to see the readership of this book universally understood data science from a socioeconomic perspective. The best way to treat this book based on my suggestion, thus, is data science materials rather than literary work in economics.

---

[39] There are many versions related to "algorithmic accountability." I only listed the newest one as the reference. ***See:*** Caplan R, Donovan J, Hanson L, Matthews J. Algorithmic Accountability: A Primer. Datasociety.Net. April 18, 2018. Retrieved from https://datasociety.net/wp-content/uploads/2018/04/Data_Society_Algorithmic_Accountability_Primer_FINAL-4.pdf.

This book is based on data sources of two main types. One is the data from the Internet. The Internet has sometimes been dubbed as the "data ocean,"[40] in which a partly indexed by search engines called the surface internet that contains at least six billion webpages. But the rest, namely the deep web, is about 400–500 times larger than the surface. My research does not include dark web. My work focuses on search engine results (e.g., Bing, Baidu, Google), government data (e.g., Data.gov), curated public sources (e.g., Wikipedia, Wikidata, Wiktionary, and Investopedia), international organization content (e.g., BIS, IMF), and the world wide web. I am particularly interested in social media sources (e.g., Facebook, Twitter, Reddit) and third-part aggregators (e.g., McKinsey, PWC, Thomson Reuter). The former represents digital interpersonal communication related to public participation, social connections, etc.; and the latter is willing to sell their available data including news stories (from large and small news organizations around the world, both global and local), company reports, annual reports, financial filings, worldwide patents, marketing and market reports, corporate communications, etc. To acquire these data, I manually identified web pages for a crawler to crawl. The crawler is for downloading lots of pages, of which the focus is scalability and volume. Other tools include scrapers, browser automation, and third-party APIs (Application Programming Interface). Scrapers are more hand-tuned than crawlers and focus instead on extracting content (e.g., numeric and metadata information). Browser automation tools can be combined with custom scripting to explore the results that might be inaccessible. Third-party APIs provided by third party content providers (e.g., Thomson Reuters, Bing, etc.) are typically more comfortable to use than the others. Of course, data dumps may be required if having some paid reports that are in the Portable Document Format (i.e., PDF files). I trend to develop custom data acquisition and ingestion tools that may be the most effective.

The other is the data from my academic research in finance and healthcare, involving the US and China. I focused on transaction sequence data at the stage of my doctoral research, which is from both supermarkets and stock markets in China. Transaction sequence data mining aims to identify the changing price pattern of commodities or securities, and its significant tasks such as detecting association rules with time constraints, analyzing similar missing data patterns. These approaches can also be used in nucleotide and protein sequence data. I have been hired by a hospital affiliated with one prestigious medical school in

[40]This figure is according to some estimates. ***See:*** (1) Retrieved from https://www.worldwideweb-size.com; (2) Thompson C. Beyond Google: Everything you need to know about the hidden interest. Business Insider. December 16, 2015. Retrieved from https://www.businessinsider.com/difference-between-dark-web-and-deep-web-2015-11; (3) Rice M. The deep web is the 99% of the Internet you can't Google. Curiosity.com. May 22, 2018. Retrieved from https://curiosity.com/topics/the-deep-web-is-the-99-of-the-internet-you-dont-see-curiosity.

Boston after finishing my doctorate. My main research area lies in electronic health records (EHRs) and other related medical data, which come from my school's two teaching hospitals, and four cooperative tertiary hospitals in Shanghai, China as well. The studies were constructed to electronic health record data mining, temporal expression extraction and representation learning in clinical notes, and clinical decision support on patient needs and hospital ranking via analyzing questionnaires or literature data. These projects involve pediatric pneumonia, chronic pulmonary obstruction, sepsis, and diabetes, coving modern Western medicine and traditional Chinese medicine. To put it bluntly, the primitive data types are mostly free text although I have some interests in imaging data.

The truth is that the two data sources contain bias despite being organically produced. So what do we mean by "data bias"? I use the definition in a broader sense. It denotes: (1) data does not include variables that properly capture the phenomenon we want to analyze; (2) data includes content produced by humans which may contain bias against groups of people. Based on my analysis, the most two common types of data bias are (a) response or activity bias and (b) omitted variable bias. The type (a) occurs in content generated by humans, such as patient comments, reviews on Amazon (4% of users produced 50% of the reviews), Facebook posts (7% of users produced 50% of the posts), Twitter tweets, and Wikipedia entries (0.04% of Wikipedia's registered editors produced 50% entries of English Wikipedia) due to the fact that "a small proportion of people contribute some content and their opinions and preference are unlikely to reflect the opinions of the population as a whole."[41] Type (b) occurs in data in which critical attributes that influence the outcome are missing. This typically happens when data generation relies on human input or the process recording the data does not have access to key attributes, EHR systems included. To exclude data bias, I sometimes appeal to comparative analysis, such as applying multiple learning models in the same dataset and using similar analytical strategies on different datasets. Besides, I welcome the readers to do the evaluation.

I should perhaps add that my purpose is to ensure this book has original content and solutions. When summarizing his book, Haskel said modestly that the rise of intangibles economy might be speculative. I think some data experiments could lead him to withdraw the words. Mayer-Schönberger envisages a tax credit—the ability to pay some taxes with data, but he didn't describe for us some processing supports. Zuboff believes surveillance capitalism is inevitable. Her book, however, does not discuss possible reverse techniques that may delay this trend. Above all, I try to show here possible technology solutions, not prophecy or fairy tale.

---

[41] ***See:*** (1) Baeza-Yates R. Bias on the Web. Communications of the ACM, June 2018;61(6):54-61; (2) Krishnamurthy P. Understanding data bias. Towardsdatascience.Com. September 11, 2019. Retrieved from https://towardsdatascience.com/survey-d4f168791e57.

The Main Results of This Study

What are the major conclusions to which these data-related studies have led me? The first is that one should be wary of any political determinism regarding inequalities of data resources. The history of resource allocation has always been deeply political. This is shaped by the way economic, social, and political actors view what is just and what is not. Due to the limited resources available for production relative to the supply and demand for social use, the social embeddedness of markets is always under government constraints. The view of data as a resource is now accepted by a public consensus so that political intervention from the governments worldwide have been increasingly prominent. One example: there may be sound antitrust reasons to break up or take a stake away from the largest data firms, but this alone will spawn more political interventions from the governments in their businesses and then more participations in the markets. Instead detecting data rights may solve the root problem by relying on the rule of "survival the fittest," to produce smaller tech firms and open the field from more competitors.

The second conclusion, which is the heart of my book, is to give the status of "data sovereign"[42] to the data producers. This will be the benign mechanisms for income and wealth distribution related to data products to push toward convergence and divergence alternatively, further, to weaken or undermine destabilizing, inegalitarian forces of long standing from traditional domains. I defined one data producer that also does not refer to an individual who provides personal data. Just like Zuboff enumerated in her *Surveillance Capitalism* (see Footnote 38):

> *The idea of "data ownership" is often championed as a solution. But what is the point of owning data that should not exist in the first place? It's like negotiating how many hours a day a seven-year-old should be allowed to work, rather than contesting the fundamental legitimacy of child labor.*

The so-called "free services" in the early digital era, in essential, is an exchange process of services and personal data. Service providers collected the data from involved individuals through the process. We are all staring wide-eyed at either Google or Facebook to complain about their data activities, but healthcare providers did in the same way. We make our personal data as currency (and pay money) to get a healthier body from hospitals. The current inequity lies in that, these service providers continue to re-/process the data they initially generated to be newer data products but did not distribute data profits to the individuals who provided the data. I adapted the definition of Piketty's convergence and divergence: convergence is "toward reduction and compression of

[42] Tang, C., Plasek, J.M., Zhu, Y. *et al.* Data sovereigns for the world economy. Humanit Soc Sci Commun 7, 184 (2020). https://doi.org/10.1057/s41599-020-00664-y.

inequalities" on data re-/processing income or wealth distribution; and divergence can exist even in a world where there is adequate investment in data skills and where all the conditions in terms of "market efficiency" mentioned by economists appear to be satisfied.

Let's consider first the mechanisms pushing toward convergence. The forces of convergence are based on the ability of "data sovereigns" to protect the data they owned or controlled. I definite the term "data sovereigns" in which the word "sovereign" is borrowed from a fundamental economic notion of "consumer sovereignty."[43] Because I believe that data producers in the emerging data economy are less vulnerable (just like the consumers in pervious industrial sectors) so that they need some powers. Max Weber defines "power" (see Footnote 23) as the ability to possess any resource. Following Weber's formulations, data sovereign as the one owned or controlled an economic resource should be preserved through "the sole grantor of the right to physical force"[44] with the legitimacy of domination. I pin my hopes on the enterprises and entrepreneurs who owned or controlled the data. The bottom of the pyramid is more unusual, where makes the butterfly effect possible to send out. I want to look at public participation (i.e., data enterprises and their alliance) serving with socioeconomic changes put into practice. Data sovereigns can increase this power they want and use to protect data resources for investing data skills, data science included. The adoption of "data sovereign" enables economic agglomeration in cyberspace without considering geopolitical borders. Put in other words; utilizing data as an economic resource usually has the threat of new entrants and substitute products. When a data sovereign cannot effectively protect their own data resources, it may be prudent to join an alliance (within or across states) of data sovereigns for added protection. Such a power related to data protection, we call data force, frequently promotes the diffusion of data knowledge and skill. Data knowledge and skill diffusion, the key to "overall productivity growth as well as the reduction of inequality both within and between countries" (see Footnote 1), is conducive for a data enterprise alliance to external market competition and internal data resource sharing. Furthermore, there exists no substantial technical gap in data science between emerging economies and developed countries. The data technological convergence is fundamentally a process of establishing global market mechanisms, rather than the diffusion and sharing of knowledge in general meaning.

---

[43] I here borrowed the word "sovereign" from "customer sovereignty." One vital issue in the current data economy needed to be addressed is data producers may be more vulnerable. ***See:*** Hutt, W. The concept of consumers' sovereignty. The Economic Journal, 1940;50 (197):66-77.

[44] ***See:*** Translated by Waters, D., Waters, T., Hahnke E., Lippke M., Ludwig-Glück, E., Mai, D., & Ritzi-Messner, N. The distribution of power within the community: Classes, Stände, Parties by Max Weber. Journal of Classical Sociology, 2010;(10):137.

We do not have to worry too much, such as Zuboff's surveillance capitalism, when data sovereigns were granted. On the contrary, more political interventions may lead to undermining individual self-determination, autonomy and decision rights that are about us but not for us. Fortunately, there's a virtually infinite sea of possibilities between now and the doomsday from a technical perspective. Each possible move represents a different game, a different universe in which we make a better move. There are more possible games of chess than there are atoms in the universe. No one could possibly predict them all, even ourselves. But it also means that if we make a mistake, there's a nearly infinite amount of ways to fix it. Meanwhile, that economic rationality would in some sense automatically give rise to democratic rationality. Someone may question that data collected by ill means do not preclude the guy from being a data sovereign. I don't think so. First and foremost, the title of "data sovereign" is granted to a data producer. I speculate the guys who obtained the data "maliciously" aims to make money fast. They were largely absent in both investing in data skills and processing the data to be a data product. They didn't meet the definition of data producers so that they cannot have data sovereignty.

The analysis of consumers' habits and preferences, as for one critical feature of surveillance capitalism criticized by Zuboff, is also no matter. This is an evolved strategy of data-rich market segmentation in nature. Mayer-Schönberger divided the markets into two types—traditional money-based markets and emerging data-rich markets, and argued in his book *Reinventing Capitalism in the Age of Big Data*:

> *Data-rich markets have the advantage of not deducing preferences from price. They offer another advantage over price, as well: not only do individuals have multiple preferences regarding a potential transaction, they also likely weigh different preferences differently.*

In segmenting money-based markets, analysts typically took for common characteristics such as similar demographic profiles, shared needs, and common interests. So, in dividing data-rich markets, the consumers' habits and preferences are included, without exception. The main difference lies in whether the relevant data exists or not. The process of segmenting the market is deceptively simple but in practice, this task can be very laborious with high cost "since it involves poring over voluminous data, and requires a great deal of skill in analysis, interpretation and some judgements."[45] The cost of market segmentation was shifted to the consumers ultimately. Current data analysts use their algorithmic capabilities and computational power based on user data logs to generate a prediction of user click-through rates as a signal of individuals' preferences. Identifying the market to be segmented via data techniques is no longer a time-

[45] I reference the entry of "Market Segmentation" on Wikipedia. ***See:*** Retrieved from https://en.wikipedia.org/wiki/Market_segmentation.

consuming and manual operation and reduce the cost; furthermore, it resulted in reducing asymmetric information to make the markets much fairer. In sum, data sovereignty indirectly guarantees consumers' sovereignty.

Another optimistic viewpoint, which supports my argument, is the governing logic about "saving over the life cycle—people accumulate wealth when young in order to provide for their old age." A pension is created according to this logic. The big development challenge behind the world's rapidly aging population; that is, all most all countries have been facing the dilemma of making ends meet when the welfare expenditure increased sharply. It is possible to be a solution if a data profit sharing mechanism exists by negotiating between data sovereigns and individual data owners because data products on the same dataset can be continuously productive. I recommend insurance companies or banks may consider designing some services related to personal data saving.

Outline of This Book

The remainder of the book consists of thirteen chapters partitioned into four parts. Part I, titled "Capital: Something to Nothing," contains two chapters and introduces the basic research content of this book—data capital derived from the notion of capital in general. Specifically, Chap. 2 presents the conclusions of previous economic studies on capitalism and capital and then demonstrates what meaning and capital characteristics are. This chapter contains nothing really new, and the reader familiar with the content may wish to skip directly to Chap. 3. The only thing to look at is brainstorming on researching capital for hospitals from hospital ranking. Chapter 3 gives a more detailed description of which the benefits data capital might have. These benefits include reclassifying the existing capitals, possibly changing the current "anchorless" currencies, and resulting in human capital sharing. I explored the possibility of data to be "anchor" from a discussion about world epidemics and global stock market performance. The original work in Chap. 3 is an experiment in social media clout to evaluate whether data capital exists. I used Facebook's posts produced by two venture-capital groups between October 2016 and September 2018. One group is private (that means only members can give a post), and the other is public (that means anyone who visits can give a post).

The purpose of Part II, titled "Data Capital as Factor of Production," which consists of three chapters, is to explain data capital by contrasting the production of Marx's *Capital* from a data science perspective. Chapter 4 looks at data capital, starting with the properties of data as a resource, and introduces two components of the data in detail: memories and time. Chapter 5 presents some potential economic scenarios when the data becomes a capital, the basic ideas of the original work that are used repeatedly in this book. In Chap. 6, I briefly described multilevel data production to indicate the role of data economically capitalized in various. The two examples and one brainstorm given in Chap. 6 are related to primitive accumulation, simple reproduction, and enlarged reproduction of data capital. These are mostly original work, in which the most interesting one is the future scenarios of tongue image learning (supported by Chinese traditional medicine), such as the catering business.

Part III, titled "Hopes and Challenges Brought from Data Capital," consists of three chapters relating to the perspectives from the market, government, and interdisciplinary research, respectively. Centered around the orders of magnitude on data capital profitability, Chap. 7 explores data-rich markets compared with traditional markets with some barriers, financial markets, and the general labor market. I also shared an idea of how to drive equitably in the profit return in Chap. 7. Specifically, bridging two mHealth products (i.e., a bone health workout and a self-governing openness based data transaction) as a brainstorm familiarizes the reader with data capital rights, income and output, and wealth distribution, in cross-border markets. Chapter 8 studies the problem of government intervention behind data capital relating to public choice. Finally, Chap. 9 looks at the prospects for of existing leading indicators generally used in economics.

The purpose of Part IV, titled "Regulating Data Capital in the Twenty-First Century" and consisting of three chapters, is to draw a solution in the policy aspect from the previous three parts. Chapter 10 reviews the current socioeconomic status and then examines what a "global society" suited to data capital might look. Chapter 11 proposes a consideration of Piketty's global tax on capital in cyberspace. I think Piketty's progressive global tax will partially work for the data-rich globalization process, especially when looking at data capital. I recommend recognizing the identity of data sovereigns for data producers and then finding out how revenue sharing works as profits distributed between data producers and individuals (or other data owners) who provided data. The data capital tax can be deducted from taxes on income when processing the final annual settlement. In the later of this chapter, I discussed the possibility of applying Mayer-Schönberger's "tax paid in data" to resolve data resource-use rents, and a recommendation of "data as fiscal" was put forward. Chapter 12 is a blueprint for the data capital tax based on the latest technique on self-governing openness of data. This idealized blueprint can solve some potential issues, such as establishing a people-oriented mechanism, identifying the transaction in the best matches, and making system transparency via a feedback mechanism.

Some final words. The history of global capitalism is always profoundly political, chaotic, and unpredictable. The sole purpose of this book is to introduce an emerging type of capital related to the data. As a relatively new topic, some recent affairs and news are my preference to cite. I hope that this will spark a broader discussion about the rise of data capital means for both academic and industrial societies. Of course, all errors and omissions are my responsibility. Since English is not my native language, it is difficult for me to modify other native speakers' words. An incorrect modification is possible to change the intent. Thus, my choice is to excerpt and keep the original parts. I hope the reader are not uncomfortable with this.

Part I
Capital: Something to Nothing

Abstract Since the early twenty-first century, the major developed economies began to invest more in intangible assets (e.g., R&D, software, and database) than in tangible assets (e.g., buildings, machinery, and computers). Increasingly investing in intangibles is, as many economists believe, an underappreciated phenomenon ranging from economic inequality to stagnating productivity. This part concludes by presenting many prior research on what meaning and the characteristics of capital are, and explores the essence of capital metamorphoses, and then proposes the concept of data capital.

Capital, in its concept, reflects the state of development and prevailing social relations of each society, so it is not immutable. Many economists acknowledge that capital has changed radically since the twentieth century. Their observation is “capitalism does not need capital ultimately” when seeing the rise of the intangible economy. However, it is only an appearance that possibly makes no sense when taken at their own level. I attempt to consider its deep issues and cause behind this observation in this part.

Chapter 2
Principles of Capital

As the discussion progressed, it became clear to me that it is at least to know the meaning and characteristics of capital. I begin by describing previous studies from some known economists; and they are Irving Fisher,[1] Thomas Piketty,[2] Ryan Doody,[3] Paul Ekins,[4] Jonathan Haskel and Stian Westlake,[5] Thomas

---

[1] I reference directly here from the original sentence published in Irving Fisher's book of *The Rate of Interest* and his summary given by Wikipedia and the Library of Economics and Liberty. ***See:*** (1) Fisher I. Theory of Interest (Reprints of economic classics). New York: Augustus m Kelley Pubs; n Reprint edition (June 1, 1956), 555 pages, 1956; (2) Retrieved from https://en.wikipedia.org/wiki/Irving_Fisher; (3) Retrieved from https://www.econlib.org/library/Enc/bios/Fisher.html.

[2] ***See:*** PiKetty, T. translated by Goldhammer A. Capital in the Twenty-First Century. Cambridge, MA: the Belknap Press of Harvard University Press; Reprint edition (August 14, 2017), 2017: 816 pages.

[3] This article summarizes Pittkey's book. In this chapter, I refer to his summary method that is concise and clear. ***See:*** Doody R. Piketty on Capital and Inequality. March 25, 2015. Retrieved from http://rdoody.com/EconJusticePIKETTY.pdf.

[4] There exist two versions of the same reference. The newly one published in 2008 was modified by Paul Ekins *et al.* based on his 1992 version for the EU's regional sustainable development evaluation. ***See:*** (1) Ekins P. 1992. A four-capital model of wealth creation. In *Real-Life Economics: Understanding Wealth Creation*, Ekins P, Max-Neef M (eds). Routledge: London; 147–155; (2) Ekins P, Dresner S, Dahlström K. The four-capital method of sustainable development evaluation. European Environment, 2008;18(2):63-80.

[5] I reference directly here from the original sentence published in Jonathan Haskei's and Stian Westiake's book of *Capitalism without Capital: The Rise of the Intangible*. ***See:*** Haskel J, Westlake S. Capitalism without Capital: The Rise of the Intangible. New Jersey, NY: Princeton University Press; Reprint edition (October 16, 2018), 2018: 296 pages.

© Springer Nature Switzerland AG 2021
C. Tang, *Data Capital*, https://doi.org/10.1007/978-3-030-60192-8_2

A. Hemphill,[6] and Carol Corrado *et al.*,[7] Pierre Bourdieu,[8] James S. Coleman[9] as well. Before presenting my results in detail, it is best to brief their conclusion, respectively, since every economist conducts research independently (i.e., there exists no comparable, such as area of research, assumptions, etc.). In addition, the concept of capital has often been intertwined vaguely by many economic thoughts, involving the concepts of "income," "resource," "asset," "investment," and so forth. It will be useful to begin with the definitions of these peripheral concepts. Of which, I referenced partially from Wikipedia, but didn't give the specific links. I hope the reader selfly finds them if interested.

Definition 2.1 *Production* Production is a process of combining various material inputs and immaterial inputs in order to make something for consumption (output).

Definition 2.2 *Producer* A producer is any organization (e.g., corporations, industry bodies, government specifies) who plan, coordinates, and performs production.

Definition 2.3 *Economic Resource* An economic resource is the (input) factors of production used in producing goods or providing services.

Definition 2.4 *Investment*[10] Investment is what happens when a producer either acquires a fixed asset or spends resources (money, effort, raw materials) to improve it.

There is plenty of economic terms (or terms used in and finance or accounting) that overlap with the coverage of the word investment. A business investment may be capital, and some people may call it a business asset. An individual may make an investment, which is very likely to be considered an asset, but not capital. The meaning of investment depends on the context but generally means spending money in order to make more money. When a business purchases

[6]This article is a recommendation for Haskel's and Westlake's book. ***See:*** Hemphill TA. Capitalism without Capital: The Rise of the Intangible Economy by Jonathan Haskel and Stian Westlake. CATO Journal. Fall 2018. Retrievedfromhttps://www.cato.org/cato-journal/fall-2018/capitalism-without-capital-rise-intangible-economy-jonathan-haskel-stian.

[7]While this article was published in 2009, there is some evidence that this study began in 2005. ***See:*** Corrado, C., Hulten, C., & Sichel, D. Intangible capital and US economic growth. Review of Income and Wealth, 2009;55(3):661-685.

[8]Pierre Bourdieu is rarely cited for his work on social capital relative to James Coleman and Robert Putnam. This may be because: on the one side, his studies are out of economic ranges; and on the other side, his approach is too intellectually demanding. ***See:*** (1) Bourdieu, P. The Forms of Capital. In: Richardson JG. *Handbook of theory and research for the sociology of education*. Westport, CT: Greenwood Press; 1986: 241-258; (2) Retrieved from https://en.wikipedia.org/wiki/Pierre_Bourdieu.

[9]James S. Coleman (May 12, 1926 – March 25, 1995) was an American sociologist based chiefly at the University of Chicago. He was elected president of the American Sociological Association. ***See:*** Coleman JS. *Foundations of Social Theory*. Cambridge, MA: Harvard University Press (August 19, 1998); 1998: 1014 pages.

[10]Note that the definition of investment used directly the one from the United Nations' (UN) System of National Accounts. ***See:*** Retrieved from https://unstats.un.org/unsd/nationalaccount/default.asp.

some new capital, that is considered an investment because it expects that capital to pay for itself in the future. An individual or financial intermediary is likely to make an investment by purchasing a financial asset (stock, bond, option, etc.). The word "investment" conveys the idea of sacrificing wealth today in order to have more wealth in the future.

Definition 2.5 *Asset* An asset is an economic resource that is expected to provide a benefit over a period of time.

Definition 2.6 *Fixed* Asset A fixed asset is an asset that results from using up resources in the process of its production.

Definition 2.7 *Rate of Profit* The profit rate is the relative profitability of a whole investment project, but sometimes expresses the share of profits that may be relevant with income from capital.

Definition 2.8 *Interest Rate* An interest rate is the amount of interest due per period, as a proportion of the amount lent, deposited, or borrowed (called the principal sum), which varies widely depending on the identity of the lender or borrower.

Fisher on Income and Capital: About Time

Irving Fisher (February 1867–April 1947) was one of American's greatest mathematical economists, had a long, distinguished, and amazingly varied academic and business career. Fisher spent most of his academic career at Yale first in the Department of Mathematics, and then in the Department of Economics. He also served as president of the American Economic Association in 1918 and as the first president of the Econometrics Society in 1931–1933. Although he damaged his reputation by insisting throughout the Great Depression (1929) that recovery was imminent, contemporary economic models of interest and capital are based on Fisherian principles. Similarly, monetarism is founded on Fisher's principles of money and prices. Fisher was also a pioneer of econometrics, including the development of index numbers. Some concepts named after him include the Fisher equation, the Fisher hypothesis, the international Fisher effect, the Fisher separation theorem, and the Fisher market.

Fisher's Main Claims

1. **The Real Interest Rate is Equal to the Nominal Interest Rate (the one we observe) minus the Expected Inflation Rate.** Fisher claims that r, the real interest rate, is equal to i, the nominal interest rate minus π as the expected inflation rate.
   Example of Fisher Equation

	Last year	Calculation process
i – Nominal interest rate	4.90%	
π – Real interest rate	3%	$r = \frac{(1 + 4.90\%)}{(1 + 3\%)} - 1$
r – Inflation rate	0.08%	

2. **Fisher's Quantity Theory of Money.** MV is identically equal to PT (i.e., MV = PT) where M equals the stock of money; V equals velocity, or how quickly money circulates in an economy; P equals the price level; and T equals the total volume of transactions.

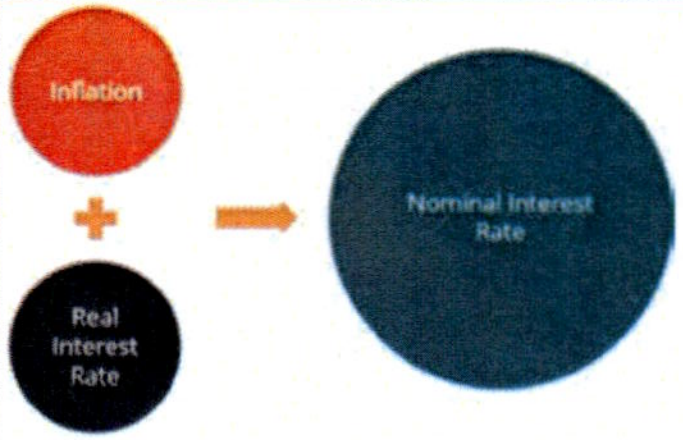

Suppose Zack owns an investment portfolio. Last year, the portfolio earned a return of 3.25%. However, last year's inflation rate was around 2%. Sam wants to determine the real return he earned from his portfolio. In order to find the real rate of return, we use the Fisher equation. The equation states that: $(1 + i) = (1 + r)\ (1 + \pi)$. We can rearrange the equation to find real interest rate:

$$r = \frac{(1+i)}{(1-\pi)} - 1$$

Therefore, the real interest that Zack's investment portfolio earned last year, after accounting for inflation, is 0.08%.

Capital and Interest

1. **What is Income?** "INCOME is a series of events." Fisher divided income into three types: real income, money income, psychic or enjoyment income.
2. **What is Capital?** "Capital, in the sense of capital value, is simply future income discounted or, in other words, capitalized."
3. **The Rate of Interest Bridges Income and Capital.** "The bridge or link between income and capital is the rate of interest. We may define the rate of interest as the percent of premium paid on money at one date in terms of money to be in hand 1 year later."
4. **What is Wealth?** "I define wealth as consisting of material objects owned by human beings (including, if you please, human beings themselves.) The ownership may be divided and parceled out among other forms of property rights."
5. **Capital-Income Relations.** "… the chief relations between capital and income are: (1) Capital value is income capitalized or discounted. (2) If the rate of interest falls, the capital value (capitalized value of expected income) rises, and vice versa. (3) This rise or fall in capital value is relatively great for durable goods like land, and relatively small for transitory goods like clothes. (4) Capital value is increased by savings, the income being decreased by the same amount that the capital is increased. (5) These savings thus diverted from income and turned back into capital will, except for mischance, be the basis for real income later." "These relations are … (1) from capital goods to their future services, that is, income; (2) from these services to their value; and (3) from their value back to capital value."

Income is a fundamental concept:
"Income is the alpha and omega of economics."
"Enjoyment income is a psychological entity and cannot be measured directly. We can approximate it indirectly, however, by going one step back of it to what is called real income. Real wages, and indeed real income in general, consist of those final physical events in the outer world which give us our inner enjoyments."
"This real income includes the shelter of a house, the music of a victrola or radio, the use of clothes, the eating of food, the reading of the newspaper and all those other innumerable events by which we make the world about us contribute to our enjoyments. Metaphorically we sometimes refer to this, our real income, as our 'bread and butter.'"
"Money income includes all money received which is not obviously, and in the nature of the case, to be devoted to reinvestment—or, as the expression is, 'earmarked' for reinvestment. For instance, if you more than "earn your living" of $6000 with a salary of $10,000, you voluntarily put by the $4000 remaining as savings. This part of your money income is saved from being turned immediately into real income. That is, instead of spending all your salary for this year's living you invest $4000 of it to help toward the cost of living of future years. And so, the $4000 is not only credited as income but debited as outgo. With it you buy durable objects such as land or buildings, or part rights in these, such as stocks or bonds. Your money income is in this case your salary (or it may be dividends, rent, interest, or profits) and it exceeds real income by the amount of your savings. On the other hand, you may be living beyond your (money) income. This means, expressed in terms of the concepts here used, that your real income for the year is greater than your money income."

Impatience and Opportunity

Interest rates, Fisher postulated, result from the interaction of two forces: the "time preference" people have for capital now, and the investment opportunity principle (that income invested now will yield greater income in the future). Fisher suggested the more popular term "impatience" in place of "time preference" and the catchword "investment opportunity" as a substitute for the inadequate term "productivity" which had come into such general use. "… investment opportunity seems to be the nearest expression in popular language to suggest or denote the technical magnitude ***r*** employed in this book. The full expression for ***r*** is the rate of return over cost, and both cost and return are differences between two optional income streams."	"This reasoning sounds very much like Eugen von Böhm-Bawerk's. Indeed, Fisher dedicated Theory of Interest to "the memory of John Rae and of Eugen von Böhm-Bawerk, who laid the foundations upon which I have endeavored to build." But Fisher objected to Böhm-Bawerk's idea that roundaboutness necessarily increases production, arguing instead that at a positive interest rate, no one would ever choose a longer period unless it was more productive. So if we look at processes selected, we do find that longer periods are more productive. But, he argued, the length of the period does not in itself contribute to productivity."

To recapitulate: Fisher's claims on capital focuses on income. He believes income is a more fundamental concept compared with capital and " interest rates bridge and link income and capital" (see Footnote 1). The concept of income means:[11] first, the past plays a vital role in income because income represents future-oriented. Secondly, income is a flow; that is, sequential events. The third is, income prediction is tough. The pain of economists, but, is not data scientists'; data scientists were, are, and have been doing this, such as describing past income characteristics with personal preferences and predicting the future occurrences of certain incomes. In data science, such data related to income can be represented as the attributes that have relationships that involve order in time or space. An attribute is a property or characteristic of an object that may vary, either from one object to another or from one time to another. Example 2.1 as shown in Table 2.1,[12] there are five different time—t_1, t_2, t_3, t_4, t_5, three different individuals—P_1, P_2, P_3, and five different incomes: wages, dividends, rent, interest, and profit—*A*, *B*, *C*, *D*, *E*. In the top table, each row corresponds to the incomes earned at a particular time by each individual. For example, individual P_2 earned incomes of *A* and *D* at time t_3. We can do variable

[11] ***See:*** Xiang, S. The New Capitalism: Rising, Crisis and Future of the Global Financial Capitalism. Beijing: China Citic Press; 1 edition, December 1, 2014, 616 pages.

[12] This is a classic textbook for data mining. I referenced and modified the original examples published in this book. ***See:*** Tan PN, Steinbach M., Kumar V. Introduction to data mining. New York City, NY: Pearson; 2 edition (January 4, 2018), 2018: 864 pages.

Table 2.1 Sequential income data

Time	Individuals	Income
t_1	P_1	*A, B*
t_2	P_3	*A, C*
t_2	P_1	*C, D*
t_3	P_2	*A, D*
t_4	P_2	*E*
t_5	P_1	*A, E*
Individual	**Time and income earned**	
P_1	(t_1: *A, B*), (t_2: *C, D*), (t_5: *A, E*)	
P_2	(t_3: *A, D*), (t_4: *E*)	
P_3	(t_2: *A, C*)	

transformation: in the bottom table, the same information indicates, but each row corresponds to a particular individual. Each row contains information about each income involving the individual, where an income is considered to be a set of items and the time at which those items were earned. Since Fisher also mentioned "opportunity" in explaining interest rates, we may retain spatial attributes (i.e., positions or areas) for the data when representing incomes. Under a premise of sufficient data, it is possible to identify "co-occurrence" relationships of a dynamic system.

Piketty on Capital and Inequality: About Classification

Speaking in 1918, at the annual meeting of the American Economic Association, Fisher warned his colleague about "a great peril."[13] The peril, which risked "perverting the democracy for which we have just been fighting," was extreme inequality. Thomas Piketty is a French economist whose work focuses on wealth and income inequality. Piketty specializes in economic inequality, taking a historic and statistical approach. His work looks at the rate of capital accumulation in relation to economic growth over a two hundred year spread from the nineteenth century to the present. His best-selling book, *Capital in the Twenty-First Century*, relies on economic data going back 250 years to show that an ever-rising concentration of wealth is not self-correcting. To address this problem, he proposes redistribution through a progressive global tax on wealth.

[13] *See:* Kahloon I. Thomas Piketty Goes Global. Newyorker.com. March 9, 2020. Retrieved from https://www.newyorker.com/magazine/2020/03/09/thomas-piketty-goes-global.

Piketty's Main Claims

1. **The Return on Capital is Greater than Growth.** Piketty claims that r, the average annual rate of return on capital, is in the long-run greater than equal to g, the growth of the economy that is the annual increase in income or output. (a) $r > g$ (b) And, "If … the rate of return on capital [r] remains significantly above the growth rate [g] for an extended period of time …, then the risk of divergence in the distribution of wealth is very high."	"In slowly growing economies, past wealth naturally takes on disproportionate importance, because it takes only a small flow of new savings to increase the stock of wealth steadily and substantially."
2. **Inherited Wealth Grows Faster than Income.** If $r \gg g$, then inherited wealth grows faster than output and income. The reason is "people with inherited wealth need save only a portion of their income from capital to see that capital grow more quickly than the economy as a whole." 3. **Piketty's Conclusion:** If the above conditions hold, then capitalism will lead to a distribution of wealth that resembles an aristocracy (see Footnote 3). Such a distribution is incompatible with the values fundamental to modern democracy.	"Under such conditions, it is almost inevitable that inherited wealth will dominate wealth amassed from a lifetime's labor by a wide margin, and the concentration of capital will attain extremely high levels—levels potentially incompatible with the meritocratic values and principles of social justice fundamental to modern democratic societies."

The First and Second Fundamental Law of Capitalism

1. **What is Income?** "National income is defined as the sum of all income available to the residents of a given country in a given year, regardless of the legal classification of that income." (a) National income = domestic output + net income from abroad (b) National income = capital income + labor income	How is national income related to GDP? National Income = GDP—the depreciation of the capital used in production + net income received from abroad Piketty refers to (GDP—the depreciation of the capital used in production) as "domestic output."
2. **What is Capital?** "Capital is defined as the sum total of nonhuman assets that can be owned and exchanged on some market."	"I speak of 'capital' without further qualification, I always exclude … 'human capital,' which consists of an individual's labor power, skills, training, and abilities."
3. Capital's Composition: Nonfinancial Assets (land, dwellings, commercial inventory, other buildings, machinery, infrastructure, patents, etc.) and Financial Assets (bank accounts, mutual funds, bonds, stocks, other financial investments, insurance policies, pension funds, etc.).	**Income is a Flow:** "… the quantity of goods produced and distributed in a given period." **Capital is a Stock:** "… the total wealth owned at a given point in The capital/income ratio β is related to the share of income from time."
4. **What is Wealth?** Piketty uses "'capital' and 'wealth' interchangeably, as if they were perfectly synonymous." 5. **The First Law $\alpha = r \times \beta$.** "It expresses a simple, transparent relationship among the three most important concepts for analyzing the capitalist system: the capital/income ration as β, the share of capital in come as α, and the rate of return on capital as r."	For example, given a developed country in 2017 with per capita national income of 50,000 dollars/year and around 30% of national income, so per capita, national income breaks down as 15,000 dollars of capital. With a capital/income ratio on the order of 600%, this meant that the rate of return on capital was around 5%. Each citizen owns an average of 300,000 dollars of capital.
6. **The Second Law $\beta = s/g$.** Piketty posits a dynamic law relating the capital/income ration β in an economy to an economy's saving rate s and growth rate g. A "country that saves a lot and grows slowly will over the long run accumulate an enormous stock of capital (relative to its income), which can in turn have a significant effect on the social structure and distribution of wealth."	Note that this law only holds "in the long run". The identity here is an equilibrium. And the equation is only valid if certain assumptions hold.

Inequality in Capital and Labor

Piketty claims that inequality with respect to capital is always greater than inequality with respect to labor. This is, in part, because $r > g$. Because Capital is often transferred via inheritance, he worries that capitalism naturally tends toward *Patrimonial Capitalism.*	Capital income	Labor income
	Profits, dividends, interest, rents, royalties, etc.	Wages, salary, honoraria, bonuses, etc.

Piketty's claims on capital as he proposed a series questions in the opening chapter of his book (see Footnote 2):

> *What is capital? What are its limits? What forms does it take? How has its composition changed over time?*

These questions happen to lay out his research focus, which is the composition of capital. He believes "'capital' and 'wealth' interchangeably, as if they were perfectly synonymous." And he "always exclude … 'human capital,' which consists of an individual's labor power, skills, training, and abilities," in the category of capital he studies; specifically (see Footnote 2),

> *In this book, capital is defined as the sum total of nonhuman assets that can be owned and exchanged on some market. Capital includes all forms of real property (including residential real estate) as well as financial and professional capital (plants, infrastructure, machinery, patents, and so on) used by firms and government agencies.*

Thus, Piketty denied psychic or enjoyment income Fisher defined. As a result, he reclassified Fisher's real and money income according to his own settings on capital to determine the rate of return on capital.

Definition 2.9 *Rate of Return on Capital* (see Footnote 2) The rate of return on capital measures the yield on capital over the course of a year regardless of its legal form (e.g., profits, interest, rents, dividends, royalties, capital gains, etc.), expressed as a percentage of the value of capital invested.

Obviously, the rate of return has a broader notion than the "rate of profit" and an "interest rate," while incorporating the latter two; different types of investment, simultaneously, resulting in the rate of return that can vary widely. But Piketty admitted the first law tells us nothing about these subtilities (see Footnote 2):

> *Some firms generate rates of return greater than 10 percent per year; others make losses (negative rates of return). The average long-run rate of return on stock is 7-8 percent in many countries. Investments in real estate and bonds frequently return 3-4 percent, while the real rate of interest on public debt is sometimes much lower.*

There already exist some classical methods to partition capital in economics, such as the four- or five-capital model. British economist Paul Ekins put forward a "four-capital model" in 1992 (see Footnote 4),

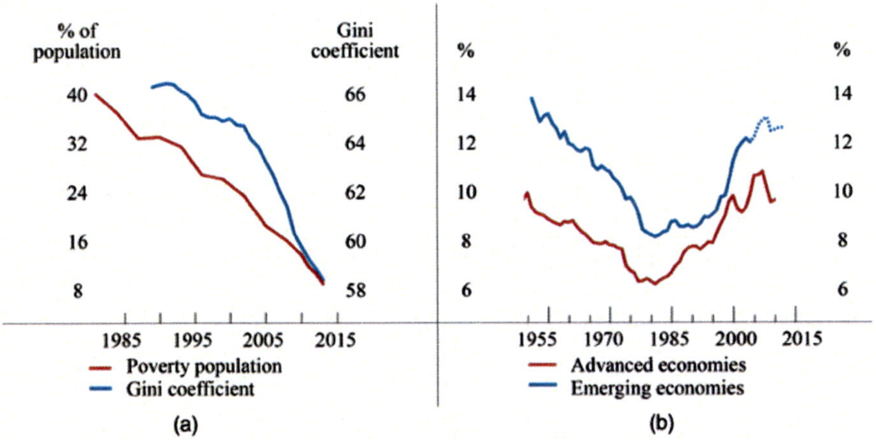

Figure 2.1 World inequality with its share of income accruing at top 1%. Notes: (1) Poverty headcount ratio at US$1.9/day; (2) 128-country world Gini coefficient estimate using a deterministic lognormal distribution. (3) Simple average across the economies listed. (4) AU, CA, CH, DE, DK, ES, FR, GB, IT, JP, NL, NO, NZ, PT, SE, and US. (5) AR, CN, ID, IN, KR, MY, SG, AND ZA. The dotted line excludes AR, ID, and IN. (Source: Author redrew the two figures using data from the World Wealth and Income database according to the BIS' 87th annual report, p. 108)

> *The four-capital model relating manufactured, human, social and natural capital to the process of production and the generation of human welfare ... represent the power to mobilize and create other kinds of capital rather than embodying real productive power themselves.*

Later scholars added the fifth type as "financial capital" to be a five-capital model, as shown in Table 2.2.

Piketty's habit in reclassification is also shown in another place. He looks at the increase in income inequality mainly at the very top end of the distribution: the top 1% and even more the top 0.1%. Example 2.2 I found to redraw a figure of showing world inequality with its share of income accruing at the top 1% (see Figure 2.1) according to the BIS's 87th Annual Report, which used the same database as Piketty's WTID. All lines in figure 2.1, interestingly, look very similar except for the differences of magnitude. We found, at first sight, they do not seem to give an adequate explanation on, for example, political institutions, environmental externalities, industrial structure in detail.

While Piketty's innovations in partitioning have trouble convincing us, newer classes sometimes lead to fresher discoveries. Human has an innate ability to classify things into categories. But manual classification often suffices for a small and simple dataset with only a few attributes. We need an automated solution. The general idea behind classification in data mining, one of the most techniques in data science, is as follows. Given the data that consists of a collection of instances. Each instance is characterized by the tuple (x, y), where x is the set of attribute values that describe the instance and y is the class label of the

Table 2.2 Common types of capital

Type	Five types of explanation
Manufactured Capital (MC)	MC is what is traditionally considered as capital: produced assets that are used to produce other goods and services. Some examples are machines, tools, buildings and infrastructure.
Natural Capital (NC)	In addition to traditional natural resources, such as timber, water, and energy and mineral reserves, natural capital includes natural assets that are not easily valued monetarily, such as biodiversity, endangered species and the ecosystems that perform ecological services (e.g. air and water filtration). NC can be considered as the components of nature that can be linked directly or indirectly with human welfare.
Human Capital (HC)	HC generally refers to the health, well-being and productive potential of individual people. Types of HC include mental and physical health, education, motivation and work skills. These elements not only contribute to a happy, healthy society, but also improve the opportunities for economic development through a productive workforce.
Social Capital (SC)	Like HC, SC is related to human well-being, but on a societal rather than individual level. It consists of the social networks that support an efficient, cohesive society, and facilitate social and intellectual interactions among its members. SC refers to those stocks of social trust, norms and networks that people can draw upon to solve common problems and create social cohesion. Examples of SC include neighborhood associations, civic organizations, and co-operatives. The political and legal structures that promote political stability, democracy, government efficiency and social justice (all of which are good for productivity as well as being desirable in themselves) are also part of SC.
Financial Capital (FC)	Unlike the other types, FC has no real value itself but is representative of manufactured, natural, human, social capital; e.g. shares, bonds or banknotes.

instance. The attribute set x can contain attributes of any type, while the class label y must be categorical. The task of classification is to calculate an instance (x, y) correctly if $f(x) = y$, in which y is given by some known rules. Another task called clustering starts from capturing the nature of the data to discovers potential classes (i.e., clusters) via automatic techniques, in which y is learned from the data. Also, the task to evaluate a specific instance whether belong to a class (cluster) or not is outlier detection. So, data science is much better at partitioning or classifying automatically.

Haskel and Westlake on Intangible Capital Accounting: About Feature

Let's go back to look at "professional capital," as Piketty did not give any definition in his book. According to some literature I searched online:[14]

[14] ***See:*** Hargreaves A, Fullan M. The power of professional capital. The Learning Professional, 2013;34(3):36.

Professional capital is a function of the interactive, multiplicative combination of the three kinds of capital: human, social, and decisional. ... We know that both human and social capital have links to ... achievement. Decisional capital, a notion that comes from the field of law, is about how you develop your capabilities over time, particularly your capacity to judge. All professions involve judgment in situations and circumstances where the evidence and the answers aren't incontrovertibly clear.

Strangely, professional capital not only is accumulated by focusing on the capital of individuals, but must be complemented by social capital to achieve substantial improvement. Piketty's "missing" definition of "professional capital" may be an intentional result, possibly, which is most likely used to avoid self-contradictions in partitioning the capital. I ruled out the idea of excluding "intangible" professional capital from capital. In contrast, any twenty-first-century corporation's intangibles (invested in "organizational capabilities," or used to develop "specific products or processes") would need to be incorporated into both these corporations' financial accounts and economists' models for calculating economic growth. In 2005, Carol Corrado *et al.* initiated their foundational research to measure intangible investment in the United States (and gradually other countries), based on what they began to recognize as the results of research and development (R&D) and the largely nonphysical ideas resulting from it (see Footnote 7):

The result was an operational definition of intangible investment divided into three broad categories each producing a different type of capital asset: (1) computerized investment, e.g., software and databases; (2) innovative property, e.g., R&D and product and service development; and (3) economic competencies, e.g., brand names, business models, and organization – level training.

Jonathan Haskel and Stian Westlake's seminal research also focus on the concept of intangible investment. Haskel is a professor of economics at Imperial College Business School. He is the son of Baron Simon Haskel, a British politician from the Labor Party. His research interests are productivity, innovation, intangible investment and growth. He constructed studies on (1) how much firms investment in "intangible" assets, such as R&D (research and development) and new business processes (2) how much intangibles (and the science budget) contributes to economic growth as whole, and so forth.

Stian Westlake is a senior fellow at Nesta, the UK's national foundation for innovation, and a policy adviser to three UK science and universities ministers. His research interests included the measurement of innovation and its effects on productivity, the role of high-growth businesses in the economy, financial innovation, and how government policy should respond to technological change.

Main Claims from Haskel and Westlake

1. **Something Fundamentally Different About Intangible Investment.** "First, conventional accounting practices ignore intangibles. Second, an intangible-rich economy behaves differently from a tangible-rich one." 2. **Intangible Investment Overtook Tangible Investment.** Haskel and Westlake present economic data for Europe and the United States that show intangible investment overtook tangible investment at the time of the Great Recession in 2008. "It did so for five reasons: First, labor-intensive services become more expensive relative to manufactured goods, and intangible investments (such as design, R&D, and software development) depend much more on labor. Second, new technology also seems to be increasing the opportunities for businesses to invest in intangibles, and, as many intangibles involve information and communications, they can be made more efficient with better information technology. Third, while the structure of the economy will affect the relative importance of intangibles, that effect will change over time. Fourth, there is some evidence that looser regulation of product markets and labor markets encourages intangible investment. Fifth, market size is critical, because intangibles (such as Facebook's software) can be 'scaled up' more or less indefinitely."	**"… Understanding the steady move to intangible investment helps us understand some of the key issues facing us today: innovation and growth, inequality, the role of management, and financial and policy reform."** **"In 2005, Carol Corrado, Charles Hulten, and Dan Sichel initiated their foundational research to measure intangible investment in the United States (and gradually other countries). The result was an operational definition of intangible investment divided into three broad categories each producing a different type of capital asset: (1) computerized investment, e.g., software and databases; (2) innovative property, e.g., R&D and product and service development; and (3) economic competencies, e.g., brand names, business models, and organization-level training."**

"Four S's" Characteristics of Intangibles

1. **Scalability.** "Scalability means intangible assets can be used repeatedly and in multiple places at the same time. The scalability of knowledge derives from a key feature of ideas; namely, that nonrivalry and powerful network effects fuel scalability."	Haskell and Westlake summarized the "unusual economic characteristics of intangibles," and they are, scalability, sunkenness, spillovers, and synergies (the "Four S's").
2. **Sunkenness.** "Sunkenness involves irrecoverable costs, and when it comes to intangible assets, unless a company has valuable intellectual property (IP), such 'knowledge' becomes almost impossible to liquidate. This creates a problem because investments with high irrecoverable costs can be difficult to finance, especially with debt, in which there is a lack of collateral available to a lender liquidation."	The "Four S's" is "useful for identifying and defining the characteristics of intangibles" and produce "two general characteristics: uncertainty and contentedness."
3. **Spillovers.** "Spillovers are often created by intangible investments, and it is relatively easy for other businesses (including competitors) to take advantage of these intangible investments. While companies may keep trade secrets, some ideas are simply nonexcludable. Such spillovers often emanate from company R&D activities, but they are also found in branding and marketing, copying of organizational innovations, and employee training." 4. **Synergies.** "Synergies are important in intangible-based economies, as ideas and other ideas (especially in technology) often work well together. The concept of 'open innovation,' defined as when a firm directly engages with and benefits from new knowledge gathered from outside the company, is a major driver of intangible synergy at both the company and industry levels, as well as for the national and local economies."	"Spillovers matter, argue Haskell and Westlake, for three reasons: First, if companies are unsure if they will obtain the benefits of their investments, they will likely invest less. Second, there is a company premium in making the most of their own intangible investments (or from other companies' intangible spillovers), and those firms will have competitive success. Third, spillovers affect the geography of modern economies, such as creative people locating to urban areas to encourage connectedness."

Inequality in Intangible Capital

Haskel and Westlake further document that the rise of synergies and spillovers from intangibles might be expected to increase wealth and income inequality, including inequality across regions, inequality in profitability between competing companies, and wealth or esteem inequality in individuals.	"Income inequality rises as synergies and spillovers increase the gap in profitability between competing companies, raising the demand for managers and leaders with coordinating skills; wealth inequality rises as cities, where spillovers and synergies abound, become increasingly attractive, driving up the property prices; esteem inequality rises as psychological traits like openness to experience become more important."

Haskel and Westlake's research on intangibles began with investment. Why investment matters? They gave the answer as follows (see Footnote 5).

> *Investment is what builds up capital, which, together with labor, constitutes the two measured inputs to production that power the economy, the sinews and joints that make the economy works.*

Haskel and Westlake took Microsoft Corporations as a stark example of "capitalism without capital." When looking at Microsoft's 2006 balance sheet that records its assets, Haskel and Westlake found that (1) Microsoft's market value was around US$250 billion, and (2) the traditional assets of plant and equipment were only US$3 billion, equivalent to 1% of its market value. They calculated via the easiest way to show that Microsoft's intangibles exceed 98% of its market value.

Measuring intangible investment, of course, is not easy and requires at least three steps (see Footnote 5):

> *First, we need to find out how much firms are spending on the intangible. Second, in some cases, not all of that spending will be creating a long-lived asset. Third, we need to adjust that investment for inflation and quality change so we can compare an investment in different periods when prices and quality are changing.*

In Haskel and Westlake's view, whatever tangible and intangible investment are things following the definition of investment; that is (see Footnote 5):

> *(a) cost money, (b) are expected to generate a longer-term return, and what's more, (c) the company making the investment has a reasonable chance of enjoying a worthwhile portion of the return itself.*

Capital accounting is about the financial management of a firm's capital assets. Capital assets are long lasting pieces of property used to generate revenue for the firm. And now, these assets can be intangibles, such as intellectual property, patents, and trademarks. Generally Accepted Accounting Principles (GAAP) provide some guidance as to reporting requirements associated with an intangible asset that may be capitalized, but the ultimate policy will be developed by each firm individually. Each firm devises its own capital accounting standards, declaring with assets will be capitalized and how these capital assets will be managed. Capitalization refers to the practice of recording a piece of property as an asset on a firm's balance sheet, versus treating the cost of the property as a deductive expense. Because the complexity involved with the capital accounting process, firms may choose to hire some professional accounting staffs or set up a special-purpose accounting department. As measuring intangibles is challenging in conventional arduous process, however, accountants (being cautious types on the precautionary principle) prefer not to do so except in limited circumstances (such as after the asset has been successfully developed and sold).

Accounting practices are typically very numbers-driven. Some might see them as too simplistic for a complex analytic task. But we can do something else before performing the task. Every intangible investment has multiple features, so the challenge is to find a way to make accountants having a truer sense of intangibles. I here use the simplest visual solution to explain. Imagine being able to view or "meet" an intangible capital asset to get a sense of its characteristics. The accountant (or other users) could form an opinion of it quickly and over time understand whether to expense or capitalize expenditures, and perhaps participate in early investment decisions or forecast future values.

There exist several ways of showing multivariate variables. First, and most common is via tables. But tables leave it up to the user to decipher and evaluate individual numerical values, making it time-consuming and hard to gather interesting insights. Second is to combine tables with simple charts to alleviate the data table overload for users, such as a line chart or line plot commonly used in economics that belongs to the category of simple charts. Example 2.3 to be in line with the first figure in Haskel and Westlake's book (see Footnote 5), I searched Elsevier's ScienceDirect database for literature that mentioned intangibles in the abstract, title, or keyword in the field "Economics, Econometrics and Finance" between 1996 and 2019 and drew two simple charts as shown in figure 2.2.

Data Source 2.1 *Literature Data* Literature data is any single part of a scientific paper (e.g., titles, abstracts, backgrounds, methods, results, conclusions, or the whole article) collected by searching of several specific databases (e.g., DBLP: Computer Science Bibliography can be accessed at https://dblp.uni-trier.de, MEDLINE/PubMed, ScienceDirect), library catalog services (e.g., Harvard's HOLLIS), or online facilities (e.g., Google Scholar) during a given period of time. Quantitative approaches to archives and texts developed by digital humanists have offered one such expansion. These approaches often treat literature as a text mine. Each paper included in the analysis can be classified as either a paper describing article type (e.g., a description, or a review of the literature) or other information: year of publication, journal title, specialty area, the field of study, and characteristics given by authors to data reuse.

Figure 2.2 Using a line (**a**) and radar (**b**) chart to show "intangibles" references in scientific journals. Note: Data are the number of mentions of the word "intangible" in the abstract, title, or keyword in academic journals in the field "Economics, Econometrics and Finance" recorded in the database ScienceDirect, a total of 1596 found between 1996 and 2019. (Source: The author drew this via Microsoft Excel)

Figure 2.2a is similar to Haskel and Westlake's "Figure 1.1 'intangibles' references in scientific journals" (see Footnote 5) by only changing the time from "1980 to 2006" to "1996 to 2019," and indicates a decline of research interest in 2017. I visualized the same information in figure 2.2b via a radar chart, which consists of a sequence of equiangular spokes, called radii, with each spoke representing one of the variables (i.e., year). The data length of a spoke is proportional to the magnitude of the variable for the data point relative to the maximum magnitude of the variable across all data points. A line is drawn connecting the data values for each spoke. This gives the plot a star-like appearance which can be used to strikingly show outliers and commonality—where each variable corresponds to "worse" or "better" in some respect, and all variables on the same scale. So, Figure 2.2b gives a gap between 2016 and 2019 clearly, which is not like figure 2.2a's specious answer that requires us a double check in the relevant table.

But what if the intangibles have numerous and complex features? Chernoff fishes can combine all variables related to a single entity into one view to help understand the full picture. Similar in concept to Chernoff faces invented by Statistician Herman Chernoff (who held teaching posts at MIT and Harvard) in 1973, Chernoff fishes display multivariate variables in the shape of a fish. Julie Rodriguez and Piotr Kaczmarek explored the idea of relying on a fish's attributes to be the vessel of fund characteristics.[15] The individual parts, such as eyes, belly, tail, and fin, represent values of the variables by their width, height, size, and position.

[15] *See:* Rodriguez J, Kaczmarek P. Visualizing Financial Data. Wiley; 1 edition (May 2, 2016), 2016: 480 pages.

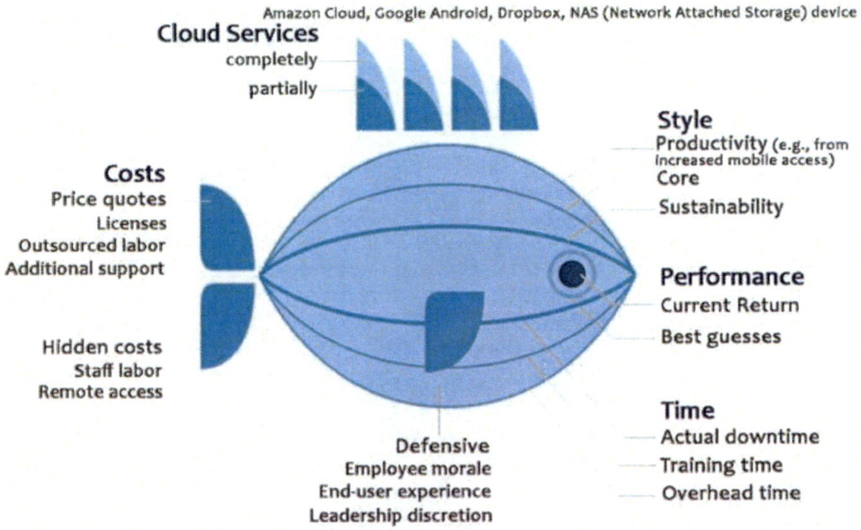

Figure 2.3 Chernoff fishes for depicting a firm's investment in moving to cloud services. (Source: Author created this by herself)

Here is an example to explain how Chernoff fishes contribute to computing investment in cloud services. Example 2.4 implementing cloud file storage and management is a more complicated capital accounting, compared with replacing a new server. Because the former is spread out over time, while the latter has a more clear front-loaded and costs. I conducted Chernoff fishes for depicting this investment as shown in figure 2.3.

Each feature could say something specific about each variable so that, collectively, this Chernoff fish represents this specific intangible investment. First is identification. The fish could embody that the investment's major traits in a way that enables the accountant to see these first and get a sense of its personality. For example, the overall width of the fish would show the investment style and total capital. A large belly would represent a big investment scale, a prominent back curvature would suggest value-based guess while a sleek-bodied fish would signal small-cap, short-term investment. These primary characteristics stand out so the accountant can quickly assess its nature, such as whether the investment is large, slow-moving and steady, or small, fast-moving and dynamic.

Second is evaluation. The fish needs to convey secondary traits; the ones that users want to understand upon closer inspection. These further explain the investment's market exposure and results. Examples include the top fin height and position to show cloud service options, the tail and lower fin to show costs (price quotes and hidden costs) and the eye size to represent performance.

Third is about insights. By illustrating these features, the fish enables the user to build a picture of its personality, including its strategy on style and time—perhaps with an indication of past performance embedded in its eyes—as well how it behaves against market conditions. The object makes all this clear to see without losing any detail. Armed with this insight, the accountant can effectively administrate, evaluate, and implement capital asset programs, leading to more proper management of schedules (e.g., before, during or after the investment) on intangibles that may be capitalized or deprecated.

A single object to convey multivariate variables has enormous potential to change how firms interact with information. First and foremost, they'll increase visibility. By replacing individual tables and charts with a single representative object, users can see the entirety of their intangible projects with just one truly holistic view. Firms can also make quicker at-a-glance assessments. The object would show major features, such as productivity, clearly so the user (e.g., leadership and end-user; mainly, accountants) sees these instantly and can make rapid initial evaluations—without having to trawl through too many numerical summaries and multiple charts.

Capital accounting is essentially a self-identification within an organization. It is possible to form social capital in this organization if capitalizable assets were approved by external social recognition. According to intangibles of "3S's (sunkenness excluded)" given by Haskel and Westlake, intangibles seem to be more natural, finishing this transformation process than tangibles. I will come back to this in the example section.

Bourdieu on Capital and Power: About Pattern

The first to extend the idea of capital to intangible categories (e.g., social, cultural, and symbolic) is probably not economists, and sociologists instead. As early as 1986, the prescient sociologist Pierre Bourdieu, also a French, observed and laid out this. For Bourdieu (see Footnote 8):

> *Each individual occupies a position in a multidimensional social space; a person is not defined by social class membership, but by every single kind of capital they are articulate through social relations. That capital includes the value of social networks, which could be used to produce or reproduce inequality.*

Bourdieu (August 1930–January 2002) was a Marxist sociologist, anthropologist, philosopher, and public intellectual. Bourdieu's work was primarily concerned with the dynamics of power in society, especially the diverse and subtle ways in which power is transferred and social order is maintained within and across generations. His key terms were capital, habitus, and field. Bourdieu inherited much of traditional sociology and anthropology to synthesize into his own theory. Bourdieu retained Max Weber's importance of domination and symbolic systems in social life, for example, as well as the idea of social orders, to transform from a sociology of religion into a theory of fields. Another example, Bourdieu stressed the role of practice and embodiment in social dynamics by having gained Karl Marx's understanding of "society" as the ensemble of social relationships.

Bourdieu's Main Claims

1. **One must understand is, what capital is and how its accumulation, its history and growth, effects all of society.** Bourdieu believed that capital "is a *vis insita*, a force inscribed in objective or subjective structures, but it is also a *lex insita*, the principle underlying the immanent regularities of the social world."	"The social world is accumulated history, and if it is not to be reduced to a discontinuous series of instantaneous mechanical equilibria between agents who are treated as interchangeable particles, one must reintroduce into it the notion of capital and with it, accumulation and all its effects."
2. **Capital and Time.** Bourdieu argued that capital accumulation, in every form, takes time. "And the structure of the distribution of the different types and subtypes of capital at a given moment in time represents the immanent structure of the social world, i.e., the set of constraints, inscribed in the very reality of that world, which govern its functioning in a durable way, determining the chances of success for practices."	"Capital, which, in its objectified or embodied forms, takes time to accumulate and which, as a potential capacity to produce profits and to reproduce itself in identical or expanded form, contains a tendency to persist in its being, is a force inscribed in the objectivity of things so that everything is not equally possible or impossible."

Capital with Its Forms

1. **What is Capital? Capital is social energy.** Capital is accumulated labor, that is, it's all the things people have made through their collective work, through their social energy, through collaborating with each other.	"Capital is accumulated labor (in its materialized form or its 'incorporated,' embodied form) which, when appropriated on a private, i.e., exclusive, basis by agents or groups of agents, enables them to appropriate social energy in the form of reified or living labor."
2. **Capital Conversion.** Economic capital is at the root of all other types of capital. Each conversion comes at the expense of one type of capital but also, more importantly, ensures that the system of capital itself remains in total power.	"Capital can present itself in three fundamental guises: as economic capital, which is immediately and directly convertible into money and may be institutionalized in the forms of property rights; as cultural capital, which is convertible, on certain conditions, into economic capital and may be institutionalized in the forms of educational qualifications; and as social capital made up of social obligations ('connections'), which is convertible, in certain conditions, into economic capital and may be institutionalized in the forms of a title of nobility."

Capital Capitalizes Capital

Bourdieu purposed capitalization based on the twin notion of dominant capital (i.e., economic capital) and differential accumulation (i.e., to give accumulation a universal form on both productive labor and utility or abstract labor). He recognized this symbolic creature. He began by assuming that capitalization is entirely fictitious, and then related to the actual, or real capita via conversions. "The conversion of capital into capital, the fact that one type of capital can be transformed into another type, is what lets the entire system of capital reproduce itself. … This transmission involves both loss and concealment. The more capital is lost in transmission, the more concealed the capital transmitted is, and vice versa. In other words, the more capital is lost in its transmission, the less people notice the capital that actually gets transmitted."[16]	"… economic capital is at the root of all the other types of capital and that these transformed, disguised forms of economic capital, never entirely reducible to that definition, produce their most specific effects only to the extent that they conceal (not least from their possessors) the fact that economic capital is at their root, in other words—but only in the last analysis—at the root of their effects. The real logic of the functioning of capital, the conversions from one type to another, and the law of conservation which governs them cannot be understood unless two opposing but equally partial views are superseded: on the one hand, economism, which, on the grounds that every type of capital is reducible in the last analysis to economic capital, ignores what makes the specific efficacy of the other types of capital, and on the other hand, semiologism (nowadays represented by structuralism, symbolic interactionism, or ethnomethodology), which reduces social exchanges to phenomena of communication and ignores the brutal fact of universal reducibility to economics."

We can begin by association analysis to perform the task, such as describing a form of capital, as well as presenting a conversion among two or more different types of capital. Association analysis is useful for discovering interesting relationships hidden in a dataset. The uncovered relationships can be represented as association rules (as known as frequent itemsets). Example 2.5 an association rule can be found from the data shown in Table 2.3:

$$\{\text{High tuition}\} \rightarrow \{\text{High salary}\}$$

[16]This article translated Pierre Bourdieu's *The Forms of Capital* in a straightforward way. ***See:*** The Dangerous Mabe. An American "Translation" of Bourdieu's "The Forms of Capital". Medium.com. April 17, 2019. Retrievedfromhttps://medium.com/@mdowns1611/an-american-translation-of-bourdieus-the-forms-of-capital-166a95f84e16.

Table 2.3 An example of education investment and economic returns

Individuals	Items
Seven	{High tuition, High salary}
Zack	{High tuition, High cost of living, High salary}
Joseph	{Low cost of living, High salary}

This rule suggests a relationship between investing in "high tuition" and getting "high salary" to return. Such a rule we called is a pattern (so that the advanced association analysis is known as pattern mining). Yet, this discovered pattern may be spurious since it might happen simply by chance (on the three data records). We thus need to do evaluation about the strength of an association rule measured by in terms of its support and confidence. Support determines how often a rule is applicable to a given dataset, while confidence's definition is below.

Definition 2.10 *Confidence* (see Footnote 12) Given an association rule that is an implication expression of the form $X \rightarrow Y$, where X and Y are disjoint itemsets, i.e., $X \cap Y = \varnothing$. Confidence determines how frequently items in Y appear in patterns that contain X. The formal metric is:

$$\text{Confidence}, c(X \rightarrow Y) = \frac{\sigma(X \cup Y)}{\sigma(X)}$$

We thus can obtain the formal metric of support is:

$$\text{Support}, s(X \rightarrow Y) = \frac{\sigma(X \cup Y)}{N}$$

In practice, one drawback of association analysis (or pattern mining) is the number of resulted patterns (e.g., structure pattern to describe a specific form of capital, transform pattern to present a conversion between two types of capital) can be very large. It is useful to identify a small representative set of patterns from which all others can be derived.

In next section, I examine the issues of capitalization and capital conversion by a form of brainstorms; I consider them beginning at the organizational level in public service areas (i.e., healthcare; actually, education is similar).

Hospital Ranking Challenges Capital for Hospitals: A Brainstorm

Patients today are enamored of hospital ranking. It is easy to see why. Many healthcare providers are uneven in specialties, and patients want to know more information about those. Hospitals usually don't like the rankings. Just recently (In February 2020), one children's hospital published a letter to the US News & World Report that releases the best hospital ranking annually, to call for the removal of traditional rankings for Pediatric Cardiology and Heart Surgery. They believed that, the rankings can be significantly reducing revenue for hospitals, dangerously misleading for patients on (1) erroneous and costly decision making, (2) unnecessary differentiation between hospitals, and (3) inaccurate representation of whether a hospital is suited to the specific needs of the patient. Here is an example in this letter:[17]

> *My family lives in Wisconsin, 90 minutes from a surgical center currently ranked 29th. Our child has a complex defect, so we are always thinking about the next possible surgery. We annually review the rankings of hospitals as we want to be sure our child goes to the best center. This year, we discovered that we have family in Florida near the hospital ranked 11th. Should our son need surgery this year, the rankings suggest that we should split up our family, take leave from our jobs, and pay out-of-pocket for travel, food and housing costs, so we can be at "the best" center for complex surgery. Only, how would we have known to look past the rankings at number and type of surgeries performed to better understand that the hospital in Florida is not the best suited for our child? That hospital, ranked 11th, performs 2/3 the number of surgeries that our nearby hospital, ranked 29th, performs. Not to mention, the new surgeon in our nearby hospital specializes in the type of defect our son has. The rankings may have led us to make a decision that could have cost us dearly, both financially and with the health of our child.*

Obviously, this letter only indicates the general attitude of healthcare providers. No even this hospital believes this little will work to make the rankings disappear.

While the newly released Honor Rolls are still over there, hospitals begin to present their concerns for the next year, and top-notch hospitals affiliated to the best medical schools are no exception. All too often, work for improving the scores of various specialties is fragmented and compartmentalized. Hospital executives find it difficult to get things done under a unifying framework. So, they cannot offer integrated dynamic ranking-related organizations and managerial behaviors. Hospital rankings, in fact, resulted in social clout; and its positive part represents all qualities of social capital, involving trust, mutual understanding, shared aims, and ethical values. One of the first formal use of social capital Pierre Bourdieu laid out is (see Footnote 8):

[17] ***See:*** Basken A. Hospital Rankings – A letter to US News and World Report. February 7, 2020. Retrieved from https://www.conqueringchd.org/hospital-rankings-a-letter-to-u-s-news-and-world-report.

Social capital is the aggregate of the actual or potential resources which are linked to possession of a durable network of more or less institutionalized relationships of mutual acquaintance and recognition – or in other words, to membership in a group – which provides each of its members with the backing of the collectively owned capital, a credential which entitles them to credit, in the various senses of the words.

American sociologist James S. Coleman described the word "collectively" as follows (see Footnote 9).

Unlike other forms of capital, social capital inheres in the structure of relations between persons and among persons. It is lodged neither in individuals nor in physical implements of production.

Given this definition, thus, it can be assumed that social capital may help not only individuals, but also complex organizations, such as hospitals, cope with stress and foster "potential strategic value."[18]

It seems that hospital rankings and social capital in hospitals are interdetermined—they are both the cause and the effect of another's being. We may draw on other suitable sources to measure social capital in a health care provider, and then to study the rankings, and vice versa. A successfully measured social capital in hospitals may make it easier for hospital executives to assess the conditions (e.g., advantages) of daily work, and to improve the rankings. The rankings may help increase social capital by reducing insecurity, uncertainty, and disorientation in professional experience and workload, which represents a significant predictor of overall job satisfaction of physicians, nurses, or other professionals working in the field of patient care. I think thus ranking methodology can be a crucial element of our analyses.

Lay Out the Context and Definitions

Example 2.6 Let's look at the US News & World Report's best hospital ranking methodology.[19] I used topic modeling on the 30-page report to extract theme-words and then generated a mind map (see Figure 2.4).

Definition 2.11 *Mind Map* A mind map is a hierarchical diagram to be an image-like aid. It is often created around a single concept, drawn as an image in the center of a blank page, to which associated representations of keywords, ideas, tasks or other items in a radial way. Mind maps are considered to be a type of spider diagram (Figure 2.5).

---

[18] *See:* Derose KP, Varda DM. Social capital and health care access: a systematic review. Medical care research and review: MCRR, 2009;66(3):272.

[19] *See:* Martin G, Majumder A, Adams Z, *et al.* Methodology: U.S. News & World Report 2019-2020 Best Hospitals Procedures & Conditions Ratings. Retrieved from https://media.beam.usnews.com/da/5b/7646c0e74 4aea61f55e5aa816870/190709-bh-procedures-conditions-methodology-2019.pdf.

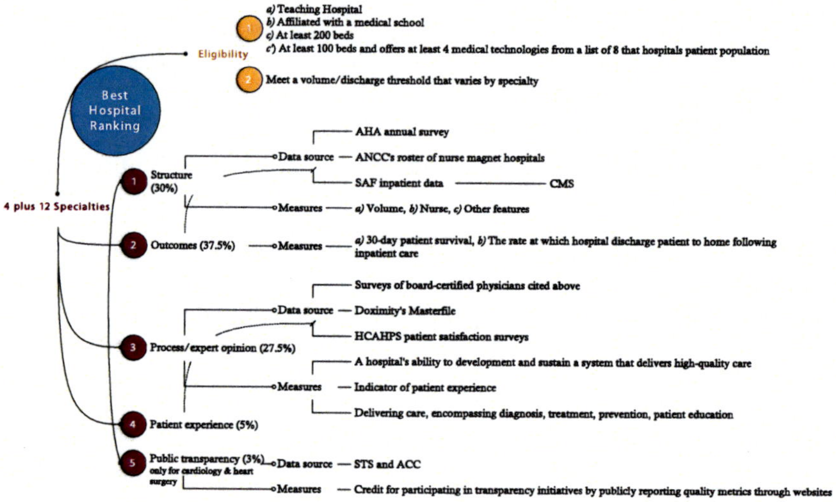

Figure 2.4 A mind map extracted from the US News & World Report's best hospital ranking methodology. Notes: (1) Abbreviations: AHA as American Heart Association; ANCC as American Nurses Credentialing Center; SAF as Standard Analytical Files (Medicare Claims); CMS as the Centers for Medicare and Medicaid Services; Doximity as partnered with U.S. News & World Report to assist with their physician survey process for the annual Best Hospitals rankings; HCAHPS as Hospital Consumer Assessment of Healthcare Providers and Systems Survey; STS and ACC as the American College of Cardiology and Society of Thoracic Surgeons. (2) 4 plus 12 specialties: 4 specialists are ophthalmology, psychiatry, rehabilitation, rheumatology, and 12 specialists are cancer, cardiology and heart surgery, diabetes and endocrinology, ear, nose and throat, gastroenterology and gastrointestinal surgery, geriatrics, gynecology, nephrology, neurology and neurosurgery, orthopedics, pulmonology and lung surgery, urology, data-driven analysis. (Source: Author's summary via a visualization method)

Also, I completed with additional relevant data. This map indicates that this ranking methodology does not straightforwardly ask for predictors, activity streams, or health outcomes; instead, it requires integrating services into a coherent whole. This ranking methodology highlights eligibility and 4 plus 12 specialties. One exciting finding is data-driven analysis is 1 of 16 specialties. Secondly, multiple data sources were found to be reusable for different assessments.

Diverging vs. Converging

A common critique of analyzing social capital is that "there is no uniform measure (see Footnote 18)."

> *Much of the social capital research has focused on structural (what people "do") and/or cognitive (what people "feel") components of social capital. ..., eight used measures*

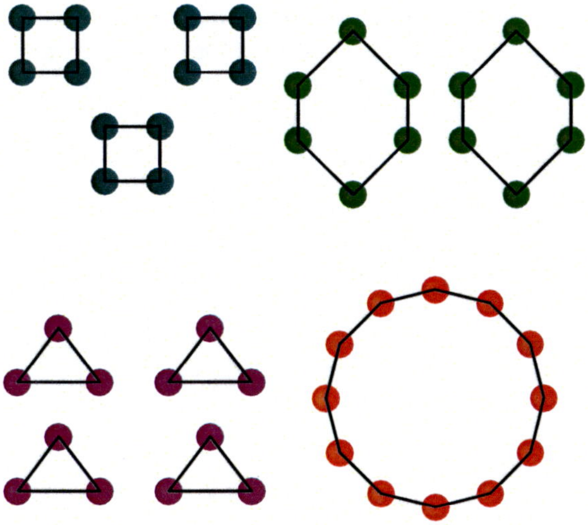

Figure 2.5 Network structures discovered by wiring pattern mining. Note: The four networks shown have exactly the same size (the same number of nodes and edges), and each node in each network has the same degree (the number of interactions with other nodes), but each network has a very different structure. (Source: The author redrew according to Footnote 20)

that included both structural and cognitive components, seven included measures that included only the structural component and the other four used measures reflecting only the cognitive component. However, the specific social capital indicators used to reflect the structural and cognitive components were diverse. Structural component measures included those relating to social and civic participation (e.g., ... volunteering), social networks (e.g., existence of a support network), number and density of community organizations (e.g., schools and ... [hospitals]), neighborhood disorganization, tenure in community, and race-ethnicity of community. Cognitive component measures included those relating to reciprocity (e.g., relationship with neighbors, helping others), trust (e.g., generalized social trust, trust of neighbors), self-esteem, social control, socil cohesion, and sense of personal safety.

Methodically, network analytics are a great alternative for uniform measurement.[20] It provides a means of interconnecting the complex world of entities. We can begin by deciphering 16 specialties; that is, with these specialties as entities being modeled as nodes and their connections as edges, comprising large networks. Each entity of specialties has descriptors in detail, such as physicians (and nurses or other professionals) and patients (and family members or friends). Also, these descriptors can be seen as entities—sub-entities with their own characteristics. According to pattern mining (i.e., advanced association analysis), thus, we may take an essential step in that direction by providing a scalable heuristic framework for grouping entities based on their wiring patterns and using the discovered patterns for revealing the higher-order organizational principles of hospital ranking networked systems. Mining such wiring patterns of

[20] *See:* Pržulj N., Malod-Dognin N. Network analytics in the age of big data. Science. 2016;353(6295):123-124.

networks to uncover the functional organization aims to discover the number of interactions that each entity (node) has with other entities (called node degree), because two networks can be identical in a similar description but have a very different connectivity structure (see figure 2.5). Of course, it is also possible to analyze a part of networks via the same approach. Anyway, I set this as an open question to allow the interested reader to go further.

Chapter 3
The Rise of Data Capital

On September 28, 2019, *Business Insider*—an online newspaper known for providing, analyzing, and aggregating news stories on various subjects from around the web, published a post titled "How WeWork spiraled from a US$47 billion valuation to talk of bankruptcy in just 6 weeks?"[1] Coincidentally, an article in the *Harvard Business Review* written by Dartmouth College's Vijay Govindarajan answered this question: No, WeWork isn't a tech company. Here's why that matters.[2] WeWork is an American commercial real estate company that provides a shared-office space for technology startups and other enterprises. Obviously, the market does not recognize WeWork's claims and in that regard, is a "SaaS (Space as a Service)"[3] tech company with the "asset-light" model where a business owns relatively fewer capital assets compared to the value of its operations. Govindarajan stated that (see Footnote 2),

> *A successful modern tech company can transform whole industries, achieve expansion of scale and scope at breakneck speeds, and make enormous profits, without requiring significant capital investments. It typically has most, if not all, of these five features: low variable costs, low capital investments, a lot of customer data and customer intimacy, network effects, and ecosystems that boost expansion with little cost. ... WeWork does not meet any of the qualifications that enable a modern tech company to achieve exponential growth as well as winner-take-all profits. In contrast, even asset- and inventory-heavy companies like Amazon, Tesla, and Apple meet three of the five criteria. ... WeWork does not deserve the large EBITDA-based (earnings before interest, depreciation, and taxes) valuation multiple that is often ascribed to tech companies.*

[1] ***See:*** Campbell D. How WeWork spiraled from a $47 billion valuation to talk of bankruptcy in just 6 weeks. Business Insider. September 28, 2019. Retrieved from https://www.businessinsider.com/weworks-nightmare-ipo.

[2] ***See:*** Govindarajan V, Srivastave A. No, WeWork isn't a tech company. Here's why that matters. Harvard Business Review. August 21, 2019. Retrieved from https://hbr.org/2019/08/no-wework-isnt-a-tech-company-heres-why-that-matters.

[3] ***See:*** WeWork's S1filing. August 14, 2019. Retrieved from https://www.sec.gov/Archives/edgar/data/1533523/000119312519220499/d781982ds1.htm.

© Springer Nature Switzerland AG 2021

C. Tang, *Data Capital*, https://doi.org/10.1007/978-3-030-60192-8_3

WeWork's self-reporting in S-1 filing (see Footnote 3) was indeed slightly dim. Although the company's rapid growth (86% year-over-year revenue growth), WeWork's hefty losses (US$1.6 billion on revenues of US$1.8 billion in 2018). EBITDA, namely earnings before interest, taxes, depreciation, and amortization, is a measure of profitability. Although no required for companies to disclose, according to the US GAAP, EBITDA can be worked out and reported using the information found in a company's financial statements. For example, the earnings, interest, and tax figures are located on the income statement, while the depreciation and amortization figures are usually on the cash flow statement.

Definition 3.1 *EBITDA* EBITDA is a metric that excludes expenses associated with debt by adding back interest expense and taxes to earnings.

$$\text{EBITDA} = \text{net income} + \text{interest} + \text{taxes} + \text{depreciation} + \text{amortization}$$

$$\text{EBITDA} = \text{operating profit} + \text{depreciation expense} + \text{amortization expense}$$

Its proponents argue that EBITDA offers a more precise reflection of operations by eliminating the effects of financing and capital expenditures. Its opponents believe that EBITDA can sometimes be used intentionally to obscure the real profit performance of a company because it is a "non-GAAP" measure. Specifically, EBITDA can serve as a distraction for investors when a company has borrowed heavily or are experiencing rising capital and development costs. Case 3.1 indicates an example of how to measure the true value of companies via EBITDA.

Case 3.1: An Example of Using EBITDA to Dig the True Value of Companies

Given two companies with different capital structures but a similar business. Company A has a current EBITDA of $20 million and Company B has EBITDA of 18 million:

There are a few possible conclusions because Company B has a smaller interest expense account.

	Company A	Company B
EBITDA	$20,000,000	$18,000,000
Depreciation amortization	-$2,000,000	-$2,500,000
Interest	-$8,000,000	-4,500,000
Taxes	-$2,000,000	-$2,000,000
Net income	$8,000,000	$8,000,000

1. Is it possible that Company B could borrow more and increase both EBITDA and net income? If the company is underutilizing its ability to borrow, this could be a source of potential growth and value.
2. If both companies have the same amount of debt, perhaps Company A has a lower credit rating and must pay a higher interest rate. This may indicate additional risk compared to Company B and a lower value.
3. Based on the amount of depreciation and amortization, Company B is generating less EBITDA with more assets than Company A. This could indicate an inefficient management team and a problem for Company B's valuation.

Govindarajan's statement represents a growing belief among business economists since the new economy by the early 2000s, in particular in the United States, that "businesses were spending significant sums of money on things that have no physical presence, but that, nevertheless, were valuable and durable."[4] They were more than brand names, software, and R&D, data also included. This change was self-evident; for example, the capitalization of software added about 1.1% to 1999s US GDP, and R&D added 2.5% to 2012s, with these figures growing all the time.[5] However, in my opinion, one of five features as "low capital investments" is much more consistent with tech companies in the big data ear than those in the information age. An eye-opening historical look is that, creatively losing money was the mark of a successful dot com. And few could make money prolifically without the venture capitalists, who backed aiming for IPOs (initial public offerings) because that's when they got paid. A tendency to sell products (or services) at a loss is to reach market share, and then to gain consumer data. In contrast, data enterprises won't require a capital investment to do so due to the fact that they owned or controlled the data.

WeWork fits into the feature of "low capital investments" actually. In addition to raise awareness on branding and advertising, WeWork invested little in both tangible and intangible capital assets. In particular, WeWork as a commercial real estate company only invested in obtaining some long-term, cheap leases via its brand effects, instead of buying building properties. WeWork, in fact, embodied what we think of the supernova development model shared some or all of traits (see Case 3.2's speech bubbles): a business plan that promised to "change the world," a Get Big Fast strategy to reach ubiquity and corner a particular market, and a willingness to spend lavishly in a move toward domination in the (construction-related) business sphere.

WeWork advertised itself as the "Google Analytics for space" (see Footnote 3), but failed to grasp Google's essentials that are to invest in the "intangible" intelligence it can extract from the vast volume of data, nevertheless. The current trend of capital assets for investment, as Govindarajan's statement, is on intangibles and further to data assets—"a lot of customer data and customer intimacy" (see Footnote 2). This is because; the return on tangible capital is relatively fixed. Take Thomas Piketty's elaboration[6] as an example,

---

[4] ***See:*** Haskel J, Westlake S. Capitalism without Capital: The Rise of the Intangible. New Jersey, NY: Princeton University Press; Reprint edition (October 16, 2018), 2018: 296 pages.

[5] ***See:*** Corrado, C., Hulten, C., & Sichel, D. Intangible capital and US economic growth. Review of Income and Wealth, 2009;55(3):661-685.

[6] ***See:*** PiKetty, T. translated by Goldhammer A. Capital in the Twenty-First Century. Cambridge, MA: the Belknap Press of Harvard University Press; Reprint edition (August 14, 2017), 2017: 816 pages.

For nineteenth-century novelists and their readers, ... land (or government bonds) yields roughly 5 percent of the amount of capital invested (or, equivalently, that the value of capital corresponds to roughly twenty years of annual rent), ... the two measuring scales [capital and annual ren] were used interchangeably, as if rent and capital were synonymous, or perfect equivalents in two different languages. ... Now, at the beginning of the twenty-first century, we find roughly the same return on real estate, 4-5 percent, sometimes a little less, especially where prices have risen rapidly without dragging rents upward at the same rate.

Haskel and Westlak's studies (see Footnote 4) highlight that some unusual characteristics (i.e., Four S's characteristics) of intangible investments, especially scalability and spillovers, could magnify profits and productivity. For example, Uber is easy to extend its offerings to Uber Eats with minimal investments. Compare with other intangibles, data asset investments could affect businesses' incentives to invest more because its output with diversity achieves higher profits. The same data product (e.g., medication use pattern mining for inpatient childhood pneumonia), for instance, can reveal different insights into different industrial sectors (e.g., for hospitals, medical equality or drug inventory management; for pharmaceutical companies, pharmaceutical production or sales) and may also indirectly promote their core businesses through better revenue and profitability margins, respectively. Example 3.1 Since WeWork's prospectus did not display such an investment for the tendency of "intangible, data" capital assets, I drew figure 3.1 in illustrator that may help convince future investors based on some highlight parts of Case 3.2. They include lead generation practices, diversified products, and shifts in internal ecosystem function.

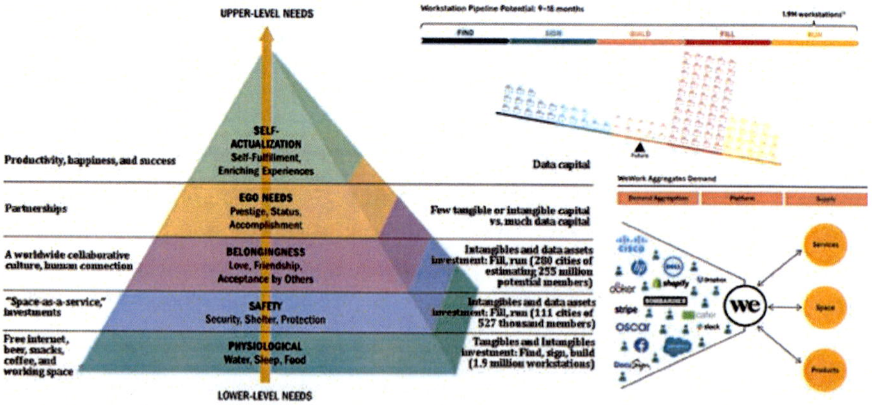

Figure 3.1 An illustrator of revised WeWork's prospectus "SUMMARY" based on Maslow's hierarchy of needs 1. Notes: The script on the left of Maslow's model is the highlighted part in Case II.2, and the script on its right is to interpret capital assets investment meets the demand. 2. source: The author drew this according to some materials in WeWork's prospectus (see Footnote 3)

Case 3.2: Some Paragraphs from WeWork's Prospectus "SUMMARY"

- Our Story

We are a community company committed to maximum global impact. Our mission is to elevate the world's consciousness. We have built a worldwide platform that supports growth, shared experiences and true success. We provide our members with flexible access to beautiful spaces, a culture of inclusivity and the energy of an inspired community, all connected by our extensive technology infrastructure. We believe our company has the power to elevate how people work, live and grow.

In early 2010, we opened our doors to our first member community at 154 Grand Street in New York City. In the beginning, our members consisted mostly of freelancers, startups and small businesses. Over the past 9 years, we have rapidly scaled our business while honoring our mission. Today, our global platform integrates space, community, services and technology in over 528 locations in 111 cities across 29 countries. Our 527,000 memberships represent global enterprises across multiple industries, including 38% of the Global Fortune 500. We are committed to providing our members around the world with a better day at work for less.

- Our Global Platform

We have proven that community, flexibility and cost-efficiency can benefit the workplace needs of everyone from global citizens to global enterprises. We pioneered a "space-as-a-service" membership model that offers the benefits of a collaborative culture, the flexibility to scale workspace up and down as needed and the power of a worldwide community, all for a lower cost. Through iterative product development at scale and significant investment in technology infrastructure, we have demonstrated that we can build better solutions for less money. We are changing the way people work globally and, in the process, we have disrupted the largest asset class in the world real estate.

We start by looking at space differently: as a place to bring people together, build community and enhance productivity. Philosophically, we believe in bringing comfort and happiness to the workplace. We employ over 500 designers and architects who work relentlessly to create spaces that are beautiful but simple, elevated but approachable, global yet locally unique, all delivered at a high quality without the associated expense. Next, we add a team of over 2500 trained community managers who foster human connection through collaboration and holistically support our members both personally and professionally. Lastly, with a persistent dedication to improving the member experience, we add products and services to our platform, either by building them ourselves, acquiring them or entering into partnerships. The entire member experience is powered by technology designed to enable our members to manage their own space, make connections among each other and access products and services, all with the goal of increasing our members' productivity, happiness and success.

- Addressable Market Size

We expect to expand aggressively in our existing cities as well as launch in up to 169 additional cities. We evaluate expansion in new cities based on multiple criteria, primarily our assessment of the potential member demand as well as the strategic value of having that city as part of our location portfolio. Based on data from Demographia and the Organization for Economic Cooperation and Development, we have identified our market opportunity to be 280 target cities with an estimated potential member population of approximately 255 million people in aggregate.

When applying our average revenue per WeWork membership for the 6 months ended June 30, 2019 to our potential member population of 149 million people in our existing 111 cities, we estimate an addressable market opportunity of $945 billion. Among our total potential member population of approximately 255 million people across our 280 target cities globally, we estimate an addressable market opportunity of $1.6 trillion.

a business plan that promised to "change the world"

a willingness to spend lavishly in a move toward domination in the (construction-related) business sphere **Finished acquisitions include a building information modeling firm *Case*, a digital sign-in system for offices called *Welkio*, and a physical data software company *Euclid*.**

a Get Big Fast strategy to reach ubiquity and corner a particular market

Figure 3.1 indicates investing capital assets is from the lower-level of tangibles, the middle-level of intangibles, to the upper-level of data assets, if following Maslow's hierarchy of needs (proposed by Abraham Maslow in 1954, i.e., physiological, safety, belongingness, ego needs, and self-actualization). Although it is unclear whether WeWork would invest in customer data and how it would use that data to develop customer-intimate solutions, this data is virtual gold for the company, "as it enables targeted ads and the sale of tailor-made products; it also increases users' switching costs (note that successful companies typically try to employ strategies that incur high switching costs on the part of consumers to dissuade them from switching to a competitor's product, brand, or services) as people use the service and get customized solutions in return" (see Footnote 2). We believe data capital assets for investment are promising. In the next sections of this chapter, I will examine various benefits from the rise of data capital.

Capital Reclassification

In Chap. 2, I introduced some classical types of capital in economics. In this part, I am going to concentrate on capital reclassification. A reclass or reclassification, in the capital, can be seen as an accounting concept; that is, a journal entry transferring an amount from one general ledger account to another. This can be done to correct a mistake; or to record that an asset is now being used for a different purpose (e.g., land's becoming investment property intended for resale, rather than as means of production). What capital reclassification solves, thus, is the measurement problem caused by the metamorphoses of capital. But first, I want to take a brief detour through economic literature and the World Wide Web, which offers a better understanding of existing capital classification.

Capital Classification Described in Economic Literature

I will begin by looking at how the economists involved in Chap. 2 classify the capital. The focus here is on Jonathan Haskel and Stian Westlake (see Footnote 4) and Thomas Piketty (see Footnote 6), besides the claims from Pierre Bourdieu.[7] Figure 3.2 is the summary I synthesized from these scholars' books. To be convenient, I also copied the corresponding content in Cases 3.3 and 3.4. As shown in figure 3.2, the scholars kept essentially the same attitude on capital classification, although having conducted their research independently.

[7] *See:* Bourdieu, P. The Forms of Capital. In: Richardson JG. *Handbook of theory and research for the sociology of education*. Westport, CT: Greenwood Press; 1986: 241-258.

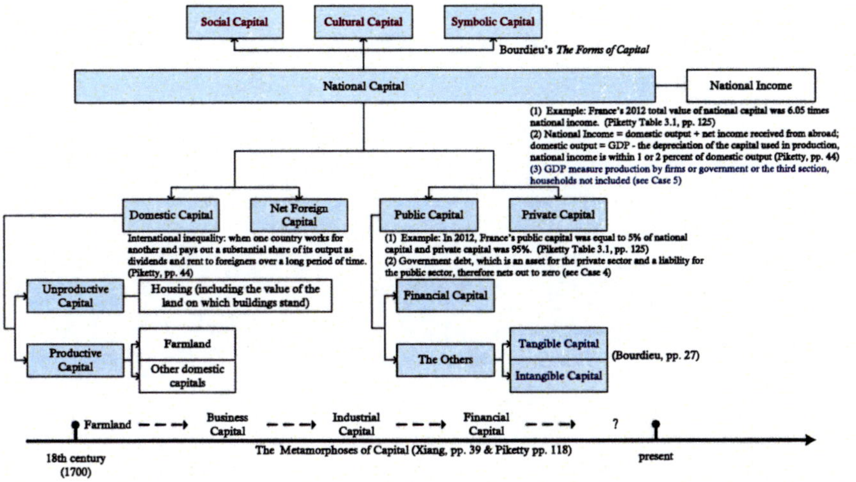

Figure 3.2 Capital classification described in economic literature. (Source: Author summarized to draw this by herself)

Case 3.3: Some Paragraphs Related to Capital Classification from Haskel and Westlake's *Capitalism Without Capital: The Rise of the Invisible Economy*

The Rise of the Invisible Economy

- What are Investment, Assets, and Capital?

… National accounts measure *production* by firms or government or the third sector. Production by households (say, doing the laundry or cooking at home) is not included, and so neither is investment by a household, say, in a washing machine or stove. This is a definitional feature of the way national account are calculated, and it is one of the reasons people criticize GDP (not least because it is large, and because it excludes from the record a part of the economy that has historically been run primarily by women). Perhaps 1day "production" will have a broader definition in national accounts; for our purposes, most of the changes we describe in this book would, we believe, apply to the household sector as well as to so-called producers.

So, in this book when we talk about "investment" we are *not* talking about the buying or selling of pieces of paper on a stock market or households paying university tuition. Rather, we are talking about spending by business, government, or the third sector that creates a fixed (i.e., nonfinancial) asset, that is, resources spent that create a long-lived stream of productive services. We shall call such a fixed asset providing these long-lived productive services "capital." Because both capital and labor produce such productive services, economists refer to them as "factors of production."

Case 3.4: Some Paragraphs Related to Capital Classification from Piketty's *Capital in the 21 Century*

- What is Capital?

Nonhuman capital, which in this book I will call simply "capital," includes all forms of wealth that individuals (or groups of individuals) can own and that can be transferred or traded through the market on a permanent basis. In practice, capital can be owned by private individuals (in which case we speak of "private capital") or by the government or government agencies (in which case we speak of "public capital"). There are also intermediate forms of collective property owned by "moral persons" (that is, entities such as foundations and churches) pursuing specific aims…

- Capital and Wealth

Some definitions of "capital" hold that the term should apply only to those components of wealth directly employed in the production process. For instance, gold might be counted as part of wealth but not of capital, because gold is said to be useful only as a store of value. Once again, this limitation strikes me as neither desirable nor practical (because gold can be a factor of production, not only in the manufacture of jewelry but also in electronics and nanotechnology). Capital in all its forms has always played a dual role, as both a store of value and a factor of production. I therefore decided that it was simpler not to impose a rigid distinction between wealth and capital.

Similarly, I ruled out the idea of excluding residential real estate from capital on the grounds that it is "unproductive," unlike the "productive capital" used by firms and government: industrial plants, office buildings, machinery, infrastructure, and so on. The truth is that all these forms of wealth are useful and productive and reflect capital's two major economic functions. Residential real estate can be seen as a capital asset that yields "housing services," whose value is measured by their rental equivalent. Other capital assets can serve as factors of production for firms and government agencies that produce goods and services (and need plants, offices, machinery, infrastructure, etc. to do so). Each of these two types of capital currently accounts for roughly half the capital stock in the developed countries.

- The Metamorphoses of Capital in Britain and France

More precisely, remember that national capital, …, is defined as the sum of private capital and public capital. Government debt, which is an asset for the private sector and a liability for the public sector, therefore nets out to zero (if each country owns its own government debt). As noted in Chap. 2, national capital, so defined, can be decomposed into domestic capital and net foreign capital. Domestic capital measures the value of the capital stock (buildings, fi rms, etc.) located within the territory of the country in question. Net foreign capital (or net foreign assets) measures the wealth of the country in question with respect to the rest of the world, that is, the difference between assets owned by residents of the country in the rest of the world and assets owned by the rest of the world in the country in question (including assets in the form of government bonds).

Domestic capital can in turn be broken down into three categories: farmland, housing (including the value of the land on which buildings stand), and other domestic capital, which covers the capital of firms and government organizations (including buildings used for business and the associated land, infrastructure, machinery, computers, patents, etc.). These assets, like any asset, are evaluated in terms of market value: for example, in the case of a corporation that issues stock, the value depends on the share price.

- Public Wealth, Private Wealth

To be sure, the distinction between public and private capital changes neither the total amount nor the composition of national capital, whose evolution I have just traced. Nevertheless, the division of property rights between the government and private individuals is of considerable political, economic, and social importance.

I will begin, then, by recalling the definitions introduced in Chap. 2. National capital (or wealth) is the sum of public capital and private capital. Public capital is the difference between the assets and liabilities of the state (including all public agencies), and private capital is of course the difference between the assets and liabilities of private individuals. Whether public or private, capital is always defined as net wealth, that is, the difference between the market value of what one owns (assets) and what one owes (liabilities, or debts).

Concretely, public assets take two forms. They can be nonfinancial (meaning essentially public buildings, used for government offices or for the provision of public services, primarily in health and education: schools, universities, hospitals, etc.) or financial. Governments can own shares in firms, in which they can have a majority or minority stake. These firms may be located within the nation's borders or abroad. In recent years, for instance, so called sovereign wealth funds have arisen to manage the substantial portfolios of foreign financial assets that some states have acquired.

In particular, for the issue on "not distinguishing productive and non-productive capital," the results are surprisingly consistent (see the underlined sentences in Cases 3.3 and 3.4). And Bourdieu as a Marxist also suggested that, capital has to do with utility or abstract labor, and it is not antithetical but essential to the material economy (i.e., productive labor for achine and production lines).

Capital Classification Learned from the Internet

A quick glance at the Internet scarcely motivates scholarly exploration, since its immediate appearance evokes billboards, video games, and hypes. Some critics purport to look for information on the Internet is like trying to "drink from a firehose." Why should a serious-minded reader learning stuff summed up from a frustrating, overloaded virtual universe?

The answers are so simple that they may seem uncompelling:[8]

> *The Internet is an enormous, in some ways unique, academic resource. In many important respects, the Internet is new kind of library, electronic library of libraries, with text, pages, books, collections of books, and collections of collections. ... [If having convenient tools for search and augmented analytics,] the Internet is immediately accessible, quite easy to search, ... For some kinds of information, the Internet is much the quickest route; for other kinds, which are on the increase, the Internet is the only source available.*

Let's see an example. There exists an online post that translates Pierre Bourdieu's The Forms of Capital in a straightforward way, whose blogger called himself as "The Dangerous Maybe." He claimed that,[9]

> *I think this essay is incredibly important and insightful, but, like so many essays by French thinkers, it can be very difficult for an American audience to read. ... Along the way, I add in examples and thoughts of my own, which is why this is more of a weird collaboration of sorts than it is a serious translation.*

I think the blogger's "examples and thoughts" are really wonderful, and this post can be a supplement to Bourdieu's original essay; specially, about Bourdieu's most classic exposition of "capitalization and capital conversion." I listed a comparison of both in Case 3.5. The green highlighted segment in Case 3.5 given by the blogger is obviously a further addition to Bourdieu's opinion.

---

[8] ***See:*** Browner S. Literature and the Internet: A Guide for Students, Teachers, and Scholars. Abingdon, Oxon: Routledge; 1 edition (July 11, 2017). 2017, 191 pages.

[9] This article translated Pierre Bourdieu's *The Forms of Capital* in a straightforward way. ***See:*** The Dangerous Mabe. An American "Translation" of Bourdieu's "The Forms of Capital". Medium.com. April 17, 2019. Retrieved from https://medium.com/@mdowns1611/an-american-translation-of-bourdieus-the-forms-of-capital-166a95f84e16.

Case 3.5: A Comparison of Translating Bourdieu's Exposition of "Capitalization and Capital Conversion"

- On the website of Medium.com

Economic capital can be transformed into other types of capital (cultural, social, symbolic). However, it takes time and energy to make these transformations. … But the point is that if you have economic capital and if you want to transform it into a different type of capital, … *Capital capitalizes capital.* … you can just immediately buy whatever goods and services you need. … But there are times when economic capital isn't enough and you also need social capital (connections) in order to get the things you want. In these scenarios, it's not enough to just have money; you also have to know the right people.

… At the end of the day, economic capital is at the root of all other types of capital. They aren't entirely reducible to it, they do have a relative autonomy, but they do stem from it and primarily serve to secure and reproduce its power. Cultural, social and symbolic capital are the little minions of economic capital. These other sorts of capital only work effectively when they appear to have nothing to do with economic capital (money, profits, self-interest, etc.). Their main function, though a secret one, is to create a capitalism with a human face. In other words, they make it seem as though capitalist society isn't all about greediness, selfishness, exploitation and inequality. They simulate friendship, recognition of human integrity, respect for others, community and meaning where there really is none (the bourgeois class has a lot of things but not these). In capitalist society, everything begins and ends with economic capital, but such a bleak, cold, depressing truth must be concealed in order for things to go on "functioning"–enter cultural, social and symbolic capital.

… The conversion of capital into capital, the fact that one type of capital can be transformed into another type, is what lets the entire system of capital reproduce itself. The convertibility of capital is the main mechanism through which capital takes over society at large and works its way into every aspect of our lives. … Each conversion comes at the expense of one type of capital but also, more importantly, ensures that the system of capital itself remains in total power. We can identify the various types of capital based on what it takes to reproduce them, that is, transmit them to the next generation of the bourgeoisie. Economic capital is passed on in ways that differ from how cultural capital is inherited. This transmission involves both loss and concealment. The more capital is lost in transmission, the more concealed the capital transmitted is, and vice versa. In other words, the more capital is lost in its transmission, the less people notice the capital that actually gets transmitted.

… The bourgeoisie is always discovering new strategies for disguising the transmission of their arbitrary power. They mainly use various conversion strategies to pull this off, that is, they are always finding more secretive, cunning, elusive and camouflaged tactics to stay in power. These tactics usually involve converting one type of capital into another. The more difficult we make it for economic capital to be passed on and the more regulations we put on capital, the more capital spreads out into other areas or fields of society. If economic capital finds it hard to be economic capital, then it will convert itself into cultural capital. That way it can't be legally and politically policed as much as it can be in the form of economic capital. Converting itself into cultural capital allows economic capital to slip through the fingers of those who try to get a firm grip on it.

- On the website of Marxists.org

The different types of capital can be derived from economic capital, but only at the cost of a more or less great effort of transformation, which is needed to produce the type of power effective in the field in question. For example, there are some goods and services to which economic capital gives immediate access, without secondary costs; others can be obtained only by virtue of a social capital of relationships (or social obligations) which cannot act instantaneously, at the appropriate moment, unless they have been established and maintained for a long time, as if for their own sake, and therefore outside their period of use, i.e., at the cost of an investment in sociability which is necessarily long-term because the time lag is one of the factors of the transmutation of a pure and simple debt into that recognition of nonspecific indebtedness which is called gratitude. In contrast to the cynical but also economical transparency of economic exchange, in which equivalents change hands in the same instant, the essential ambiguity of social exchange, which presupposes misrecognition, in other words, a form of faith and of bad faith (in the sense of self-deception), presupposes a much more subtle economy of time.

In accordance with a principle which is the equivalent of the principle of the conservation of energy, profits in one area are necessarily paid for by costs in another (so that a concept like wastage has no meaning in a general science of the economy of practices). The universal equivalent, the measure of all equivalences, is nothing other than labor-time (in the widest sense); and the conservation of social energy through all its conversions is verified if, in each case, one takes into account both the labor-time accumulated in the form of capital and the labor-time needed to transform it from one type into another.

When searching the Internet,[10] I found that the online primary interest in capital classification lies in being intangible (e.g., intellectual capital), nonfinancial (e.g., meaning capital), and nondomestic (e.g., cross-border capital flows) (see Footnote 13). I listed these definitions as follows.

Definition 3.2 *Intellectual Capital* Intellectual capital is the intangible value of a business, covering its people (human capital), the value relating to its relationships (relational capital), and everything that is left when the employees go home (structural capital), of which intellectual property is but one component. It is the sum of everything everybody in a company knows that gives it a competitive edge.

Definition 3.3 *Meaning Capital* Meaning capital involves several forms of capital (e.g., spiritual, psychological, and experiential capital), which more directly connect to an individual or a collective sense of purpose and what emotions are.

Definition 3.4 *Cross-Border Capital Flows* Cross-border capital flows, at the global level, can be calculated as the sum of capital inflows or the sum of capital outflows, excluding financial derivatives but including flows related to international banking transactions.

Of course, the three categories of capital with their segments are all relevant to data capital I want to purpose. Some can be measured, and others may be replaced. In the following section, I will discuss what data capital is, and furthermore, what the emergence of data capital helps.

What Is Data Capital?

When researching the four S's of intangibles, Haskel and Westlake investigated "knowledge, data, information, and ideas" (see Footnote 4). They believe intangibles are more inclined to knowledge. However, in my opinion, this might not be correct. Coincidentally, I explained the similar problem in the first chapter of my prior book, which is published earlier than theirs—*The Data Industry: The Business and Economics of Information and Big Data*. So, I compared the content from two sources in Case 3.6.

In fact, as shown in the green highlighted segment in Case 3.6, Haskel and Westlake understood that intangibles are based on "analyzable data." But they put intangibles' definition at the knowledge level resulting in a difficult challenge they repeatedly emphasized in their book; that is, how to measure the

[10] There are several online links. ***See:*** (1) Esbjorn-Hargens S. Using the ten forms of capital for impact. January 20, 2019. Retrieved from https://thetransformseries.net/2019/01/20/ten-forms-of-capital; (2) Eight forms of capital and why they matter. August 11, 2016. The GoodLifeRevival. Com. Retrieved from http://thegoodliferevival.com/blog/eight-forms-of-capital; (3) The entry of "Intellectual Capital" on Wikipedia. Retrieved from https://en.wikipedia.org/wiki/Intellectual_capital; (4) James E, McLoughlin K, Rankin E. Cross-border capital flows since the global financial crisis. Retrieved from https://rba.gov.au/publications/bulletin/2014/jun/pdf/bu-0614-8.pdf.

Case 3.6: A Comparison of Describing Data Assets and Intangibles

- My book *The Data Industry: The Business and Economics of Information and Big Data* on data asset in 2016

As defined in academic, a standard asset has four characteristics: (1) it should have unexpired value, (2) it should be a debit balance, (3) it should be an economic resource, and (4) it should be future economic benefits. The US Financial Accounting Standards Board expends on this definition: "[assets are] probable future economic benefits obtained or controlled by a particular entity as a result of past transactions or events." Basically, by this definition, assets have two properties: (1) an economic property, in that an asset must be about to produce an economic benefit, and (2) a legal property, in that an asset must be controllable.

Our new common understanding is that the intellectual asset, as one of the three key components of intellectual capital, is a "special asset." This is based on the concept of intellectual capital introduced in 1969 by John Galbraith, … and later expanded by deductive argument due to Annie Brooking, Thomas Stewart, and Patrick Sullivan. In more recent years, the concept of intellectual asset was further refined to a stepwise process by … Max Boisot (1999) … theorized on the "knowledge asset;" by … George Stigler (2003) … added an "information asset;" and by … Tony Fisher (2009) suggested a "data asset" specification process that would closely follow the rules presented in the DIKW (data, information, knowledge, and wisdom) pyramid [appears in a 1988 address to the International Society for General Systems Research, by American organizational theorist Russell Ackoff] …

According to the ISO 27001:2005 standard, data assets as an important component of information assets , in that they contain source code, applications, development tools, operational software, database information, technical proposals and reports, job records, configuration files, topological graphs, system message lists, and statistical data.

We therefore want to treat data asset in the broadest sense of the term. … we … redefine the data asset as data exceeding a certain scale that is owned and controlled by a specific agent, collected from theagent's past transactions involved in information processes, and capable of bringing future economic benefits to the agent.

- Haskel and Westlake's book *Capitalism without Capital: The Rise of the Invisible Economy* on intangibles in 2018

The words "data," "information," and "knowledge" seem interchangeable. As Goodridge and Haskel (2016) point out, the UK Data Protection Act "controls how your personal information is used," that the UK Information Commissioner "promotes data privacy for individuals," and the Freedom of Information Act allows citizen to request publicly held datasets (all our italics). Romer (1991), when talking about intangibles, uses terms like "idea," "blueprints," and "instructions." The OECD talks about the "knowledge economy," while economists typically refer to "knowledge" that is embodied or disembodied. Meanwhile, in his masterful work on the Industrial Revolution, the economic historian Joel Mokyr divides "knowledge" into propositional and prescriptive (Mokyr 2002). How does all this fit together?

Let's start with data. Define two kind of data: raw records and transformed data. Raw records are raw data not yet cleaned up, formatted, or transformed not ready for analysis. … Transformed data has been cleaned up, formatted, combined, and/or structured such that it is suitable for some form of data analytics.

Turning to information, we can think of information as synonymous with transformed data. Shapiro and Varian (1998) take information to mean anything that can be digitized, thereby implicitly defining information as digitized data.

We define knowledge as connections made between pieces of information, supported by evidence, to form a coherent understanding, and knowledge is required to fully understand and interpret information. Knowledge can, therefore, include theories, hypotheses, correlations, or causal relationships observed from information constituted by analyzable data.

Joel Mokyr (2002) introduces a distinction between different types of knowledge, "propositional" and "prescriptive." Propositional knowledge includes science and discoveries: knowledge of nature and its properties. Prescriptive knowledge prescribes actions for the purposes of production, such as "recipes," "blueprints," or "techniques." … Mokyr's argument is that preindustrial, stop-and-start growth was founded on chance discoveries. Postindustrial steady growth was only possible due to discoveries being founded on propositional knowledge.

Tangible assets … consist of metal but also lots of knowledge, … It's helpful to think of "embodied" and "disembodied" knowledge. …

Other classifications of knowledge are "tacit" as opposed to "codified," meaning whether the knowledge is directed primarily toward a specific, practical objective or whether it is theoretical, with no particular application in mind. … Finally, "commercialized" knowledge is knowledge applied to a particular business end.

intangibles (because pure value of knowledge is hard to distinguish from the value-added assets). Meanwhile, such a definition cannot attribute financial capital to intangibles (see figure 3.2).

"A theory is the more impressive," Albert Einstein once said, "the greater the simplicity of its premises is, the more different kinds of things it relates, and the more extended is its area of applicability." This is also my original purpose of coining the term "data capital." For now, I will give some definitions based on the prior given meaning of data asset.

Definition 3.5 *Data Asset* A Data asset is the data asset as data exceeding a certain scale that is owned and controlled by a specific agent, collected from the agent's past transactions involved in information processes, and capable of bringing future economic benefits to the agent.

My prior assets' definition is based on a consensus in the data science community (see the introduction); that is, "accessing data trumps owning it" (see Footnote 22). According to this, the data asset is defined to belong to data producers. Thus, the definitions of data production and data producer are given as below.

Definition 3.6 *Data Production* Data Production is a process of combining various analyzable data inputs in order to make something for consumption (output). Something for consumption (output) here means data product and the others.

Definition 3.7 *Data Producer* A data producer is any agent (e.g., corporations, industry bodies, government specifies, individuals) who has the ability to pursue adding value to row data or other primary data products.

Definition 3.8 *Data Capital* Data capital, for a specific agent, can be calculated as the sum of add-valued data asset inflows or outflows.

I would say that not both concepts of the asset and capital are equivalent. Due to the special nature of the data, data capital must be accumulated by data asset flows.

As mentioned by Haskel and Westlake, tangibles also consist of metal but also lots of data; for example, manufacturing decision-making with digital twin technology. Manufacturing processes are becoming increasingly digital. Within some smart factories, cyber-physical systems monitor physical processes, create a virtual copy of the physical world, and make decentralized decisions. Digital twins are core to the operation of these systems (note that a digital twin is a digital replica of a physical entity whatever living or non-living). With the help of the digital twin, management and specialists can get detailed and demonstrative information about the economic and production consequences for different maintenance policy configurations. This is done by running various what-if scenarios where the user can vary different core parameters (e.g. maintenance policy, production plan, working schedule, etc.). It is also possible to change the characteristics of the line, if needed. Such services can be considered as data production and the value-added part of data assets that comes from the next manufacturing process and investment can attribute to data capital. In the same way, the financial category of capital also belongs to data capital because of the fully digital status in this field.

The data capital I introduced will cover almost all current types of capital, from tangibles' digital part to the financial category, and to intangibles. Given relevant analyzable data, the pure value of various economic resources is easy to distinguish from the value added by any production or investment to be calculated. One significance to propose data capital is to help reclassify the capital using scientific measurement.

A Nominal "Anchor" for Future International Monetary Coordination?

The pursuits of monetary economics are mostly a stable high-value currency. And for monetary stability to prevail, it is vital to examine the credibility of the monetary framework, including money and its functions (e.g., medium of exchange, store of value, and unit of account), regulation of money and associated financial institutions, and international aspects. We should know, there exists a great monetary stability period in the history of capitalism; that is, in the eighteenth and nineteenth centuries. And even, up to World War I, inflation (see Irving Fisher's definition; inflation is nominal growth with a real component taken out) was zero or close to it. This is because; money at that time didn't lose monetary bearings as an "anchor"—the gold standard. As Karl Polanyi wrote in his renowned book *The Great Transformation: The Political and Economic Origins*:[11]

> *Nineteenth-century civilization rested on four institutions. The first was the balance-of-power system which for a century prevented the occurrence of any long and devastating war between the Great Powers. The second was the international gold standard which symbolized a unique organization of world economy. The third was the self-regulating market which produced an unheard-of material welfare. The fourth was the liberal state. ..., two of these institutions were economic, two political. ..., two of them were national, two international. Between them they determined the characteristic outlines of the history of our civilization.*
>
> *Of these institutions the gold standard proved crucial; its fall was the proximate cause of the catastrophe. By the time it failed, most of the other institutions had been sacrificed in a vain effort to save it. But the fount and matrix of the system was the self-regulating market. It was this innovation which gave rise to a specific civilization.*
>
> *The gold standard was merely an attempt to extend the domestic market system to the international field; the balance-of-power system was a superstructure erected upon and, partly, worked through the gold standard; the liberal state was itself a creation of the self-regulating market. The key to the institutional system of the nineteenth century lay in the laws governing market economy.*

[11] I reference directly here from the original sentence published in Karl Polanyi's book. ***See:*** Polanyi K. The Great Transformation: The Political and Economic Origins of Our Time. Boston, MA: Beacon Press; 2 edition (March 28, 2001). 2001: 364 pages.

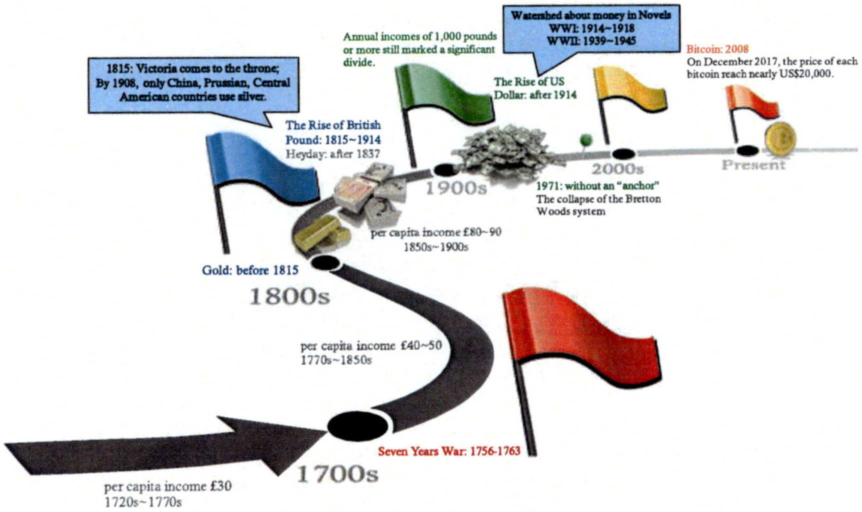

Figure 3.3 Monetary Bearings: From the gold standard to unanchored monetary era, and to bitcoin. (Source: The author drew by herself according to Piketty's book and other materials)

I will begin by reviewing the history of money from the eighteenth century to the era of unanchored monetary (or fiat money). Then I am going to concentrate on the disadvantages of anchorless.

Past, Present, and Future of Monetary Bearings

According to Piketty's conclusion (see Footnote 6) (on monetary from European and American literary classics in the eighteenth and nineteenth centuries) and other materials, I drew figure 3.3 along a timeline. I will have a lot more to say about this figure.

Let's look at the colored parts in figure 3.3. A purple highlighted line locates at a time point of August 15, 1971. The date was the collapse of the Bretton Woods system, of which a brief history is below.[12]

> *Established in 1944 and named after the New Hampshire town where the agreements were drawn up, the Bretton Woods system created an international basis for exchanging one currency for another. It also led to the creation of the International Monetary Fund (IMF) and the International Back for Reconstruction and Development, now known as the World Bank. The former was designed to monitor exchange rates and lend reserve currencies to nations with trade deficits, the latter*

[12] I reference directly here from an article in the Time. ***See:*** Stephey MJ. A brief history of Bretton Woods system. Time.com October 21, 2008. Retrieved from http://content.time.com/time/business/article/0,8599,1852254,00.html.

to provide underdeveloped nations with needed capital, ... Each of the 44 nations who joined the discussions contributed a membership fee, of sorts, to fund these institutions; the amount of each contribution designated a country's economic ability and dictated its number of votes.

In an effort to free international trade and fund postwar reconstruction, the member states agreed to fix their exchange rates by tying their currencies to the U.S. dollar. American politicians, meanwhile, assured the rest of the world that its currency was dependable by linking the U.S. dollar to gold; US$1 equaled 35 oz. of bullion. Nations also agreed to buy and sell U.S. dollars to keep their currencies within 1% of the fixed rate. And thus the golden age of the U.S. dollar began.

Though it came on the heels of the Great Depression and the beginning of the end of World War II, the Bretton Woods system addressed global ills that began as early as the first World War, when governments (including the U.S.) began controlling imports and exports to offset wartime blockades. This, in turn, led to the manipulation of currencies to shape foreign trade. Currency warfare and restrictive market practices helped spark the devaluation, deflation and depression that defined the economy of the 1930s.

The Bretton Woods system itself collapsed in 1971, when President Richard Nixon severed the link between the dollar and gold — a decision made to prevent a run on Fort Knox, which contained only a third of the gold bullion necessary to cover the amount of dollars in foreign hands.

The collapse of the Bretton Woods system led the last connection between monetary and material commodities also destroyed. As a result, most major world economies had allowed their currencies to float freely against the dollar. With increased flexibility of exchange rates, currency competition has become increasingly fierce, and sovereign, regional, and international currencies coexist. More importantly, unanchored monetary or fiat money has paved the way for financial capital to replace industrial capital.

Second is the rise and fall of the British pound and the rise of the US dollar. Someone comments on the economic history of European that is "currency war history." The answer to this neither yes, nor no. This is what the "balance-of-power system" mentioned by Polanyi is. In other words, if a superpower emerges in a region or the world, its currency must play a dominant role in that place. The French Empire, for example, was the dominant power in much of continental Europe before the middle of the eighteenth century. But after the Seven Years War, British replaced it.[13]

The Seven Years War was the nearest thing the eighteenth century had to a world war. Like the global conflicts of the twentieth century, it was at toot a European war. Britain, France, Prussia, Austria, Portugal, Spain, Saxony, Hanover, Russia and Sweden were all combatants. But the fighting raged from Coromandel to Canada, from Guinea to Guadeloupe, from Madras to Manila. Indians, native Americans, African slaves and American colonists all became involved. At stake was the future of empire itself. The question was simply this: Would the world be French or British?

... By 1743 it had been possible to speak of "the British Empire, taking together as one body, via. Great Britain, Ireland, the Plantations and Fishery in America, besides its Possessions in the East Indies and Africa." ... "this vast empire on which the sun never sets and whose bounds nature has not yet ascertained."

[13] ***See:*** Ferguson N. Empire: The Rise and Demise of the British World Order and the Lessons for Global Power. New York, NY: Basic Books; Reprint edition (April 14, 2004). 2004: 400 pages.

The international monetary system, with the British pound sterling at the core, began to take shape. In 1868, only British with its few dependent counties (e.g., Portugal, Egypt, Canada, Intellect, and Australia) adopted the gold standard. France with the "Latin Monetary Union," as well as Russia, Prussia, and all the Latin American nations used bimetallism or the double standard; and the rest is the sliver standard. By 1908, in addition to China, Prussia, and a few Central American countries continued to implement the silver standard, the gold standard dominated the monetary system which is actually to use British pound.

As Piketty's conclusion based on European and American literary classics, in the eighteenth century, per capita income grew very slowly; that is (see Footnote 6):

> *... money was everywhere, not only as an abstract force ... Writers frequently described the income and wealth of their characters in francs or pounds, not to overwhelm us with numbers but because these quantities established a character's social status in the mind of the reader. Everyone knew what standard of living these numbers represented.*
>
> *... In Great Britain, the average income was on the order of 30 pounds a year in the early 1800s, when Jane Austen wrote her novels. The same average income could have been observed in 1720 or 1770. Hence ... to live comfortably and elegantly, one needed ... at least twenty to thirty times that much. ... in the 1850s, the average income was barely 40~50 pounds a year. ... By the turn of the twentieth century, the average income in Great Britain had risen to 80~90 pounds a year. The increase was noticeable, but annual incomes of 1,000 pounds or more – the kind that Austen talked about – still marked a significant divide.*
>
> *... One could easily multiply examples by drawing on [French,] American, German, and Italian novels, as well as on the literature of all the other countries that experienced this long period of monetary stability. Until World War I, money had meaning, and ...*
>
> *This world collapsed for good with World War I. To pay for this war of extraordinary violence and intensity, to pay for soldiers and for the ever more costly and sophisticated weapons they used, governments went deeply into debt. As early as August 1914, the principal belligerents ended the convertibility of their currency into gold. After the war, all countries resorted to one degree or another to the printing press to deal with their enormous public debts. Attempts to reintroduce the gold standard in the 1920s did not survive the crisis of the 1930s: Britain abandoned the gold standard in 1931, the United States in 1933, France in 1936. The post– World War II gold standard would prove to be barely more robust: established in 1946, it ended in 1971 when the dollar ceased to be convertible into gold.*
>
> *It is surely no accident that money – at least in the form of specific amounts – virtually disappeared from literature after the shocks of 1914~1945. Specific references to wealth and income were omnipresent in the literature of all countries before 1914; these references gradually dropped out of sight between 1914 and 1945 and never truly reemerged. ... The novels of Orhan Pamuk, set in Istanbul in the 1970s, that is, in a period during which inflation had long since rendered the meaning of money ambiguous, omit mention of any specific sums. In Snow, Pamuk even has his hero, a novelist like himself, say that there is nothing more tiresome for a novelist than to speak about money or discuss last year's prices and incomes. The world has clearly changed a great deal since the nineteenth century.*

After the outbreak of World War I in 1914, the British pound began to decline significantly whose status as an international currency was quickly replaced by the US dollar. While arguing in the exact time point of this replacement, we consider 1914 as a watershed year. Put in another word, we believe that the period from 1815 to 1914 is the British pound era, in which the heyday is from 1837 Victoria comes to the throne after the death of William IV.

By the end of World War II, the US dollar had undoubtedly replaced the British pound. One reason lies in multiple aspects of the US's institutional innovation. This has dramatically improved the productivity and per capita income leaving the UK and other countries far behind. The US's roads to the institutionalization of modern science and technology the UK never tried. By 1946, there were four scientific jobs for every 1000 employees in the US manufacturing sector, four times as many as in the UK.[14] Besides, American corporates are particularly close to universities and research intuitions.

The third is about currencies in cyberspace. Rife with complicated terminology, the money in the virtual world of can be very confusing. It will be useful, to begin with, the clarified concepts, and then case study.

Definition 3.9 *Digital Currency* The digital currency is the blanket term used to describe all digital or electronic, and intangible money, which includes both virtual currency and cryptocurrency. It can be regulated or unregulated and is used online.

In February 2020, the head of IMF highlighted digital currencies as a top priority for its 189 member countries, imploring central banks or big tech firms to embrace the emerging technology.[15] For me, I am more interested in Libra proposed by Facebook.[16] While no being decentralized, it may help data productivity.

Definition 3.10 *Virtual Currency* Virtual currencies are a type of digital currency, typically controlled by its creators and used and accepted among the members of specific virtual community.

Virtual currencies are sometimes represented in terms of tokens and may be unregulated without a legal tender such as coins or banknotes. A famous virtual currency called the Linden dollar within a three-dimensional (3D) online game *Second Life*[17] was developed by Linden Labs. Users of the game, known as residents, can pay real money (e.g., the US dollar) to acquire Linden dollars. In 2015 alone, Second Life users had cashed out approximately US$ 60 million and Second Life had an approximate GDP of US$ 500 million, higher than some small countries.

Definition 3.11 *Cryptocurrency* "Crypto" in cryptocurrency refers to the fact that many encryption algorithms and cryptographic techniques are used to ensure

---

[14] ***See:*** Xiang, S. The New Capitalism: Rising, Crisis and Future of the Global Financial Capitalism. Beijing: China Citic Press; 1 edition, December 1, 2014, 616 pages.

[15] ***See:*** Head of IMF Highlights Digital Currencies As Top Priority in 2020 for Its 189 Member Countries. TheDailyHodel.com. January 16, 2020. Retrieved from https://dailyhodl.com/2020/01/16/head-of-imf-highlights-digital-currencies-as-top-priority-in-2020-for-its-189-member-countries.

[16] ***See:*** The entry of Libra (digital currency) on Wikipedia. Retrieved from https://en.wikipedia.org/wiki/Libra_(digital_currency).

[17] ***See:*** Maiberg E. Why Is Second Life Still a Thing? Motherboard. April 29, 2016. Retrieved from https://www.vice.com/en_us/article/z43mwj/why-is-second-life-still-a-thing-gaming-virtual-reality.

security across the network. This level of security also makes cryptocurrencies hard to counterfeit.

Bitcoin and Ethereum (both considered as virtual currencies) are the two most well-known and widely used blockchain-based (decentralized) cryptocurrencies. Unlike cash that is entirely anonymous, the two can be traced on the blockchain while initially no knowing participants' identities.

One another note, all world economies just treat cryptocurrencies as financial products like commodities, without any monetary property. Cryptocurrencies thus tend to be characterized by price volatility because their value is solely on supply and demand. A developer on May 22, 2010, for example, bought two pizzas using 10,000 bitcoins, which once reached US$20 million at the end of 2018.[18]

There exists an online post from bitcoin's proponents.[19] The proponent claimed that bitcoin is in fact backed something. These include scarcity (given by bitcoin's finite supply of 21 million), divisibility (the same as the properties of any form of other money), and credibility (from a pre-existing form of real money). He also listed an example of the US dollar as a contradiction. I will discuss it as Case 3.7.

Despite claiming to aware of "first principle" (see the green highlighted segment), the proponent gave some inductive inferences; that is, his expectations go beyond the observed he made. For example, the observation that bread of a certain appearance has thus far been nourishing seems to justify the expectation that the next similar piece of bread he eats will also be nourishing, as well as the claim that bread of this sort is generally nourishing. His three points are shown as the yellow highlighted segments. Next, I am going to talk about the underlined section. Regarding scarcity, a criticism of bitcoin[20] is "the top 100 bitcoin addresses control 17.3% of all the issued currency," or "1000 people who own 40% of the market;" meanwhile, "a limit and finite supply of bitcoin of 21 million" may be a lie. According to the Marxist labor theory of value, the dollar is credible relative to productivity, rather than the amount of dollar-denominated debt that exists. Decentralization means that the economy cannot produce efficiently and is not equivalent to be non-manipulable. The statistics succeed in showing a cryptocurrency worth US$2 billion can be hacked (by 51% attacks)[21] profitably with as little as US$1.5 million. In addition, as Friedrich F. von

[18] ***See:*** Wong JL. Eight years ago today, someone bought two pizzas with bitcoins now worth $82 million.Quartz.May22,2018.Retrievedfromhttps://qz.com/1285209/bitcoin-pizza-day-2018-eight-years-ago-someone-bought-two-pizzas-with-bitcoins-now-worth-82-million.

[19] ***See:*** Lewis P. Bitcoin is not backed by nothing. Unchained-Capital.com. September 27, 2019. Retrieved from https://unchained-capital.com/blog/bitcoin-is-not-backed-by-nothing.

[20] ***See:*** Kharif O. The bitcoin whales: 1,000 people who own 40 percent of the market. Bloomberg Businessweek. December 8, 2017. Retrieved from https://www.bloomberg.com/news/articles/2017-12-08/the-bitcoin-whales-1-000-people-who-own-40-percent-of-the-market;.

[21] ***See:*** Varshney N. Here's how much it costs to launch a 51% attack on PoW cryptocurrencies. TheNextWeb.com. May 30, 2018. Retrieved from https://thenextweb.com/hardfork/2018/05/30/heres-how-much-it-costs-to-launch-a-51-attack-on-pow-cryptocurrencies.

Case 3.7: An Online Post to Support "Bitcoin Is Not Backed by Nothing" on Unchained-Capital.com

The value of the dollar did not emerge on the free market. Instead, it emerged as a fractional representation of gold (and silver initially). Essentially, the dollar was a solution to the inherent limitations in the convertibility and transferability of gold; its inception was dependent on the monetary properties of base metals, rather than properties inherent in the dollar itself. It was also initially a system based on trust: accept dollars and trust that it could be converted back to gold at a fixed amount in the future. … without gold, the dollar would have never existed in its current construct.

Over the course of the twentieth century, the dollar transitioned from a reserve-backed currency to a debt-backed currency. While most people never stop to consider why the dollar has value in the post gold era, the most common explanation remains that it is either a collective hallucination (i.e. the dollar has value simply because we all believe it does), or that it is a function of the government, the military, and taxes. Neither explanation has any basis in first principles, nor is it the fundamental reason why the dollar retains value. Instead, today, the dollar maintains its value as a function of debt and the relative scarcity of dollars to dollar-denominated debt. In the dollar world, everything is a function of the credit system. Nominal GDP is functionally dependent on the size, and growth of the credit system, and taxes are a derivative of nominal GDP. The mechanisms that fund the government (taxes and deficit spending) are both dependent on the credit system, and it is the credit system that allows the dollar to function in its current construct.

The problem for the Fed's economy (and the dollar) is that it depends on the functioning of a highly leveraged credit system. And in order to sustain it, the Fed must increase the amount of base dollars. This is what quantitative easing is and why it exists. In order to sustain the amount of debt in the system, the Fed has to systematically increase the supply of actual dollars, otherwise the credit system would collapse. Increasing the amount of base dollars has the immediate effect of deleveraging the credit system, but it has the longer-term effect of inducing more credit. It also has the effect of devaluing the dollar gradually over time. This is all by design. Credit is ultimately what backs the dollar because what the credit actually represents is claims on real assets, and consequently, people's livelihoods. Come with dollars in the future or risk losing your house is an incredible incentive to work for dollars.

The relationship between dollars and dollar credit keeps the Fed's game in play, and … this can go on forever. And while the dynamics of the credit system create relative scarcity of the dollar, it is also what ensures dollars will become less and less scarce on an absolute basis.

As is the case with any monetary asset, scarcity is the monetary property that backs the dollar, but the dollar is only scarce relative to the amount of dollar-denominated debt that exists. And it now has real competition in the form of bitcoin. … Dollar scarcity is relative; bitcoin scarcity is absolute. The dollar system is based on trust; bitcoin is not. The dollar's supply is governed by a central bank, whereas bitcoin's supply is governed by a consensus of market participants. The supply of dollars will always be wed to the size of its credit system, whereas the supply of bitcoin is entirely divorced from the function of credit. … Ultimately, bitcoin's monetary properties are emergent and increasingly unmanipulable, whereas the dollar is inherently and increasingly manipulable.

Wieser showed in his opportunity cost theory, the cost of a commodity also depended on its subjective, or psychological, value. Such reasoning brought into playing psychological rather than only material and monetary factors in economic discussions. This is why, *Second Life* still has dedicated users in 2020.[22]

Of course, this proponent's post has some merit. He raised a question about the "anchor" of stability confidence related to price (or currency).

[22] ***See:*** Buscemi J. Second Life still has dedicated users in 2020. Here's what keeps them sticking around. Mic.com. February 16, 2020. Retrieved from https://www.mic.com/p/second-life-still-has-dedicated-users-in-2020-heres-what-keeps-them-sticking-around-18693758.

A book review of *Reinventing Capitalism in the Age of Big Data* from MIT concluded that:[23]

> *... Authors Viktor Mayer-Schönberger and Thomas Ramge argue that the data's rise means money's decline, that meaningful economic growth overwhelmingly depends on data innovation, and that regulating market competition requires rethinking data access.*
>
> *"In data-rich markets," they write, "participants no longer use price as the primary conveyor of information ... one of the central tasks that money has played in the economy will be gone." Money's role will erode to the point where "(r)ather than equating markets with money and the economy with finance capitalism in which money rules supreme, markets will be understood to surge because of rich data flows (not money). Finance capitalism will be as old-fashioned as Flower Power."*

While agreeing with this, I don't think the next scene is like this. I envision there may exist a transition period before the future scenarios. So, lowering a “data” anchor at the sea of “currencies” may be more realistic. I will give an example from the stock market to explore this.

Stock Market Is Not Just Backed by Business Achievement

Some “chicken soup” articles often admonish investors to identify “good companies” built for growth as one popular strategy that can be set to make money in the stock market. These companies are expected to increase their profits (or revenues) at a faster-than-average pace, enabling the investor to earn big returns through capital appreciation. They use the story we heard growing up about the tortoise and the hare to give an example. The hare is a typically supercharged growth stock that has soaring earnings, even faster price gains, and can surge with a master move in just several months. The slow and steady tortoise, given enough time, can outperform even the quickest hares. Which better? People are suggested to consider a blended approach from two noted investors: Warren Buffett (Buffett's investment methodology falling within “founder centrism”) and William J. O'Neil (O'Neil invented the CAN SLIM strategy, a formula for determining which stocks were likely to grow). But earlier, when researching transaction data in my Ph.D. phase, I realized that watching a stock not only is about such a definition of “good companies” but also needs another two: information (e.g., news, gossip) and players' behavior (i.e., actor behavior such as buyers, sellers, dealers, brokers, and market makers).

Data Source 3.1 *Market Data* Market data is price and trade-related data for a financial instrument (e.g., equities, fixed-income products, derivatives, and currencies) reported by a trading venue such as a stock exchange.

---

[23] This article is a recommendation for *Reinventing Capitalism in the Age of Big Data*. ***See:*** Schrage MD. From money to data. Finance & Development, 2018;55(1):58.

Table 3.1 An illustration of Day-by-Day data

Day / Behavior object	D_1	D_1	…	D_i	…	D_n
O_1	$\{b_1, b_3, b_7\}$	$\{b_1, b_3\}$	…	$\{b_2, b_6\}$	…	$\{b_7\}$
O_2	$\{b_6, b_8\}$	$\{b_1, b_3\}$	…	$\{b_1, b_4\}$	…	$\{b_5, b_9\}$
…	…	…	…		…	…
O_j	$\{b_1, b_3\}$	$\{b_3\}$	…	$\{b_5, b_9\}$	…	$\{b_1, b_3, b_7\}$
…	…	…	…		…	…
O_m	$\{b_2\}$	$\{b_1, b_3\}$	…	$\{b_6\}$	…	$\{b_2, b_7\}$

Data Source 3.2 *Day-by-Day Data*[24] Day-by-Day data is recorded in the computer systems according to human routine rhythm, such as stock trading on each closing price in a week as $\{C_1, C_2, \ldots, C_7\}$. Table 3.1 shows the day-by-day behavior data, namely a collection of behaviors which include each person's daily behavior. Trade-related data for a financial instrument reported by a trading venue such as a stock exchange. Market data allows traders and investors to know the latest price and see historical trends for instruments such as equities, fixed-income products, derivatives, and currencies.

There exists a mysterious connection between news and stocks. A previous example was "Japan's Fukushima Daiichi nuclear disaster (in 2011)" caused the US stock market to climb 18% while insurance companies declined sharply in value, and the recent one results from the COVID-19 (novel coronavirus) epidemic.[25] In general, traditional news organizations more reliable because their business model is to paint the clearest picture of the world that they can manage. But in the post-trust era, we know that any news source can steer us wrong at times, and they're likely all jumbled together in our new feed anyway.

Example 3.2 On *MarketWatch*, a subsidiary of Dow Jones & Company along with *The Wall Street Journal* and *Barron's*,[26] produced two inconsistent opinions from

[24] ***See:*** Xiong Y, Zhu Y. Mining peculiarity groups in day-by-day behavioral datasets. In 2009 Ninth IEEE International Conference on Data Mining. December 2009, pp. 578-587.

[25] I integrated all references related to the COVID-19 epidemic. ***See:*** (1) The entry of "2020 Stock Market Crash" on Wikipedia. Retrieved from https://en.wikipedia.org/wiki/2020_stock_market_crash; (2) The entry of " Socio-Economic Impact of the 2019~2020 Coronavirus Pandemic" on Wikipedia. Retrieved from https://en.wikipedia.org/wiki/Socio-economic_impact_of_the_2019%E2%80%9320_coronavirus_pandemic; (3) Reinicke C. Dow slips 2,000 points as Trump coronavirus response disappoints, Fed funding boost fades. BusinessInsider.com, May 12, 2020. Retrieved from https://markets.businessinsider.com/news/stocks/stock-market-news-today-index-reaction-coronavirus-trump-travel-ban-2020-3-1028990017.

[26] ***See:*** (1) Maierhofer S. Opinion: The odds of a 2008-like stock-market meltdown are low. MarketWatch.com. March 6, 2020. Retrieved from https://www.marketwatch.com/story/the-odds-of-a-2008-like-stock-market-meltdown-are-low-2020-03-06; (2) Maierhofer S. Opinion: This is how big bear-market rallies have been in the past nine decades. MarketWatch.com. April 1, 2020. Retrieved from https://www.marketwatch.com/story/heres-how-big-of-a-bear-market-rally-in-stocks-you-can-expect-according-to-nine-decades-of-history-2020-04-01?mod=home-page.

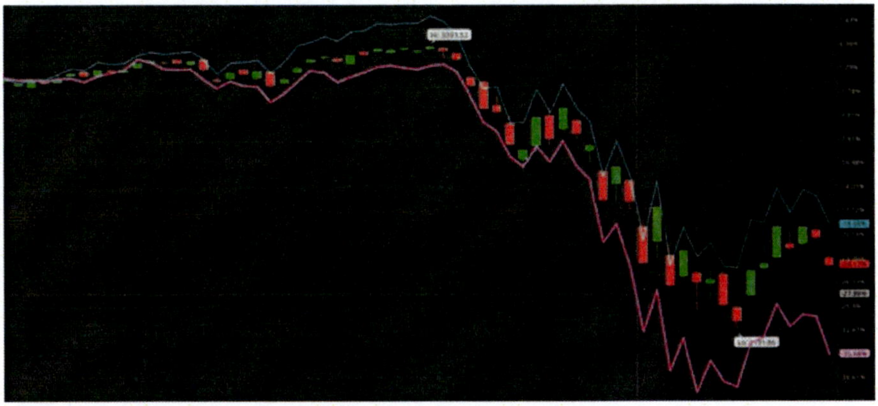

Figure 3.4 The US stock market's four meltdowns within 10 days during the COVID-19 epidemic. (Source: Author obtained this screenshot by herself)

the same stock analyst in 1 month. The new one is expected to change because of showing no sign of a rebound for a while (see figure 3.4).

March 6, 2020	April 1, 2020
Over the past two weeks, the stock market has had some of its biggest drops since the 2008 financial crisis. Investors are asking whether this decline will turn into a [a bear market].	From top to bottom, the S&P 500 Index (Standard & Poor's 500) has fallen 35% in this bear market. Here are the two most common questions I receive in my inbox: Is the bear market over? If it is not, can there be a bear market rally?
[No], I think that while the stock market could remain volatile for the next month, it will be the time to buy for those with a 12-month horizon.	In my humble opinion, the bear market may already be over, ...

The news to calculate the patterns and cycles of the past is much better. One and a half months before outbreaks reported in the US, for example,[27] a *CNBC* (Consumer News and Business Channel) investigation based on past virus scares estimates "stocks may have more to lose." This estimation was given via looking back 20 years, as shown in Table 3.2, previous epidemics from SARS (Severe acute respiratory syndrome) in 2003 to the Ebola scare 6 years ago shaved 6–13% off the S&P 500 over different lengths of time. There was a lot decline in the S&P 500 index from the day of news released (February 18, 2020) to of opinion given (April 1, 2020).

[27] ***See:*** Li Y. Market reactions to past virus scares show stocks may have more to lose. CNBC: Consumer News and Business Channel. January 28 2020. Retrieved from https://www.cnbc.com/2020/01/28/market-reactions-to-major-virus-scares-show-stocks-have-more-to-lose.html.

Table 3.2 Market reactions during virus emergencies: S&P 500 performance over the course of each outbreak

Virus	Data range	Trading days	S&P 500 change
SARS	January 2003–March 2003	38	−12.8%
Avian Influenza	January 2004–August 2004	141	−6.9%
MERS	September 2012–November 2012	43	−7.3%
Ebola	December 2013–February 2014	23	−5.8%
Zika	November 2015–February 2016	66	−12.9%
Coronavirus	January 2020–Present (02/28/2020)	5	−2.6%

As claimed by Stephen A. "Steve" Ross (February 1944–March 2017, best known for the development of the arbitrage pricing theory that is a general theory of asset pricing), detecting such a mysterious connection between news and stocks sometimes requires an intuition. The intuition is the combination of guesswork (often based on experience) and imagination (or wishful thinking). He even said:[28]

> *The intuition and the theory of finance are coconspirators. The theory is less a formal mathematical structure driven by its own imperatives – although that is always a danger – as it is a handmaiden that attempts to bridge the gap between the intuitions and the data.*

He also harshly criticized that a hypothesize-and-test paradigm greatly diluted this intuition. History as a giant dataset could lead us to a better understanding of the relationships between entities, such as news and stocks, and finance and economics as well. General data mining tools are previously used to discover all correlations (e.g., dependency, association, aggregation, composition, inheritance); and now, Bayesian can help analyze causations based on previous correlation findings.

Analyzing the behavior of the actor is a part of behavioral economics. Behavioral economics is the study of finance and psychology, as it related to the decision-making process behind an economic outcome of individuals and institutions. Nobel prize in economics has already been awarded to some behavioral finance theorists, such as Daniel Kahneman, Richard H. Thaler, and so forth. For example, in Belief in the Law of Small Numbers, Kahneman and his longtime collaborator Amos Tversky laid out "eleven cognitive illusions" affecting the human judgment, which uses small-scale empirical experiments to demonstrate how subjects make irrational decisions under uncertain conditions. This work was highly influential in the field of economics and can be used to presume largely rationality of all actors. On the contrary, a growing interest of actor behavior in data science is due to the availability of vast amounts of data. Behavior analysis aims to detect common behavioral patterns in multivariate temporal stock data (e.g., day-by-day data), which can be dramatically different

[28] ***See:*** Ross SA. The interrelations of finance and economics: Theoretical perspectives. The American Economic Review, 1987;77(2), 29-34.

from the others and show their peculiarities. For instance, we try to figure out how actors are operating in the stock market; these behaviors include, but not limited to, opening a stock portfolio, pushing or pulling on stocks, and short selling for a profit or loss. Some play as the academicism, who has a solid theoretical foundation and classic trader tips. Some act in a pirate style, featuring as quick buy-in and forcible pull-high. Some seem to spend a long-time latency time between buying and selling. All players' operations have accidentally left clues in massive stock data. The similar trends among different stocks in various time periods can indicate similar behaviors of an abnormal group of actors determined by their psychologies.

Despite recent progress in analyzing news and actor behavior, I think it is not enough to only use the two types of data for understanding the stock market, whether using traditional mathematical models or data mining approaches. Therefore, the stock market is repeatedly "pricing in a bad scenario" when meeting disease outbreak news. With the popularity of social media as a news source, information overload caused by fake news and tricks is a problem apart from regular hype. In addition, some inside information-based transaction to profit is a big problem. After the US stock market four-time meltdown in March 2020, *The New York Times* exposed an incident that several senators sold major holdings in stocks when Trump administration was playing down coronavirus treat.[29]

> *The record of Richard Burr's stock transaction shows he and his wife sold 33 different stocks on Feb. 13 that were collectively worth US$628,000 to US$1.7 million, according to the disclosures filed with the secretary of the Senate. Those sales include as much as US$150,000 worth of stock in two hotel chains, Wyndham Hotels and Resorts and Extended Stay America. ... He also sold as much as US$65,000 worth of stock in Park Hotels & Resorts.*
>
> *Dianne Feinstein [Democrat of California, who is also a member of the Intelligence Committee] and her husband sold US$1.5 million to US$6 million worth of stock in Allogene Therapeutics, a California-based biotech company, in transactions that took place on Jan. 31 and Feb. 18.*
>
> *James M. Inhofe [Republican of Oklahoma] sold a large amount of stock — all on Jan. 27 — including holdings in PayPal, Apple and Brookfield Asset Management, a real estate company, with the overall value of the sales totaling as much as US$400,000, a disclosure report shows.*
>
> *Kelly Loeffler [Republican of Georgia] and her husband, Jeffrey C. Sprecher, who is the chairman of the New York Stock Exchange, reported 27 stock sales worth millions of dollars starting on Jan. 24. ... The stocks the couple sold were in companies including Exxon Mobil, Ross Stores and AutoZone.*

Hussein Sayed, the chief market strategist at FXTM, told *Business Insider* in an email when commenting this disclosure report:[30]

---

[29] ***See:*** Lipton E, Fandos N. Senator Richard Burr Sold a Fortune in Stocks as G.O.P. Played Down Coronavirus Threat. The New York Times. March 19, 2020. Retrieved from https://www.nytimes.com/2020/03/19/us/politics/richard-burr-stocks-sold-coronavirus.html.

[30] ***See:*** Reinicke C. Dow slips 2,000 points as Trump coronavirus response disappoints, Fed funding boost fades. BusinessInsider.com, May 12, 2020. Retrieved from https://markets.businessinsider.com/news/stocks/stock-market-news-today-index-reaction-coronavirus-trump-travel-ban-2020-3-1028990017.

Markets are in complete crisis mode, past economic data has zero influence on investors' decision, central bank emergency easing policies are not being effective and politicians' actions are only adding more confusion.

The one thing that investors are monitoring is how fast the coronavirus is spreading, how many lives it is taking and the number of countries and cities in complete lockdown. It literally feels like we are living in a science fiction movie.

These puzzles concerned by financial advisers or economists must approach them in multiple ways; specifically reaching across disciplinary divides;[31] for example,[32]

... In 2011, not long after the global financial crisis of 2008, Bank of England chief economist Andy Haldane worked with theoretical ecologist Robert May to harness infectious-disease modelling techniques to investigate risks and vulnerabilities in the global financial system. Their resulting Nature paper 2 showed researchers how they could collaborate to understand other complex networks, such as those involving trade or information. "I am continually surprised at how much impact that paper has had," says Haldane.

According to Google Scholar, this article has more than 1200 academic citations. Thus, I suggest anchoring the stock market with all available data, coving literature, healthcare, environmental, and other fields, and cross-border data flows included.

For the stock market under the COVID-19 epidemic:[33] global data on disease trajectories and the effectiveness and economic impact of different prevention and mitigation measures is needed to accurately simulate local scenarios. Meanwhile, data analysis based on multiple data sources is better than a single

[31] ***See:*** Economists and scientists: solve big societal problems by working together. February 26, 2020. Retrieved from https://www.nature.com/articles/d41586-020-00532-4.

[32] ***See:*** Haldane AG, May RM. Systemic risk in banking ecosystems. Nature, 2011;469(7330), 351-355.

[33] This is mainly the content from one of my own paper, so I integrated all references as follows. ***See:*** (1) Niiler E. An AI epidemiologist sent the first warnings of the Wuhan virus. Wired.Com. January 25, 2020. Retrieved from https://www.wired.com/story/ai-epidemiologist-wuhan-public-health-warnings; (2) Calling all coronavirus researchers: keep sharing, stay open. Nature. 2020;578:7. DOI: 10.1038/d41586-020-00307-x; (3) Chinazzi M, Davis JT, Ajelli M, *et al.* The effect of travel restrictions on the spread of the 2019 novel coronavirus (COVID-19) outbreak. Science. March 6, 2020:eaba9757. DOI: 10.1126/science.aba9757; (4) Apuzzo M, Rich M. Yaffe-Bellany D. Failures on the Diamond Princess Shadow Another Cruise Ship Outbreak. The New York Times. Retrieved from https://www.nytimes.com/2020/03/08/world/asia/coronavirus-cruise-ship.html; (5) Zhu N, Zhang D, Wang W, *et al.* A novel coronavirus from patients with pneumonia in China, 2019. New England Journal of Medicine. 2020;382:727-733. DOI: 10.1056/NEJMoa2001017; (6) HealthMap. Boston Children's hospital. Retrieved from https://www.healthmap.org/covid-19; (7) Donovan D. Map tracks coronavirus outbreak in near real time. JHU.edu. Retrieved from https://hub.jhu.edu/2020/01/23/coronavirus-outbreak-mapping-tool-649-em1-art1-dtd-health; (8) Miliard M. Epic pushes out software update to help spot coronavirus. Healthcare IT News. January 24, 2020. Retrieved from https://www.healthcareitnews.com/news/epic-pushes-out-software-update-help-spot-coronavirus; (9) Wang CJ, Ng CY, Brook RH. Response to COVID-19 in Taiwan: big data analytics, new technology, and proactive testing. JAMA, March 3, 2020, DOI:10.1001/jama.2020.3151; (10) Niehus R, Salazar PMD, Taylor A. *et al.* Quantifying bias of COVID-19 prevalence and severity estimates in Wuhan. China that depend on reported cases in international travelers. medRxiv, 2020, 13:2020.

source. News reports and other biosurveillance related data pointed to a cluster of pneumonia cases that a Canada firm called BlueDot identified as being an outbreak on December 31st, 2019 via an algorithm, a week before other global public health officials notified the public. To facilitate the dissemination of knowledge regarding COVID-19 during the outbreak, publishers are prioritizing review of and offering free, open access to relevant research findings. The global health security index (GHSI) encompasses disease prevention, detection, reporting, and response capabilities for each country. Countries with a higher GHSI like Singapore are capable of identifying undetected cases through increased epidemiological surveillance and contact-tracing, leading to a more accurate disease prevalence picture. Transportation data from airlines, and car traffic has been used to simulate the spread of a disease and estimate the effect of local and intercontinental travel restrictions. Outbreaks, such as on the Diamond Princess cruise ship, provided valuable information about how the disease is spread and its incubation period. HealthMap and online dashboard (produced by Boston Children Hospital and the Johns Hopkins Center for Systems Science and Engineering, respectively) continue to allow visualization of the global spread of COVID-19 over time. Electronic health record systems have augmented their self-reported travel screening questionnaires to help identify patients who have recently visited areas where community spread is present. Chin has integrated its health insurance database with immigration and customs enforcement data to generate real-time alerts based on travel history and clinical symptoms to aid in case identification and has used this data to decide whom to quarantine and track at the border. Policymakers, based on this, then, can make their decisions regarding travel bans, quarantines, and economic stimulus. These examples of data flow across geographic borders are extremely useful to not only public health professionals, such as for accelerating the pharmaceutical development pipeline, for triaging resources to a locale, and for making decisions about intensive care unit rooms, where to build temporary hospitals, or where to boost supplies of personal protection equipment or diagnostic tests, but also investors and financial practitioners—at least, for alleviating panic in an epidemic situation.

Leading Human Capital to Be Shareable

An economic study hosted by Capital Economics shows that, the pandemic may come for economies across the developing world after slamming developed economies in Asia, Europe, and North America, specifically,[34]

> *Economic output in emerging markets is forecast to fall 1.5% this year, the even if some developing nations manage to avoid catastrophic virus infection rates, the*

[34] ***See:*** Luhnow D and Pérez S. Santiago Emerging-Market Economies Brace for Coronavirus Hit. The Wall Street Journal. April 5, 2020. Retrieved from https://www.wsj.com/articles/emerging-market-economies-brace-for-coronavirus-hit-11586112224?mod=searchresults&page=1&pos=2.

lockdowns and expected recessions in industrialized countries will take a heavy economic toll. They likely will dent demand for beach holidays in Thailand, clothing stitched together in Bangladesh and auto parts and avocados from Mexico. ... Unemployment may become an even bigger problem than the virus.

In China, some companies start hiring "shared employees" to cope with labor market power imbalances during the outbreak of COVID-19. With more and more consumers take advantage of online shopping due to social distancing, most industry services suspended but essential businesses such as supermarkets, restaurants, and other establishments that sell ingredients or food and beverage products to the public are experiencing staff shortages. The development of such an employment model depends to a great extent on market demand. A shared employee is someone who works for more than one business at the same time. The businesses jointly coordinate, hiring, firing, salary determination, supervision, and other employer responsibilities. This idea of "shared employees" bates back to the 2008 crisis. Many developed countries encouraged "work sharing"[35] to reduce unemployment by having incorporated it into their unemployment insurance system during a recession. This permitted the payment of prorated benefits to workers who are kept on the job with reduced hours because of slack demand. By adopting work sharing, employers also can avoid the adverts effects that layoffs have on employee morale and productivity. Within actual activities and operations of the enterprise, however, hiring "shared employees" can cause some risks in human capital management.

Emerging data-driven human capital management (HCM) may help overcome challenges around stakeholder engagement and talent acquisition and management. According to MarketsandMarkets, which is a market research and consulting firm specializing in high growth markets, an estimation of global HCM market size grows from US$16.7 billion in 2019 to US$26.5 billion by 2024,[36] at a compound annual growth rate of 9.7% during the forecast period. When introducing the Hype Cycle for Human Capital Management Technology, 2019, Gartner's report identifies four key trends:[37]

---

[35] ***See:*** Abraham KG, Houseman SN. Proposal 12: Encouraging work sharing to reduce unemployment. Policies to Address Poverty in America, 2014:129. Retrieved from https://www.hamiltonproject.org/assets/legacy/files/downloads_and_links/work_sharing_abraham_houseman.pdf.

[36] ***See:*** Human Capital Management Market by Software (Core HR, Applicant Tracking System, HR Analytics, Workforce Management), Service, Deployment Type (cloud and on-premises), Organization Size, and Region - Global Forecast to 2024. October 25, 2019. Retrieved from https://www.marketsandmarkets.com/Market-Reports/human-capital-management-market-193746782.html.

[37] Since Gartner didn't permit any graphics from its branded research, interested readers can find the Hype Cycle for Human Capital Management Technology report, 2019 from primary sources. ***See:*** (1) Hype Cycle for Human Capital Management Technology, 2019. Garnter.com. July 23, 2019. Retrieved from https://www.gartner.com/en/documents/3953514/hype-cycle-for-human-capital-management-technology-2019; (2) Wiles J. 4 key trends in the Gartner Hype Cycle for Human Capital Management Technology, 2019. SmarterWithGartner.com. October 25, 2019. Retrieved from https://www.gartner.com/smarterwithgartner/4-key-trends-gartner-hype-cycle-human-capital-management-technology-2019.

Automation of the manager experience *"Importantly, these technologies are meant to reduce the amount of time managers spend on administrative tasks, not replace managers' role/authority by automation. One example is workforce management (WFM), a suite of functions designed to help manage hourly paid workers. Core functions of WFM include time and attendance, scheduling, absence management and task management. Next-generation WFM is the result of the following trends impacting the market: automation of the manager experience; employee experience; virtual assistants; new platforms; the flexible workforce."*

Voice of the employee (VoE) *"To measure and monitor employee engagement, an increasing number of organizations are using engagement data beyond formal large-scale surveys to create a more holistic 'voice of the employee.' VoE technologies, such as employee sentiment and social analytics tools, collect and analyze opinions, perceptions and feelings of employees and workers, providing a way to harness multiple sources of information to understand the dynamics of the employee experience; they are not meant to enable covert surveillance of employees." Further, "nevertheless, the potential benefit is high, and over the longer term, a VoE approach can contribute to improving an organization's overall employee experience and employment value proposition. Potential employees are more likely to be attracted to a work environment in which their voice is heard. VoE also has the potential to improve engagement, performance and productivity among existing employees over time."*

Analytics and AI in talent acquisition *"Gartner research shows steady demand from organizations to leverage these technologies, including capabilities across front-end talent processes such as sourcing, job targeting and distribution, and predictive analytics to improve time-to-hire and quality-of-hire metrics." Further, "As interest grows and the market matures, solutions are moving deeper into the hiring process to support both recruiters and applicants in more advanced activities. By expanding sourcing and screening capacity, as well as increasing levels of initial front-line responsiveness, talent acquisition teams can gain a broader reach and higher levels of candidate interaction."*

Rise of the internal talent marketplace *"The gig economy and the need for business agility have opened up new ways of working, challenging established notions of employment and changing worker expectations, and preferences around how individuals manage their careers and build their portfolio of work experiences." Further, "Internal talent marketplaces take advantage of the increased flexibility of the gig economy and marketplace-based platforms without requiring changes to employment categories. They match internal employees and, in some cases, a pool of contingent workers, to short-term project and work."*

Among these trends, hiring "shared employees" is the one; and starting from internal employees may be a good option.

As mentioned above, theoretically, data can be used to not only estimate but replace capital. Human capital should be no exception. Before presenting the specifically analytic process, it is best to proceed by stages. In the remainder of this section, I will begin by understanding the concept of human capital.

Is Human Capital Illusory?

Before discussing human capital, I will talk about what capital is used for. According to Piketty's simplest assumption, in a society in which capital is of no use in the production process, all of national income and output would go to labor; specifically (see Footnote 6),

> *no investment can increase the productivity of farmland, no tool or machine can increase output, and having a roof over one's head adds nothing to well- being compared with sleeping outdoors. Yet capital might still play an important role in such a society as a pure store of value: for example, people might choose to accumulate piles of food (assuming that conditions allow for such storage) in anticipation of a possible future famine or perhaps for purely aesthetic reasons (adding piles of jewels and other ornaments to the food piles, perhaps). ... the capital/income ratio β is quite high but the return on capital r is strictly zero. In that case, the share of capital in national income, $\alpha = r \times \beta$, would also be zero.*

Suppose further that in such a society where slaves represent a large share of the population. Put in another words, take an extreme case in which virtually an entire population is owned by a tiny minority. The sake of argument that the income from labor (that is, the yield to slave owners on the labor of their slaves) represents 100% of national income. Then, all labor in this society is related to slave capital. To explain this, Piketty claimed that (see Footnote 6),

> *There are many reasons for excluding human capital from our definition of capital. The most obvious is that human capital cannot be owned by another person or traded on a market (not permanently, at any rate). ... In any case, it is clear that this type of calculation makes sense only in a slave society, where human capital can be sold on the market, permanently and irrevocably.*

Interestingly, the simplest society assumed by Piketty in which capital is of no use in the production process seems to a portrayal of our modern society during the COVID-19 epidemic, if considering cross-border data flows via high-speed Internet service. That is, "no tool or machine can increase output" (face mask and ventilator no included), "having a roof over one's head adds nothing to well-being compared with sleeping outdoors," and "people might choose to accumulate piles of food" (with toilet paper). In that sense all of national income and output would go to labor and data capital within this society (because Piketty excluded human capital from the definition of capital).

Close to home, we will look at Marx's view on human capital:[38]

> *Marx adapted a distinction, in Hegel's Elements of the Philosophy of Right between labor power and labor and gave this distinction a new meaning. For Marx, labor power refers to ...: the physical ability of human beings and other living things to perform work, including mental labor and skills such as manual dexterity, in addition to sheer physical exertion. ... Labor power ... is ... the aspect of labor (Marx's usage of labor is somewhat similar to the later concept of "labor services" in ... [labor economics]), which becomes a commodity within capitalist society and is alienated from workers when it is sold to capitalists.*
>
> *The distinction between labor and labor power, according to Marx, helped to solved a problem that David Ricardo had failed to solve, i.e. explaining why the surplus value resulting from profit normally arises out of the process of production itself–rather than in the investment of capital (e.g. the advance of money-capital in the form of wages) in labor power (acquired from laborer).*

---

[38] ***See:*** (1) The entry of " Labour Power" on Wikipedia. Retrieved from https://en.wikipedia.org/wiki/Labour_power; (2) The entry of "Human Capital" on Wikipedia. Retrieved from https://en.wikipedia.org/wiki/Human_capital.

When comparing Marx's concept of labor power to that of human capital, we found that Marx himself may have considered a concept of human capital to be a "reification," the purpose of which was to imply that workers were a kind of capitalist. For instance, in Capital Vol. 2, Marx states:[39]

> *... economists ... say: ... [the worker's] labor power, then, represents his capital in commodity-form, which yields him a continuous revenue. Labor power is indeed his property (ever self-renewing, reproductive), not his capital. It is the only commodity which he can and must sell continually in order to live, and which acts as capital (variable) only in the hands of the buyer, the capitalist. The fact that a man is continually compelled to sell his labor power, i.e., himself, to another man proves, according to those economists, that he is a capitalist, because he constantly has "commodities" (himself) for sale. In that sense a slave is also a capitalist, although he is sold by another once and for all as a commodity; for it is in the nature of this commodity, a laboring slave, that its buyer does not only make it work anew every day, but also provides it with the means of subsistence that enable it to work ever anew.*

Piketty's view on human capital (that is equivalent to labor power) is obviously consistent with Marx. But labor economists do not think so.[40]

> *One of the most important ideas in labor economics is to think of the set of marketable skills of workers as a form of capital in which workers make a variety of investments. This perspective is important in understanding both investment incentives, and the structure of wages and earnings.*
>
> *Employee share ownership can mitigate this hold-up problem [(whereby both employers and employees exploit the bargaining weaknesses of the other)] because it can align interests, develop loyalty, signal good-will and lock in employees.*

The original idea of human capital can be traced back at least to Adam Smith in the eighteenth century, as shown in Figure 3.5. The modern theory was popularized by two Nobel Prize winners Gary Becker and Paul Romer. Becker's approach included altruistic behavior of human behavior by defining individuals' utility appropriately. Based on this, he realized the investment in workers was no different than investing in capital equipment. Romer founded the modern innovation-driven approach to understanding economic growth, in which his conceptualization and modeling working using human capital as a key factor. These theories explained how investing in education benefited people,

[39]I reference directly here from the original sentence published in Karl Marx's *Capital*. ***See:*** (1) Marx K (Author), Engels F (Introduction), Fowkes B (Translator). Capital Volume 1 (Das Kapital series Book 1). City of Westminster, London: Penguin; New Ed edition (February 5, 2004); 2004: 1134 pages; (2) Marx K (Author), Engels F (Introduction), Fowkes B (Translator). Capital Volume II (Das Kapital series Book 1). City of Westminster, London: Penguin; 2 edition (May 5, 2006); 2004: 601 pages; (3) Marx K (Author), Engels F (Introduction), Fowkes B (Translator). Capital Volume III (Das Kapital series Book 3). City of Westminster, London: Penguin; 3 edition (August 27, 1992); 1992: 1092 pages.

[40]There exists two references. ***See:*** (1) Acemoglu D, Autor D. Lectures in labor economics. Manuscript. 2011; 299 pages. Retrieved from https://economics.mit.edu/files/4689; (2) Whitfield K, Pendleton A, Sengupta S, Huxley K. Employee share ownership and organisational performance: A tentative opening of the black box. Personnel Review. 2017.

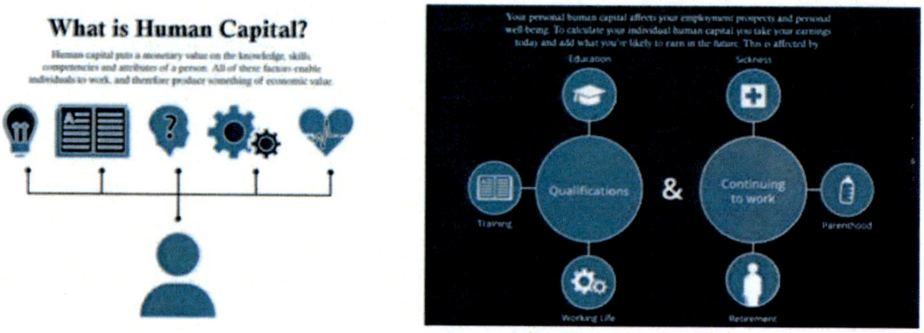

Figure 3.5 What is human capital. (Source: Author modified them according to some materials on the Internet)

companies, and countries. Piketty's recent research seems to interest in education, but he used it to study the ideology of inequality.

Finally, let me answer the question asked in the title of this section. Human capital is not illusory with the apparently growing importance. One reasonable interpretation is that technology has changed in a fundamental way that the labor factor now plays a greater role.

What Can Big Data do? Workforce Implications

Speaking of fundamental technological changes, digitalization and the Internet have dramatically revolutionized every aspect of our lives. I once described such changes in my prior book; that is, digitalization is a process that stores "information" that is generated in the real world from the human mind in digital form as "data" into cyberspace. No matter what types of new technologies emerge (e.g., the next generation mobile networks like 5G, Internet of things, cloud service, quantum computing, etc.), the data will stay the same. The data, or more fashionably, big data, is not merely about its *n*Vs (i.e., 5Vs as volume, velocity, variety, veracity, and value), but comprised of three critical elements as "data, technology, and application." Big data is first and foremost a technical terminology. It is superficial to argue whether a dataset is big enough to reach a petabyte (PB) or not. The essence of big data meaning is to investigate whether a data is valuable to analytics, or to obtain the value at the desired time; and how important a value is, depends on the decision-making applications required by groups or individuals. For example, if a dataset is not valued for a specific application, we don't need to pay it attention. By starting with human capital as an application, in this section, I will first outline how some available data sources are opening new avenues for research in the analysis of labor market activity. Secondly, I will discuss what I see to be key implications for the workforce and for the labor economics, regarding the tasks where data technology excels. Given that most big data problems require human analysis and interventions and based on human-curated datasets, I limit my discussion without autonomous robots (as the substitution of labor). Note that an autonomous robot is a robot that performs behaviors or tasks with a high degree of autonomy (without external influence).

In the past, labor economics relies heavily on administratively collected data sources. These data sources are notable for their breadth and their quality, including, but not limited to, the National Longitudinal Study of Youth (NLSY), the Current Population Survey (CPS), the Panel Study of Income Dynamics (PSID), or the Job Openings and Labor Turnover Survey (JOLTS). When evaluating these administratively collected data sources, MIT's labor economist John J. Horton commented:[41]

> *They are expensive to generate and therefore only infrequently collected and limited in their sampling and in the scope of the questions that they can be used to answer.*

Horton emphasized the benefits of the Internet data sources that are (see Footnote 41):

> *The emergence of new Internet data sources is transforming both the scale and granularity at which researchers can examine economic phenomena, ranging from ecommerce transactions to online search behavior to consumer decision-making. ... the capture of massive volumes of fine-grained data on labor activity and its analysis have many new implications, especially for research into labor-based phenomena. ... Along some dimensions, including both the granularity of observation as well as sample sizes, the gaps in these administrative data sources can be filled by new data sources that are being generated daily by the activity conducted through Internet labor market intermediaries.*

Definition 3.12 *Spatio-Temporal Visualization* The spatio-temporal visualization incorporates the legacy from the conventional cartography to track the kind of changes occurring over time, including: existential changes (e.g., appearance and disappearance), changes of spatial properties (e.g., location, sharp or/and size, orientation, altitude, height, gradient, and volume), and changes of thematic properties expressed through values of attributes (e.g., qualitative changes and changes of ordinal or numeric characteristics: increase and decrease). Of which, timeline is a graphical way of displaying a list of events in chronological order. Some timelines display as a straight (e.g., horizontal or vertical) line, while others can be a curve (e.g., a spire).

But, to my knowledge, these carefully collected data can also be better understood if developing more advanced methods. We start putting all free-text answers to similar open-ended questions extracted from surveys into a topic modeling, for example, to indicate theme evolution. I previously developed a

[41] ***See:*** Horton JJ, Tambe P. Labor economists get their microscope: big data and labor market analysis. Big data, 2015;3(3):130-137.

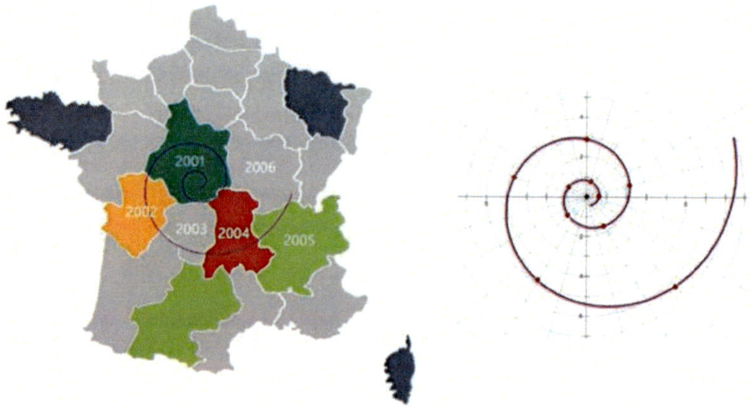

Figure 3.6 When spiral timelines combined with a geographical map. (Source: The author drew by herself according to some materials on the Internet)

classifier, we call the regional classifier, to visualize theme words under different themes in a timeline map to show chronological development of the focused topics. When combined with a geographical map (see figure 3.6), timelines can depict temporal patterns of events with respect to their spatial attributes. The classifier partitions theme words pretreated by semantic analysis tools into a real map such that words appearing in the same time period appear in the same region.

Intuitively, there exist two types of data sources related to labor market activity on the Internet. One is from employment websites; the other is from professional social networks. Employment websites, such as CareerBuilder and Monster, are online labor market intermediaries focusing on job boards, in terms of the populations they cover. Some websites serve niche markets; for example, Dice.com is a job board oriented specifically towards IT workers. There exists a growing number of employment websites in emerging economies, such as Naukri (in India) or Zhaopin (in China). LinkedIn as a social network, which is a largely successful attempt to digitize the supply side of the labor market, reaches almost 300 million users globally.

When further interpreting these data from employment websites, Horton said (see Footnote 41):

> *Job sites capture a number of different types of valuable information that are useful for labor market research, including a) workers' prior employment histories, b) employers' job openings, and c) about the matching activity that takes place to connect these two sides of the market.*
>
> *Much of the data available through the employment histories that workers post on job boards, or resumes, would not otherwise be available through government surveys. ... Employment histories generated through job boards contain rich information about prior employers, job titles, skills, and dates of exit and entry, as well as details about education. They also occasionally contain information about geographic mobility (i.e. city to city moves).*

Job listings contain information about offered wages, the skills that are required to fill a particular job opening, and other details valuable for understanding labor demand as well as perhaps related to broader questions around unemployment trends.

To obtain the recruiter's checklist, I tend to use other candidate data sources instead of the above-mentioned sources. The reason is quite simple—data credibility: the data posted job boards are possible to carefully processed by individuals, who often try to hide about themselves for a variety of reasons. Besides the preference for maintaining privacy, citing the previous literature,[42] researchers list some of the examples of stigmatized and concealable workplace identities, involving hidden medical conditions, mental illness, sexual orientation, pregnancy, minority religious group membership, non-traditional family structures, having a criminal record, and surviving sexual abuse. Under what circumstances would employers believe in recruiters, and what the costs might be?

Definition 3.13 *Data credibility* Data credibility is the extent to which the good faith of a provider of data or source of data can be relied upon to ensure that the data really represents is what the data is supposed to represent, and that there is no intent to misrepresent what the data is supposed to represent.

Assume that an engineering team has an open position that is critical to its ability to deliver on yearly goals, and the team can handle the data of the recruiter's checklist. When a candidate is identified as a lack of a necessary skill or inconsistency with company value in the latter half of the recruitment process, the team may waste a lot of time and source or reason could be really hard to find due to the mismanagement of the data in the initial stages. The importance of data management in the initial stages leads to mismanagement of it can have repercussions that the team should look at avoiding. If other candidate data sources are available (see figure 3.7), the team could score candidates based on the number or quality of projects or contributions in engineering communities like Github or Stackoverflow. The team can identify the ones who have put time and energy into sharing knowledge with the wider public through blogs or Quora answers. Also, this team can target candidates who seem to embrace values that are in line with those of the team, or who have wider interests that might make them an interesting addition to the team.

Also, these candidate data sources can be used to examine the mechanics of how professional relations form a network structure, and which network structure best encourages information (e.g., about employer) cascades or idea contagion. While also offering professional social network data, LinkedIn is not free. Specifically, for individuals: weak ties beyond three degrees are charged with a fee, i.e., users need to pay certain amounts if they want to view a network

[42] ***See:*** (1) Chaudoir SR, Quinn DM. Revealing concealable stigmatized identities: The impact of disclosure motivations and positive first-disclosure experiences on fear of disclosure and well-being. Journal of Social Issues. 2010;66(3):570-584; (2) Berkley RA, Beard R, Daus CS. The emotional context of disclosing a concealable stigmatized identity: A conceptual model. Human Resource Management Review. 2019;29(3), 428-445.

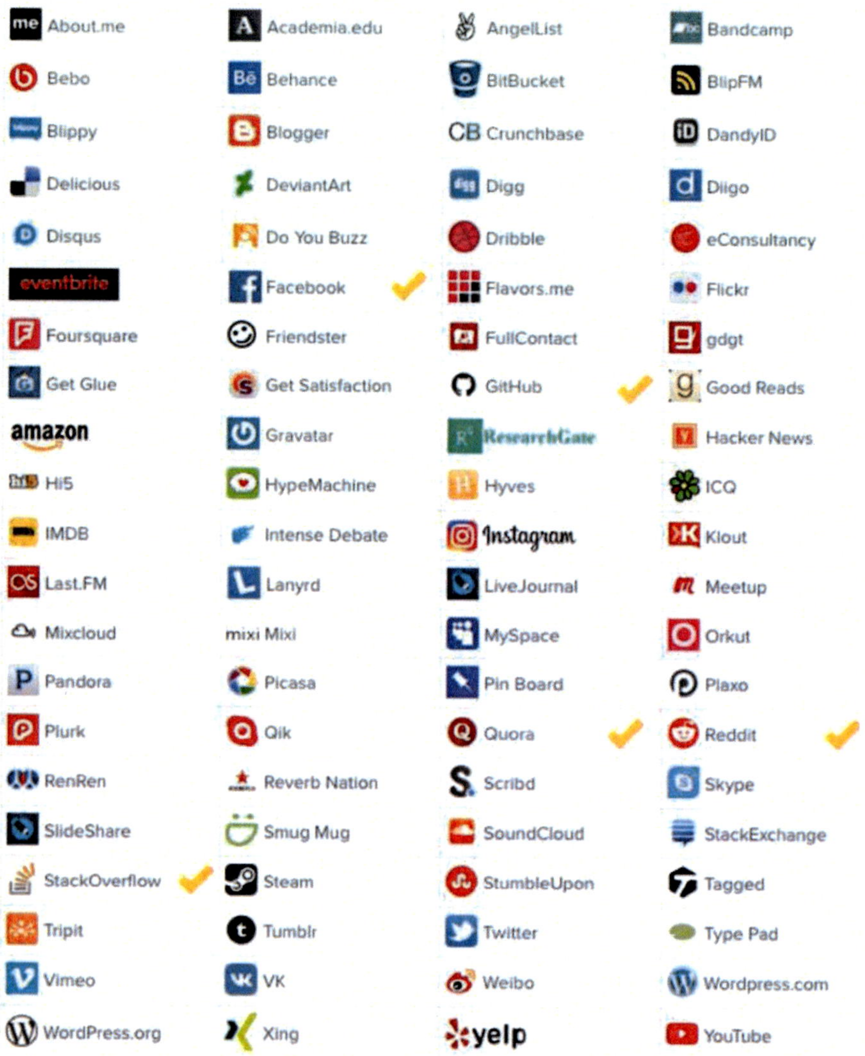

Figure 3.7 Some available candidate data sources (LinkedIn not included). Note: highlighted ones (✓) are valuable which author believed. (Source: Author drew by herself according to some materials on the Internet)

beyond three degrees (see the notion of six degrees of separation grew out of work conducted by the social psychologist Stanley Milgram in the 1960s).

Next to data sources, let's see the impact of data technology on workforce. I mentioned in the introduction, an Oxford's high-citation study[43] demonstrated

[43] ***See:*** Frey C. B., Osborne M. A. The future of employment: How susceptible are jobs to computerisation? Technological Forecasting and Social Change, 2017;114:254-280.

that many demanded jobs nowadays jobs (about 47% of 702 detailed occupations listed by the BLS in 2010) are most likely to be replaced. One of the most striking among those positions is computer programmers with a 48% high-risk because "algorithms have made it possible to train computer systems to be more accurate and more capable than those that we can manually program."[44]

With the sheer volume of training data available for some tasks, algorithms and data models may be large enough to capture highly valuable and previously unnoticed regularities. This contrasts with a person's capability of examining or comprehending to the processing algorithmic ability. For example (see Footnote 44), DNNs (deep neural networks) for image recognition have driven error rates on ImageNet,[45] a large data set of more than 10,000 labeled images, down from more than 30% in 2010 to less than 3% today. Similarly, DNNs have helped improve error rates from 8.4% to 4.9% in voice recognition since July 2016. The 5% threshold for image recognition and speech is important because that is roughly the error rate of humans when given similar data. For these tasks, a temporarily rough standard, based on the fourth paradigm put forwarded by Jim Gray, defines five criteria to help determine whether data technology more likely to be successful or not (see Footnote 44):

> *[First,] learning a function that maps well-defined inputs to well-defined outputs. Among others, these include classification (e.g., labeling images of dog breeds or labeling medical records according to the likelihood of cancer) and prediction (e.g., analyzing a loan application to predict the likelihood of future default). Although models may learn to predict the Y value associated with any given input X, this is a learned statistical correlation that might not capture causal effects.*
>
> *[Secondly,] large data sets exist or can be created containing input-output pairs. The more training examples are available, the more accurate the learning. ... Digital data can often be created by monitoring existing processes and customer interactions, by hiring humans to explicitly tag or label portions of the data or create entirely new data sets, or by simulating the relevant problem setting.*
>
> *[The Third is] no need for detailed explanation of how the decision was made. Large neural nets learn to make decisions by subtly adjusting up to hundreds of millions of numerical weights that interconnect their artificial neurons. Explaining the reasoning for such decisions to humans can be difficult because DNNs often do not make use of the same intermediate abstractions that humans do. ... For example, whereas computers can diagnose certain types of cancer or pneumonia as well as or better than expert doctors, their ability to explain why or how they came up with the diagnosis is poor when compared with human doctors.*
>
> *[Fourthly,] a tolerance for error and no need for provably correct or optimal solutions. Nearly all algorithms derive their solutions statistically and probabilistically. As a result, it is rarely possible to train them to 100% accuracy. Even the best speech, object recognition, and clinical diagnosis computer systems make errors (as do the best humans). Therefore, tolerance to errors of the learned system is an important criterion constraining adoption.*

---

[44] ***See:*** Brynjolfsson, E., & Mitchell, T. What can machine learning do? Workforce implications. Science, 2017;358(6370), 1530-1534.

[45] ***See:*** Deng J, Dong W, Socher R, *et al.* Imagenet: A large-scale hierarchical image database. Computer Vision and Pattern Recognition, 2009. IEEE Conference on. IEEE, 2009:248-255. Retrieved from http://image-net.org/challenges/LSVRC/2017/results.

> *[Finally,] the phenomenon or function being learned should not change rapidly over time. In general, algorithms work well only when the distribution of future test examples is similar to the distribution of training examples. If these distributions change over time, then retraining is typically required, and success therefore depends on the rate of change, relative to the rate of acquisition of new training data.*

As more data come online and are pooled and as we discover which tasks should be automated by data technology, we will collect data even more rapidly to create even more capable systems. Unlike solutions to tasks mastered by humans, many solutions by algorithms can be disseminated almost instantly worldwide. Historically, as some tasks have been automated, the freed-up labor has been redeployed to producing new goods and services or new, more effective production processes. Such innovations are equitant to increased capital, labor, or resource inputs as a force for raising overall incomes and living standards.

In terms of workforce impaction, the current effects of data technology lie on a relatively narrow swath of routine, highly structured and repetitive tasks. This is the reason that labor demand has fallen for jobs in the middle of the skill and wage spectrum (e.g., clerks and factory workers), whereas demand at the bottom (e.g., janitor or home health aide) and top (e.g., physicians) has held up in most advanced countries. For instance, transcripts from large sets of online chats between salespeople and potential customers can be used as training data for a simple chatbot that recognizes which answers to certain common queries are most likely to lead to sales. But a much broader set of tasks will be automated or augmented by algorithms over the coming years, including: "tasks for which humans are unable to articulate a strategy but where statistics in data reveal regularities that entail a strategy (see Footnote 44)." Take an example of "creative;" designing a complex new device has historically been a task where humans are more capable than machines. But generative design software can come up with new designs for objects like the heat exchanger (see photo) that meet all the requirements (e.g., weight, strength, and cooling rate) more effectively than anything designed by a human, and with a very different look and feel.

There are many non-technological implications for the workforce concerning labor demand and wages, involving five aspects (see Footnote 44): substitution, complementarities, the elasticity of labor supply, income elasticity, and business process redesign. First and foremost, big data will directly substitute for some tasks, replacing the human and reducing labor demand for any given level of output. Next, as calculation became automated, the demand for human programmers increased. Skills can also be complementary to other skills. For instance, interpersonal skills are increasingly complementary to analytical skills. Then, automation many change the total income for some individuals or the broader population. If income elasticity for a good is nonzero, this will in turn change demand for some types of goods and the derived demand for the tasks needed to produce those goods. Fourth is related to elasticity of labor supply. As wages change, the number of people working on the task will respond.

If there are many people who already have the requisite skills (for example, driving a car for a ride-hailing service), then supply will be fairly elastic and wages will not rise (or fall) much, if at all, even if demand increases (or falls) a lot. In contrast, if skills are more difficult to acquire, such as becoming a data scientist, then changes in demand will mainly be reflected in wages, not employment. Lastly, the business process may redesign. The production function that relates any given set of different types and quantities of labor, capital, and other inputs to output is not fixed. When faced with new technologies, they will change the production process, by design or through luck, and find more efficient ways to produce output. For example, integration of autonomous trucks onto city streets might require changes in traffic laws, liability rules, insurance regulations, traffic flow, and the like, whereas the switch from talking to a human assistant to a virtual assistant in a call center might require the relatively little redesign of other aspects of the business process or customer experience.

An Experiment of Quantifying Data Capital in Social Media Clout

Harvard's Robert Putnam identified "social capital" as the missing ingredient in addressing many key national challenges in his seminal nonfiction book *Bowling Alone: The Collapse and Revival of American Community* was published in 2000. Social capital is understood as the capacity for interconnection achieved through participation in churches, civic groups, associations, and organizations. Political polarization, weakened family structures, lagging economic development, fraying societal cohesion, and many other contemporary troubles have been said to result from a lack of such capital. Commenting on this point, Princeton's Patrick T. Brown wrote in an article:[46]

> *Without question, a renewed spirit of civic engagement, neighborliness, and social trust would be a welcome contribution toward re-knitting the frayed fabric of American society. But the forces that bind communities together also have a dark side. ...*
>
> *Social networks have always involved some kind of trade-off between inclusion and exclusivity. But as the decline in associational life has been very uneven across classes, and as political tensions have come to define group identities and lifestyles, the benefits of social capital for some often look like disadvantages of exclusion for others.*

[46] There are several guys named Patrick T. Brown when searching on the Internet. This serves as a senior policy advisor at the US Congress joint economic committee. When this paper being published, Mr. Brown is studying for his masters in public affairs, with a concentration in domestic policy, at Princeton University's Woodrow Wilson School of International and Public Affairs. ***See:*** Brown PT. The dark side of social capital. National Affairs. Summer 2019. Retrieved from https://www.nationalaffairs.com/publications/detail/the-dark-side-of-social-capital.

But what I'm trying to say is: we live in a data-rich world, and global flows of data may help solve complex real-world problems from different and complementary viewpoints. With the social data stream, for example, a more rigid class structure formed by many aspects of the physical and social environment (e.g., incomes and social status, education, physical environment, social support networks) is broken. This may engage in dialogue among people with different socioeconomic backgrounds, and even a phenomenon of "Trump's Twitter Presidency"[47] appeared.

In this section, I introduce a novel idea for detecting the ways and means to gain potential benefits via social media clout formed by digital human relations along the time; that is, the analysis of topic evolution among the post categories by considering interpersonal recommendations represented by data at that time. I then compare social structures of relationships generalized from two venture-capital social media groups on Facebook™, wherein nodes are people (the size depends on the number of online discussion topics), and two people are connected if one introduced another for networking. The findings show some evidence on potential benefits based on intangible social capital formed by interpersonal communication, which can be subdivided into public participation, social connections, trust, and social norms.

Preliminary Knowledge

This section introduces the basic concepts, setting and corpus, and methods briefly.

Data Source 3.3 *Social Media Data* Social media data refers to all of the raw insights and information collected from individuals (e.g. prospects and customers) social media activity. Social media data tracks how individuals engage with a specific content or channels like Facebook, Twitter, Instagram, and LinkedIn.

Social Media is HUGE.[48]

Reports show that 26 billion texts were sent each day by 27 million people in the US. That is 94 texts per day per person in the US in 2017. (a) Every minute on Facebook: 510,000 comments are posted, 293,000 statuses are updated, and 136,000 photos are uploaded. This number has increased more than 300 percent, from around 650,000 posts per minute in 2011. (b) Since 2013, the number of Tweets each minute has increased by 58% to more than 474,000 Tweets PER MINUTE in 2019. (c) Instagram users upload over 100 million photos and videos every day. That is 69,444 million posts every minute.

[47] McIntire M, Confessore N. Trump's Twitter Presidency: 9 Key Takeaways. The New York Times. November 2, 2019. Retrieved from https://www.nytimes.com/2019/11/02/us/trump-twitter-takeaways.html.

[48] Schultz J. How much data is created on the Internet each day? Microfocus.com. August 6, 2019. Retrieved from https://blog.microfocus.com/how-much-data-is-created-on-the-internet-each-day.

Figure 3.8 Corpus description of analyzing social clout capital. Notes: (**a**) shows the same or similar posts may be published by different authors, (**b**) shows the same post may be published in both two groups, (c) shows a real post in the full size (**c-1**) and in brief (**c-2**) that was crawled and cleaned as free-text format (**c-3**), and (**d**) and (**e**) are the descriptions of the two groups as of December 2019. (Source: The author drew by herself according to some materials on the Internet)

Example 3.3 As free and public social media data sources, Facebook's pages are easy to be crawled from its page and to be evaluated. We specifically differentiate between comments and posts as these contain fundamentally different data characteristics (i.e., they are written for different purposes, posts tend to be longer in length, posts tend to contain both free-text and pictures whereas comments rarely contain pictures). Posts are composed by a Facebook user (e.g., a public/private group member, or anyone who visits public group's Facebook page). As shown in figure 3.8, we collected the data posts from two venture-capital groups' Facebook pages from 2016 to 2018. One is a private group with 18,946 members created on December 2006 with 25 moderators. The other is a public group with 11,999 members created on May 2008 with three admins. Three thousand nine hundred fifty-two people participated in both groups. It should be noted that anyone doesn't need to apply to get approval when giving posts or comments in a public group, but permission required in the private one.

Network analytics provides a means of interconnecting the complex world of entities.[49] A qualified network for representing contact capital should reflect at least three aspects, involving outreach, interaction, and income. In a social media context, outreach is the activity of detecting any person who didn't know before but might be interested in services (i.e., potential customers). Interaction,

---

[49] ***See:*** Pržulj N., Malod-Dognin N. Network analytics in the age of big data. Science. 2016;353(6295):123-124.

here, is conversation-driven via the content of posts and comments; for example, accepting a new membership application after viewing posts corresponding to his/her profit. Income is possibly come from social media clout, such as brand awareness, leads, and gaining followers. We also envision timeline evolution may help to discover opportunity costs among such a network. Thus, nodes in the networks are people, and two people are connected if one introduced another for networking. We next weight each people in the network to reflect the number among the post categories individually, which will be represented in a different size. Then we adapt Latent Dirichlet allocation of topic modeling to derive similar discussions among the post categories globally, which will be represented in a different color. The nodes hence have differences in size and color.

The networks leverage the force-directed layout based on the Barnes-Hut algorithm for interacting, which provides a flexible method for node arrangements that use local relationships in an attempt to reveal the global shape of the graph. Specifically, the force-directed layout simulating a charge-spring model adds a repulsive force between two nodes and a tractive force among the links. Thus, the nodes (as many-body) can move under the influence of various forces when iterating and will have a balance after multiple iterations.

A Force-Directed Network of Social Clout Capital

We constructed two interactive networks for the private and public groups shown in figure 3.9. Note that we excluded one of admins' 512 posts (i.e., the largest node) for the public group due to imbalance in the number of the posts. It is easy to see that the two groups indicate themselves as a hierarchical organization (both more than five levels). An interesting thing is that, the private group's network is closed, and the public one is open. Also, the open network shows a distinct subclass having lots of links that mean frequent interactions. Table 3.3 demonstrates post topic classification of two groups, respectively, corresponding to figure 3.9 two color ones.

As shown in Table 3.3, the results are comparable among two different groups. We can easily classify the supply and demand side in venture capital networks. Obviously, more suppliers are in the private group and more demanders in the public group.

A standard definition of income is to obtain money or some equivalent value that an individual or business receives in exchange for providing goods or services or through investing capital. Essentially a people's net worth to be the sum of how often people post, obtain like and feedback and how many people tweet, get retweeted and how many others follow. Then someone applies this to various marketing initiatives. So according to the two networks, it is easy to do

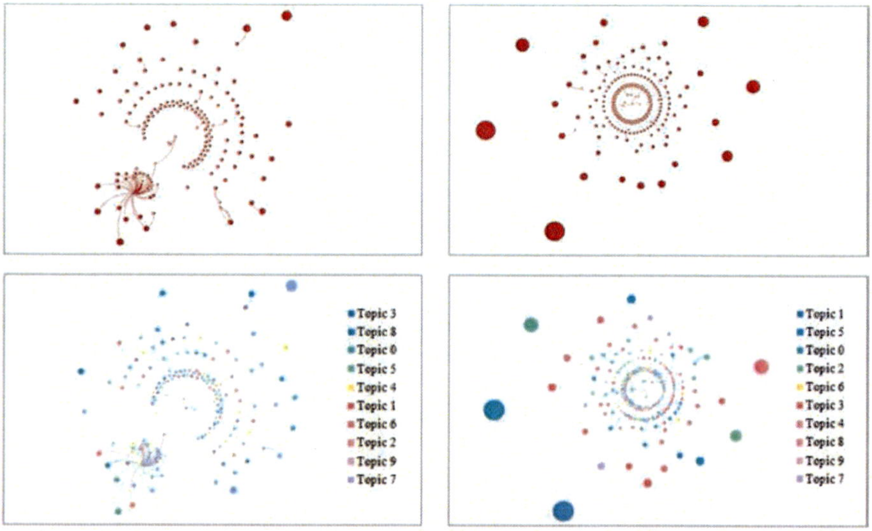

(a) (b)

Figure 3.9 Two venture-capital groups' networks with/without topic classification. Notes: **(a)** is the public group and **(b)** is the private group; (2) Networks may be slightly different when opening each time due to the force-directed layout. (Source: The author drew by herself according to some materials on the Internet)

Table 3.3 Post topic classification among two groups

The public venture-capital group		The private venture-capital group	
Topic	Theme	Topic	Theme
3	Characteristics of funded projects	1	Characteristics of investable projects
8	Describing funded projects at the industry level	5	Project selection and identification
0	Seeking financing via the Internet	0	Seeking projects via the Internet
5	Project self-promotion	2	Good project promotion
4	Existing investment hotspots	6	Existing investment hotspots
1	Conference advertising	3	Introduction to investment
6	Introduction to VC firms	4	Post-investment management
2	Business plan pitching	8	Investment income and growth
9	Summit advertising	9	Project optimization
7	Introduction to fundraising methods	7	Blockchain

further work on how social media clout capital resulted in income at the present time of the micro-influencer; for example, connecting topics or relationships with marketing strategies for businesses, such as training, conference promotion, and book sales.

Part II
Data Capital as Factor of Production

Abstract Exploring the role of data capital in data production covers multiple aspects. First is to clarify what is the difference between data and data capital, and which can be a factor of data production. Secondly, introducing two components of the data (e.g., memories and time) is vital for understanding data as capital. The third is about particularities in data capital, which is to produce; namely, the premise of being a data sovereign is the data producer. The last involves many examples of coving various metamorphoses of data capital during the production process.

In Part I, I introduced data capital by exploring economists' observation of "the rise of intangibles." Data, as a human-created resource, is naturally one capital. Data Capital not only can replace almost all current types of capital, such as financial or intangible capital, and the digital part of tangibles as well, but also measures them (and itself included).

In this part, I am going to analyze data capital from a perspective of production factor, looking at both its compositions and particularities, as interpreted by the basic concepts in data science, and its metamorphoses during the production process. Given the title of this part, moreover, someone may be natural to ask why my preference is to see data capital as a factor of production, rather than the data. I will first answer this at the beginning of Chap. 4.

Chapter 4
The Composition of Data Capital

The generally accepted view, which follows the popular phrase "data is the new oil"[1] apparently coined the British mathematician Clive Humby in 2006 but which possibly stems from an earlier article on the MIT Sloan Management Review wrote by Anany V. Levitin and Thomas C. Redman in 1998, is "data as a resource."[2] Of course, someone's opinion on data is the new "nuclear power" (see Footnote 1) with the capacity for harm. Our discussion is limited to data (digital information) measured in bytes (not stored in paper files and other physical manifestations considered by Levitin *et al.*). The IEEE 1541-2002 standard specifies the upper-case letter *B* as the symbol for byte. Historically, a byte was the number of bits used to encode a single character of text in the computer, which is the smallest addressable unit of memory depending on computer hardware architecture; but today it almost always means eight bits—that is, an octet.

Many scholars have noted the distinctions between data and other economic resources (see Definition 2.3). Before proceeding, I want to clarify our use of the term "data resource." According to *The American Heritage Dictionary* (at https://ahdictionary.com), often resources are "an available supply that can be drawn upon when needed" and "the total means available for economic … development or increasing production or profit, which can be used to advantage." As these definition's suggestions revealed by Levitin and Redman (see Footnote 2),

> *A resource is both necessary for an enterprise and a potential source of yet unrealized benefits. With these definitions in mind, we can list several categories of resources used by modern enterprises. There are the traditional resources – financial, human, plant*

[1] ***See:*** Bridle J. Opinion: Data isn't the new oil – it's the new nuclear power. Ted.com. July 17, 2018. Retrieved from https://ideas.ted.com/opinion-data-isnt-the-new-oil-its-the-new-nuclear-power/?fbclid=IwAR1gculxTbrpyMns-prhlhfORQvX2y_uoa6V03NeuJnCXU_hwojnoOdZglU.

[2] The reference may be an early one in this research topic. ***See:*** Levitin A.V., Redman T.C. Data as a resource: Properties, implications, and prescriptions. MIT Sloan Management Review. October 15,1998.Retrievedfromhttps://sloanreview.mit.edu/article/data-as-a-resource-properties-implications-and-prescriptions.

© Springer Nature Switzerland AG 2021

C. Tang, *Data Capital*, https://doi.org/10.1007/978-3-030-60192-8_4

and equipment, raw materials, and energy – and the so-called information age resources – data, information, and knowledge.

Russell Ackoff's hierarchy (i.e., the DIKW hierarchy, see Case 3.6) is with data at the bottom, information and knowledge above, and wisdom at the top. Haskel and Westlake I explained in Chap. 2, agreed that their understanding of knowledge-based intangibles is on "analyzable data." So, the "so-call information age resources" mentioned by Levitin and Redman here, ultimately, refer to only data resources.

The basic concepts of set theory seem to help us understand data resources in cyberspace. Set theory, a branch of mathematics that investigates sets and their properties (to put homogenous objects together, e.g., plants and animals in mother nature), appears to be self-evident. The relationship of the data and data resources, from a perspective of set theory, is analogous to gathering some phone numbers in a city—the data is a single resident's postal code, and a data resource refers to postal codes of all residents in the city. However, despite its apparent simplicity, studying data resources from a perspective of set theory turns out to be a very sophisticated subject. In particular, data science has shown that similarities among data objects are hard to be formalized within the theory of set. For example, image similarity is not amenable to formal specification but can be easily expressed visually. Fortunately, data science involved concepts can be organized by set theory as follows.

Definition 4.1 *Atomic Data* Atomic data is an atomic state of information that cannot be broken down into smaller pieces or sources, such as unit sales (e.g., for a particular model of the washer), gross revenue (that cannot be calculated from anything even if being broken down by transaction; note that net revenue is based on multiple inputs and is not considered atomic data), and base salary (as a hard number that is not typically calculated from the others).

Definition 4.2 *Data Item* A data item describes an atomic data concerning a specific property at a certain time point. Values of a data item are not necessarily atomic. The complexity of a value (represented by symbols like numbers, texts, images, sounds, videos or others) depends on the complexity of the property and time component.

Definition 4.3 *Data Object* A data object is a unique location in a computer memory (i.e., a specific memory address) that contains the value of *something*. This *something* can be accessed by giving a unique name that is known as the identifier of the variable that contains the data object. Data objects are often composed of a collection of limited data items, which are identified by object (o), property (p) and time (t), while the value (v) is a function of o, p, and t, $v = \boldsymbol{F}(o, p, t)$.

Definition 4.4 *Dataset* A dataset (or a data set) is a collection of data objects. Generally, data objects in a specific dataset are limited. Even though building a dataset that contains an infinite stream, there exist limited data objects at a certain (arbitrary) time point.

Definition 4.5 *Data* Data as a general concept refers to the fact that some existing information or knowledge is digitally coded in some form suitable for better usage or processing. The data can be represented as an atomic data, a data item, a data object, or a dataset arbitrarily. The size of the data is measured in bytes (i.e., *B*). Think of *NULL* data is with an empty value, of which the size is zero *B*.

Definition 4.6 *Metadata* Metadata is the data that provides information about other data. There exist multiple distinct types of metadata, including descriptive metadata, structural metadata, administrative metadata, reference metadata, and statistical metadata. For example, the American Standard for Information Interchange (ASCII) can be used to represent atomic data, elements' title (e.g., NAME) is to describe data item, EMPLOYEE (ID, NAME, RANK) is for employee data, and DATASET #1 is a dataset.

Definition 4.7 *Data Tool* A data tool involves algorithms, data models, programs, and software, which can run in a computer system. Data tools belong to a specific data object designed for some purposes (e.g., an analytical purpose). The data tool itself is also data, which is usually used to process data and can also be processed by other data tools. For example, antivirus software can process against viruses, while a virus program may replicate itself—it processes itself.

Definition 4.8 *Data Resource* As data organizations connected by logics and data areas assembled by data volumes reach a certain scale, the massive distinct datasets become a data resource.

Currently, as for scientific fields, almost all disciplines such as astronomy, geography, geochemistry, geology, oceanography, atmosphere, biology, and medical science, have their own data resources. Moreover, various large-scale transnational collaborative experiments continuously provide big data that can be generated, stored, communicated, aggregated, and analyzed, such as CERN's LHC (Large Hadron Collider) in Europe, Pan-STARRS (Panoramic Survey Telescope and Rapid Response System) to preform all-sky space astrometry in America, radio telescope SKA (Square Kilometer Array) in Austria, and the INSDC (International Nucleotide Sequence Database Collaboration; that aims to capture, preserve, and present globally comprehensive public domain biological data) and the OHDSI network (Observational Health Data Sciences and Informatics) with data product sharing for a collective research gain in health informatics. As for economic areas, data resources are constructed by financial and economic data, social behavior data, personal identity data, and Internet and social media data. Obviously, as shown in figure 4.1, the artificial ecosystem in cyberspace has been evolving into the data "nature."[3]

[3]The concept of "data nature" was coined by my doctoral advisor Yongyong Zhu in his book *Dataology* published in 2009. ***See:*** Zhu Y., Xiong Y. Dataology. Fudan University Press; 1 edition (December 1, 2019, in Chinese). 2009, 131 pages.

Before defining data nature, let's see what meanings and ideologies we have been infusing in the concept of "cyberspace."[4] While being an internet metaphor, cyberspace should not be confused with the Internet because the term cyberspace often used to refer to objects and identities that exist largely with the communication network itself. The most recent draft definition (compared with the previous 28 versions) (see Footnote 4):

> *Cyberspace is a global and dynamic domain (subject to constant change) characterized by the combined use of electrons and the electromagnetic spectrum, whose purpose is to create, store, modify, exchange, share, and extract, use, eliminate information and disrupt physical resources.*
>
> *Cyberspace includes: a) physical infrastructures and telecommunications devices that allow for the connection of technological and communication system networks, understood in the broadest sense (SCADA [i.e., supervisory control and data acquisition] devices, smartphones/tablets, computers, servers, etc.); b) computer systems (see point a) and the related (sometimes embedded) software that guarantee the domain's basic operational functioning and connectivity; c) networks between computer systems; d) networks of networks that connect computer systems (the distinction between networks and networks of networks is mainly organizational); e) the access nodes of users and intermediaries routing nodes; f) constituent data (or resident data).*
>
> *Often, in common parlance (and sometimes in commercial language), networks of networks are called the Internet (with a lowercase i), while networks between computers are called intranet. Internet (with a capital I, in journalistic language sometimes called the Net) can be considered a part of the system a). A distinctive and constitutive feature of cyberspace is that no central entity exercises control over all the networks that make up this new domain.*
>
> *... This does not mean that the dimension of power in cyberspace is absent, nor that power is dispersed and scattered into a thousand invisible streams, nor that it is evenly spread across myriad people and organizations, as some scholars had predicted. On the contrary, cyberspace is characterized by a precise structuring of hierarchies of power.*

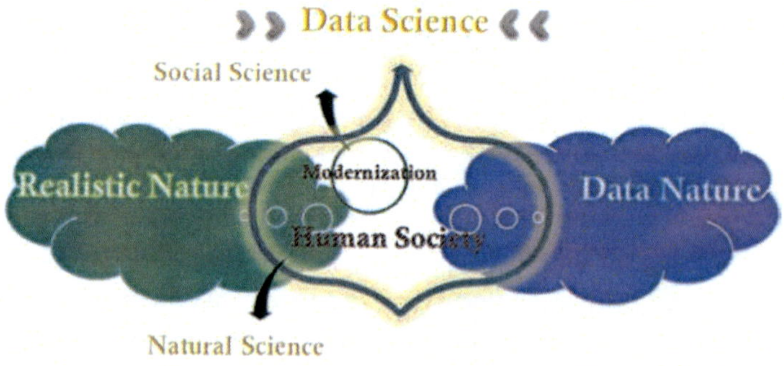

Figure 4.1 Data nature and data science. (Source: Author drew it herself)

[4]I use two references for defining the term "cyberspace." ***See:*** (1) Mayer M, Martino L, Mazurier P, *et al.* How would you define Cyberspace? Academia.edu. May 19, 2014. Retrieved from https://www.academia.edu/7096442/How_would_you_define_Cyberspace; (2) Morningstar C, Farmer, FR. The lessons of Lucasfilm's Habitat. Journal For Virtual Worlds Research. 2008;1(1).

Also, the term cyberspace often describes "alternate realities in philosophy and art" due to the fact that it entered the popular culture from science fiction and the arts. According to earlier writers' view (see Footnote 4),

> *cyberspace is defined more by the social interactions involved rather than its technical implementation, ... and the core characteristic of cyberspace is to offer an environment that consists of many participants with the ability to affect and influence each other.*

So, cyberspace is not enough to describe data nature used in data science. I give its definition separately as follow.

Definition 4.9 *Data Nature* (see Footnote 3) All data in cyberspace constitutes data nature, but the computer systems (or other IT physical facilities) as carriers do not belong to the part of data nature.

It is very difficult to clearly define when data nature has been formed, or what are the sings that data resources are reaching a peak to form data nature in cyberspace. But we can take a sign of forming data nature, which features sporadic, unordered, uncontrolled, and unknown parts of data existence. In the initial stage of digitalization, the data only represents stuff existed in reality. Digitizing library materials, for example, is to preserve scanned collections, always with consideration to the index, format, and character of the items being digitized. When characterizing to be unrecognizable, or the complexity of the issues involved in recognizing the diversity of ways, the data may correspond to something (e.g., the game *Second Life*'s Linden dollar) that cannot be found in reality. Therefore, my preference is to see data capital as a factor of production because of being mastered easily. Just as natural resources cannot be calculated to natural capital, unknown, uncontrolled data resources with diver complexity in data nature may be hard to be factors of production directly.

After having massively preserved almost all findings of realistic nature in the form of data in cyberspace, our scientific studies and social activities will evolve into an exploration in data nature. That is; we shall decide on whether we use data science to assist studies of natural science (and/or social science) obtaining more advanced discoveries or not. For instance, digital humanities may cultivate a two-way relationship between the humanities and data science: the field both employs data technology in the pursuit of humanities research and subjects data technology to humanistic questioning and interrogation, often simultaneously. Another example is to perform real experiments about herd immunity in an epidemic, which may reach extremely high risk in the population. Given 60% of the population must be infected for the whole population to acquire herd immunity. With a 0.3–1% mortality, that meant a plan in British that would accept somewhere between 117,000 and 390,000 deaths. A data experiment can be a little safer at a very low cost.

Definition 4.10 *Data Experiment* Data experiment is a scientific method to discovering how things (e.g., specific natural phenomena, or any rules in a designated dataset and/or data tool) in the universe work though the direct intervention and control of datasets and/or data tools. Three major types of data experiments include taking datasets, data tools, or simulated natural phenomena (e.g., natural disasters, urban or business operations) as the experimental object, respectively. Data experiment is characteristic that the experimental objects and tools of such a research approach are both composed of data: data verifies data through data.

To solve the terminology confusion, some relevant definitions are as below.[5]

Definition 4.11 *Repeatability* The measurement can be obtained with stated precision by the same team using the *same* measurement procedure, the same measuring system, under the same operating conditions, in the same location on multiple trials. For data experiments, this means that a researcher can reliably repeat her own computation.

Definition 4.12 *Replicability* The measurement can be obtained with stated precision by a different team using the *same* measurement procedure, the same measuring system, under the same operating conditions, in the same or a different location on multiple trials. For data experiments, this means that an independent group can obtain the same result using the author's own artifacts.

Definition 4.13 *Reproducibility* The measurement can be obtained with stated precision by a different team, a *different* measuring system, in a different location on multiple trials. For data experiments, this means that an independent group can obtain the same result using artifacts which they develop completely independently.

Once we made our decision on tending to more data experiments, all things will come to us. First is how we understand data nature. Methods of logical reasoning (e.g., deductive, inductive, and abductive reasoning) from observations have been important to scientific practice for us to understand the real world. These methods aim to define or set up a logical system, which is a mathematical model of some pre-theoretic notion of consequence coupled with an inferential practice of some sort. But, can we apply those for the data to establish that some intuitive notion of logic for data nature? The data does not imply a mathematical foundation despite the computer systems as its carriers were built on Boolean algebra.

Secondly, we may ask what interpretations of mother nature are corresponding to particular understandings of data nature. This issue could include two aspects. One is with the data presented by realistic nature, we may ask whether the rules discovered in data are equivalent to the laws of the real world or not. The other is without the data presented by realistic nature, we may ask how we deal with phenomena on a law of only existing in data nature but not in reality.

The third is about the relationships between realistic nature and data nature. That is to say, the data is possible to fundamentally change our daily life. This involves almost all real life events, covering Third, how to change the life in real

[5] *See:* Association for Computing Machinery. Artifact Review and Badging. 2016. Retrieved from https://www.acm.org/publications/policies/artifact-review-badging.

nature by understanding the nature of data. This involves social and human issues, covering ways of earning a living, systems of trade and government, social roles, religions, the tradition in clothing and foods and arts, expectations for behavior, attitudes toward other cultures, and beliefs and values about all of these activities.

The above-mentioned issues obviously are difficult to answer. I will begin by discussing two components of the data as memories and time.

Memories: Stored, Hidden, But Never Erased

We all have the feeling like this: meeting someone familiar at a party whose name we cannot recall. Or, we leave our office cubicle to reach a colleague but wander through because we cannot remember what exactly we should do, and after having come back later we "realize" suddenly it. I have been scolding myself in such situations for my feeble remembering skills. We may not like it, but "forgetting is something very human."[6] It is part of how our mind works and even goes in depth for example, on designing the Long Short Term Memory recurrent neural networks used in machine learning algorithms—we usually called LSTMs, capable of learning how to forget.

Just as Daniel L. Schacter, who is William R. Kenan, Jr. Professor of Psychology at Harvard University, augured:[7]

> *Our brain constantly reconfigures our memory – what we remember, based at least in part on our present preferences and needs. Our memory is a living evolving construct. Human (re)construction is not so much a deficiency as a benefit. Using generalizations, relying on conjecture, emphasizing the present, and respecting subsequent experiences, helps us to reason swiftly and economically, to abstract and generalize, and to act in time, rather than remain caught up in conflicting recollections.*

The human brain is superbly complex (see figure 4.2). It consists of a hundred billion neurons or nerve cells specialized in processing information received from our senses. Many such nerve cells are arranged together with each that has thousands of connections in our brain to form a network of nerves. Specifically,[8] the dendrites receive the impulse from the terminal button or synapse of an adjoining neuron. Dendrites carry the impulse to the nucleus of the nerve cell we called soma. Here, the electrical impulse (i.e., stimuli) is processed and then passed on to the axon. The axon is longer branch among the

---

[6] ***See:*** Mayer-Schönberger V. Delete: The Virtue of Forgetting in the Digital Age. Princeton University Press; Revised edition (July 5, 2011). 2011; 253 pages.

[7] ***See:*** Schacter DL. The seven sins of memory: How the mind forgets and remembers. New York, NY: Mariner Books; 1st edition (May 7, 2002). 2002; 1,318 pages.

[8] ***See:*** Panchal S. Artificial Neural Networks – Mapping the Human Brain. Medium.com. May 6, 2018.Retrievedfromhttps://medium.com/predict/artificial-neural-networks-mapping-the-human-brain-2e0bd4a93160.

dendrites which carries the impulse from the soma to the synapse. The synapse then passes the impulse to dendrites of the second neuron. Thus, a complex network of neurons is created in the human brain. The same concept of the network of neurons is used artificially on a computer. In this case, an artificial neuron is created as a squashing function that outputs to a range between 0 and 1 (i.e., a type of activation function); such as, the sigmoid function: its name comes from the Greek letter Sigma, and when graphed, it appears as a sloping "*S*" across the *y*-axis. The sigmoidal functions may attempt to help an artificial neural network find a desired solution given a set of inputs. That is; to determine the output and that output will be used as the input for the following node, and such a process will repeat until the solution to the original problem is found.

Contrary to popular belief that we only use a small fraction of our brain's power, the entire network of nervous and synapse is active in healthy human beings. But all the incredible processing and storage capacity of this vast network would be overwhelmed quickly if we committed to memory every sensual stimulus we receive. For instance,[9] the human ear has about 12,000 sensory cells. Nerve cells can fire up to 500 times a second. Thus, doing a simplistic calculation, the ear could produce as much as 6 million bits of information every second. With 50,000 s in the conscious hours of every day, even a very large storage system would quickly fill up. And if we add in human's other five senses, the total amount of incoming information would be much higher still. To cope with the sea of stimuli, our brain uses multiple levels of processing and filtering before committing information to long-term memory. For example (see Footnote 6), from millions of individual impulses our brain constructs a picture, perhaps of something pink and moving, and a split second later we "realize" there is a little pink balloon floating towards us. That more abstract information—a pink

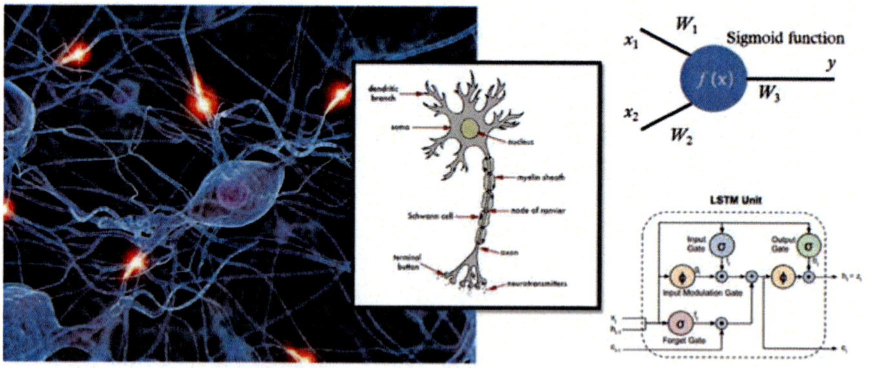

Figure 4.2 Human brain and artificial neural networks (mapping human brain). (Source: Author drew it herself)

[9] *See:* Smith SW. The Scientist and Engineer's Guide to Digital Signal Processing. Burlington, MA: Newnes; 1 edition (October 22, 2013). 2013; 650 pages.

balloon approaching us—is more usable than millions of individual bits of information contained in a series of snapshots of something pink getting closer and closer. More importantly, it is also much more compact. As our nerve cells process the incoming information, from simple stimuli to pattern recognition, a tremendous amount of information is deliberately lost. It is the first layer of unconscious biological forgetting—and one we rarely realize. Once an external stimulus makes it through that initial layer or after we have formed a thought, it is then usually stored in what's called short-term memory. In this state, we can easily retrieve it, but information in short-term memory fades very quickly—in a matter of seconds, eliminating most of the information. Short-term memory actually consists of three distinct parts that are subordinate to central executive—a special control unit. The psychologists Alan Baddely and Graham Hitch explained their model of working memory that (see Footnote 7),

> *The first part acts a short-term mental sketchpad for visual-spatial impression. Researchers have shown that it can hold a few (perhaps on average between four and seven) "chunks" (digits, letters) of information. The second part functions as a short audio-loop, recording about two seconds of sound and speech. If not refreshed (for example, through constant repetition), that audio information is quickly lost. That's why we can hold onto phone number for a litter while by saying them again and again. The third part is a short-term buffer of episodic memory, of things we have experienced.*

Short-term memory is the information-processing bottleneck through which almost all information must pass before reaching long-term memory. Once a piece of information is transformed into long-term storage, it is stored in what we call "human memory."[10] Actually, remembering is a two-step process.[11] The first is successfully committing information to long-term storage. The second is recalling that information from memory. Scientists classify human memory into two types: procedural memory and declarative memory. Procedural memory is part of implicit memory—memory that we acquire and recall without realizing it. Riding a bike, hit the keyboard to input text, or doing somersaults are similar to such routines. If recall from procedural memory is automatic—we do not have to remember how to hit the keyboard, we just hit it—declarative memory requires a conscious act of recall. We have to think about it, and actively "search" our mind to recall that experience or event in our past. Because these are abstract knowledge (i.e., abstract memory) or specific episodes of our lives we have experienced (i.e., episodic memory). Think, for example, of hearing Ben Lewis' (act as the Phantom) words during a performance of Love Never Dies: remembering words of a play is abstract memory, while recalling the context in which we heard about Love Never Dies; for example, on Broadway—is episodic memory.

[10] ***See:*** Baddeley AD. Human memory: Theory and practice. Boston, MA: Allyn & Bacon (April 1990). 1990: 515 pages.

[11] ***See:*** Baddeley AD, Hitch G. Working memory. In G.H. Bower (Ed.), The psychology of learning and motivation: Advances in research and theory. New York, NY: Academic Press. 1974; vol. 8, pp. 47-89.

Recurrent neural networks (RNNs) are networks with loops in them, allowing information to persist.[12] A recurrent neural network can be thought of as multiple copies of the same network, each passing a message to a successor. Similar to human-like reading, our thoughts have persistence—we don't throw every away and start thinking from scratch again. So, RNNs can easily help us to predict the last word in "the clouds are in the *sky*," because it doesn't need any further context—it's pretty obvious the next word is going to be "sky." But RNNs seem like a major shortcoming that is not able to learn long-term dependencies. Consider trying to predict the last word in the text "I grew up in China … I speak fluent *Mandarin*." Recent information suggests that the next word is probably the name of a language, but if we want to narrow down which language, we need the context of "China," from further back. It's entirely possible for the gap between the relevant information and the point where it is needed to become very large. To solve the problem of learning long-term dependencies, we should introduce the LSTM proposed by Sepp Hochreiter and Jürgen Schmidhuber in 1997.[13] Our LSTM at first step is to decide what information we're going to throw away from the cell state. This decision is made by a sigmoid layer called the "forget gate layer." The next step is to decide what new information we're going to store in the cell state. This includes two parts. One is to use a sigmoid layer called the "input gate layer" to decide which values we'll update. Next, a tanh layer creates a vector of new candidate values that could be added to the state. We then combine these two to create an update to the state. For the language model example, since it just saw a subject, it might want to output information relevant to a verb, in case that's what is coming next. For example, it might output whether the subject is singular or plural, so that we know what form a verb should be conjugated into if that's what follows next. What I've described so far is a pretty normal LSTM. In fact, it seems like almost every paper involving LSTMs uses a slightly different version as a LSTM variant. I've been designed a model related to LSTM in case the original one cannot deal with irregular time-lapse segments. Unlike other approaches, my modification was to add two layers: a flatten layer followed by a dense layer, to utilize LSTM output matrices to generate irregularities, which represent a period of similar disease conditions learned from clinical notes.

However, in data nature "comprehensive digital memory can remember things for you" (see Footnote 6) despite the model can be designed for forgetting, such as Facebook's "memories to look back on today."

> *It counters the annoying side of human forgetfulness, reminding us of birthdays and wedding anniversaries as well as pledges and promises. … And while remembering has its indisputable benefits, too much of it may lead to terrible consequences. …*

---

[12] ***See:*** Olah C. Understanding LSTM Networks. Github.io. August 27, 2015. Retrieved from https://colah.github.io/posts/2015-08-Understanding-LSTMs.

[13] ***See:*** Hochreiter S, Schmidhuber J. Long short-term memory. Neural computation. 1997;9(8), 1735-1780.

Privacy experts have been warning ... present behavior: I act as if I am watched even if I am not.

To appreciate the concrete from that digital remembering acts in today's world, it is good to borrow Viktor Mayer-Schönberger's known research in this aspect to do an extension. This interesting part is also related to time so that it easily brings us going to the next section.

Mayer-Schönberger on Perfect Digital Remembering

Viktor Mayer-Schönberger's *Delete: The Virtue of Forgetting in the Digital Age* is an award-winning book, and the awards include the 2010 Marshall McLuhan Award for Outstanding Book and the 2010 Don K. Price Award for Best Book in Science and Technology Politics. In *Delete*, Mayer-Schönberger traces the important role that forgetting has played throughout human history, from the ability to make sound decisions unencumbered by the past to the possibility of second chances. He argued (see Footnote 7):

Seen this way, forgetting is not an annoying flaw but a life-saving advantage. As we forget, we regain the freedom to generalize, conceptualize, and most importantly to act. ... Without some form of forgetting, forgiving becomes a difficult undertaking.

Mayer-Schönberger gave digital memories' three features with their four dangers. Based on six potential responses, he proposed a simple solution—expiration dates on information—that may help to reduce inequality and lead to "a redistribution of information power" when forgetting becomes the exception. Although an eye-opening book he offered, there is much room for improvement in terms of a comparison between now and a decade ago. I will explain it as a case study (i.e., Case 4.1), and my answer will be more nuanced.

Before proceeding, it may be useful to say a little more about the expression of digital memories. The term "information" frequently shown in Mayer-Schönberger's book is representatives "before digital memories were filed, indexed, and numbered," or "after the process of discovering from digital memories." To simplify the text, I accept this usage and use the words "data" and "digital memory" interchangeably, as if they were perfectly synonymous, including raw data and data products.

Definition 4.14 *Raw Data* Raw data (sometimes called source data or atomic data) is data that has not been processed for use. Raw data that has undergone processing is sometimes referred to as cooked data.

Definition 4.15 *Data Products* Data products (i.e., cooked data), such as models, intermediate results, and annotated training corpora, are the outcomes from data preparation, processing, and analysis (e.g., statistical analysis, data mining, and machine or deep learning). Data products also include visualizations and dashboards created by the artistic manual work of data scientists to assist in the interpretation of the analysis in an actionable way.

I return now to examine the features of digital memories. Mayer-Schönberger summed up three features of digital memories, namely accessibility, durability, and comprehensiveness. He argued that they are all about "the power associated with information, and its relative nature" (see Footnote 7), and

> *As others gain access to our information (especially when we do not approve or even know of it), we lose power and control. Because of the accessibility and durability of digital memory, information power not only shifts from the individual to some known transactional party, but to unknown other as well. This solidifies and deepens existing power differentials between the information rich and the information poor, and may even deny the latter their own conception of their past.*

Myself included, we all easy to accept this viewpoint nowadays. This is also Shoshana Zuboff of the main opinion in her book *The Age of Surveillance Capitalism: The Fight for a Human Future at the New Frontier of Power*. Capitalism transforms digital memories of "human experience"[14] as the "free factors of production" to do nearly all business activities. However, just as Mayer-Schönberger mentioned in the first potential response of "digital abstinence," sharing personal information offers users value which they may not want to forego. Put in other words; it was fairly fair from the beginning, if no considering any data brokers' over illegal or unlawful behaviors.

> *In exchange for receiving personal information, ecommerce vendors provide their customers with customized (and thus more valuable) transactional experiences: Amazon offers book recommendations, and Google serves up more targeted lists of search results.*

The inequality has risen sharply in deriving profits from subsequent data products since the recent decade when rapid advances in data collection and storage technology that enable the vendors to handle vast amounts of data. Also, personal data we worry to be misused merely accounts for a very little portion of big data. A real data scientist, in general, might cull or filter this part before dealing with the data. What I would like to add here in features of digital memories, thus, is shareability. Human has been used to sharing memories since ancient times. This is because; the fundamental quality of human memory remained unaltered: that remembering is weak. The written word (or drawn picture, etc.) made it possible for the human to remember across generations and time, so digital memories no exception. Accessibility does not simply mean shareability; meanwhile, it is distinct differences in the content and degree of shareability. Among them, sharing data products with their profits should be included.

[14] Shoshana Zuboff's Surveillance Capitalism has some commentaries. ***See:*** (1) Zuboff S. The Age of Surveillance Capitalism: The Fight for a Human Future at the New Frontier of Power. New York, NY: PublicAffairs; 1 edition (January 15, 2019). 2019; 704 pages; (2) Naughton J. "The goal is to automate us": Welcome to the age of surveillance capitalism. The Guardian.Com. January 20, 2019. Retrieved from https://www.theguardian.com/technology/2019/jan/20/shoshana-zuboff-age-of-surveillance-capitalism-google-facebook.

Case 4.1: Some Paragraphs Related to Digital Memories from Mayer-Schönberger's *Delete*

- Three features of digital memory.

Accessibility: One way we believe we control the use of our personal information is to decide whether to share it based on recipient and circumstances. … Put a bit more abstractly, we link our willingness to share information with others to a particular context and purpose – for which we have weighed the pros and cons of a relative loss in information power. … This works well if information is stored in separate disconnected containers, … however, is undermined by today's widespread ability of others to collect, access, and disseminate the information they have …

Durability: The second feature of digital remembering ... leads to redistribution of information power … Before digital memory most information, even incriminating items, were forgotten relatively quickly. Special efforts were necessary to ensure remembering, which proved costly and employed only in special cases. … In digital age, the effort to remember has diminished while value has soared.

Comprehensiveness: The third feature … is … with accessibility and durability … can get to personal information stored in different databases that have accumulated over time. [So,] … information providers can (and do) go one step further by combining such personal information from different sources and building information-rich dossiers of people, their activities and friend, their preferences, and their beliefs in a matter of seconds.

- Four dangers on digital remembering

… as we expand the use of external memory through digital remembering, we endanger human reasoning a number of ways. … First, external memory may act as a memory cue, causing us to recall events we thought we had forgotten. If human forgetting is at least in part a constructive process of filtering information based on relevance, a recall triggered by digital memory of an event that our brain has "forgotten" may undermine human reasoning. Second, comprehensive digital memory may exacerbate the human difficulty of putting past events in proper temporal sequence. Third, digital remembering may confront us with too much of our past and thus impede our ability to decide and act in time, as well as to learn. The fourth danger is that when confronted with digital memory that conflicts with our human recollection of events, we may lose trust in our own remembering.

- Six potential responses

Digital Abstinence: This approach rests on the view that individuals play a central role in responding to digital remembering. The fundamental idea is that once people understand the implications of abandoning forgetting, they will stop providing their personal information to others, and digital memory will cease to exist … The problem is that sharing personal information offers users value, which individuals may not want to forego. … Instead of expecting individuals to keep most information to themselves, people would just be expected to be more cautious when releasing information.

Information Privacy Rights: A viable and accepted mechanism of enforcement is a defining quality of information privacy rights. Such rights come in vastly different forms and shapes, … in its most basic form, information privacy rights give individuals a right to choose whether or not to share information. … Unfortunately, such legal privacy protection mechanisms [a combination of broad information privacy rights with purpose limitation rules] also come with three substantial drawbacks: they are notoriously difficult to enact, at least in the United States; they are shockingly underutilized, at least in Europe; and they depend on our society's willingness to abide by present information privacy principle in the future.

Digital Privacy Rights Infrastructure: With such a system, individuals would add metadata to their personal information detailing who can use it, for what purpose and price. Rules embedded in our computer's operating system would check this metadata and ensure that a specific usage is permissible before processing an individual's information. … one important caveat, however, … requires at least some support of the legal system, in particular in restricting attempts to circumvent the technical control mechanisms.

Cognitive Adjustment: … permanent presence of all information would lose much of its sting if we could consciously disregard old facts and accept that human are ever-changing beings, never locked into a particular state.

Information Ecology: … a deliberate regulatory constraint of what information can be collected, stored, and thus remembered, by whom and for how long. … But the prevailing political climate has not allowed rules to get enacted.

Perfect Contextualization: What if digital memory does not comprise too much information, but too little? … digital memory [is] only problematic … [because] it selectively remembers information from the past, and thus provides its users with an incomplete picture. … surveillance is about the asymmetric power to watch, a power that necessarily benefits the surveyor over the surveilled. In a world where all information is transparent, …, surveillance loses that power.

The four dangers of digital memories mentioned by Mayer-Schönberger are mostly about human beings. He solved these four well via two potential responses "cognitive adjustment" and "perfect contextualization." I particularly appreciate the latter one and agree that, a transparent society might reduce information inequality; and in such a society, comprehensive information about everything is (at least in principle) available to everyone. Also, I believe that data and technology can contribute to establishing such a society. For example, given a corpus of scientific articles and a claim about a scientific finding, a fact-checking data model can identify abstracts that support or refute the claim, specifically, it can be used to verify claims following data during global pandemics (e.g., COVID-19).

When reviewing the remaining potential responses proposed by Mayer-Schönberger, I found my evaluation of them won't be speculative. Two of them (i.e., "information privacy rights" and "digital privacy rights infrastructure") have been implemented in many societies, which is the result of the study, activism, debate, and enactment of measures to combat the perceived privacy loss in the last decade. Many societies, such as the EU, China, and Brazil,[15] successively promulgated similar regulations to become active in 2018 alone (the former two came into effect both in May, only 25 days apart), respectively. New governance of data and data product sharing can match design concepts of digital privacy rights infrastructure (see Case 4.1), which may take the form of a self-governance model like DataBox,[16] where profit sharing from the data transfer can be realized by the data owners. The last one—information ecology—has evolved into data nature in cyberspace, which will no longer require any mandated changes through correcting the political weakness, adopting new laws, or developing a novel technical architecture.

According to the EMC-sponsored IDC Digital Universe study done in 2012[17] it estimated that today's total data (in 2020) reaches 50 times that of 200, amounting to 40 ZB (zettabyte, 2^{20}). While agreeing with Mayer-Schönberger's simple solution—expiration dates on information, constructing a "rusting" memory might be unrealistic. To cope with new challenges from digital memories in the next decade, I'm going to propose some recommendations to update old ones.

---

[15] I listed the three references together to reduce space. ***See:*** (1) European Union. General Data Protection Regulation. 2016, April 27. Retrieved from https://gdpr-info.eu. (2) China. Information security technology – Personal information security specification. December 29, 2017. Retrieved from https://www.tc260.org.cn/upload/2018-01-24/1516799764389090333.pdf. (3) Monteiro, R. International Association of Privacy Professionals. The new Brazilian General Data Protection Law – A detailed analysis. August 15, 2018. Retrieved from https://iapp.org/news/a/the-new-brazilian-general-data-protection-law-a-detailed-analysis.

[16] ***See:*** Zhu Y, Xiong Y, Liao Z, *et al*. Self-governing openness of data. Big Data Research. 2018;4(2):3-14, in Chinese.

[17] ***See:*** New Digital Universe Study Reveals Big Data Gap: Less Than 1% Of World's Data Is Analyzed; Less Than 20% Is Protected. December 11,2020. PRNewswire.com. Retrieved from https://www.prnewswire.com/news-releases/new-digital-universe-study-reveals-big-data-gap-less-than-1-of-worlds-data-is-analyzed-less-than-20-is-protected-183025311.html.

In one important aspect, I envision that setting expiration dates on data products rather than raw data is an alternative. This is in line with Mayer-Schönberger's intention of setting on "information." Then, we might not do any changes to raw data. Data products are processed on raw data, and initially have a period of decline in demand; meanwhile, the continuous improvement of data tools tends to be available in the same raw data, resulting in similar data products at scale in a period. Also, I believe future data technology can produce similar but much better data products in the same raw data. For instance, a paper published on the Arxiv e-print repository often has multiple versions. There exist some "one-line" data tools[18] on GitHub which can help detect the authors' revisions between two versions. We prefer to add validity to the detection result rather than the original multiple versions of this paper.

Secondly, I envisage that hiding but no deleting any older data (e.g., comments, posts, financial information) is a better way. It is possible to squeeze the outdated (or negative) data out of people's search horizon (normally 2–3 pages), and even bury it more deeply. This is also an alternative to solve the time threat in digital memories which Mayer-Schönberger didn't solve essentially. Since this aspect may take up a lot of space, I will come back to this point by elaborating in the next section about time—another component of data.

Big Data Brain

The human brain is a complex system on multiple scales of space and time that can be decomposed into subcomponents and functional interactions between them.[19] The reason for such a complex is that the brain remembers or forgets memories, and then implements the mind. Unfortunately, the nature of the relationship between the mind and the brain is far from understood. Brain science and cognitive psychology, thus, are one of the most versatile specialty areas today.[20] All professions have a compelling interest in how memories are created, stored, and recalled in part of the brain called the hippocampus. Besides cognitive psychologists, judges, public health and safety officials, educators, curriculum designers, architects, and engineers all want to know more about the processes at work in the brain. In clinical settings, they seek to treat human mental diseases, including Alzheimer's disease, speech issues, memory loss, and sensory or perception difficulties; and in legal settings, they attempt to distinguish real memories from false ones on the witness stand.[21]

[18] ***See:*** Emken, T. Retrieved from https://github.com/temken/comparxiv.

[19] ***See:*** Turk-Browne NB. Functional interactions as big data in the human brain. Science, 2013;342(6158):580-584.

[20] ***See:*** American Psychological Association. Understanding Brain Science and Cognitive Psychology. Retrieved from https://www.apa.org/action/science/brain-science.

[21] ***See:*** Ledford H. False memories show up in the brain. Nature, 2007. DOI: 10.1038/news.2007.220.

Noninvasive studies of human brain function hold great potential to unlock mysteries of the human brain. Imaging technologies like electroencephalography (EEG) and functional magnetic resonance imaging (fMRI) allow researchers to use brain scans to read people's memories and to help uncover fundamental principles of cognition and behavior. But a single type of brain data is insufficient to analyze such a complex system. To detect metabolic activity by predicting the blood oxygenated level dependent (BOLD) fMRI signals, and by inference, neuronal activity, noninvasively throughout the brain, for example,[22] the simultaneous acquisition and analysis of EEG with fMRI is the minimum requirement. This study not only generates complex data sets: ~100,000 locations, measured simultaneously hundreds of times, resulting in billions of pairwise relations, but also remains challenging for a standard data analysis methodology to be established. As a consequence, our studies of the brain are the heterogeneity of multimodal data from different areas; for example, for some images contain free-form annotations, cross modal retrieval or multimedia event detection are required. Multimodal machine learning, which aims to build models that can process and relate information from multiple modalities, has grown in popularity in studies of the brain but provides great challenges during the phase of data integration. The better initial step of studying the memories (and even the mind) built on the physical scaffolding of the brain is to construct a "data brain" by integrating almost all available data related to the brain. In this section, I propose a data integration architecture for the future big data brain which is based on an initial research from the Shanghai Key Laboratory of Data Science, Fudan University. After explaining this architecture, I am going to provide an overview of approaches employed for integrating EEG and fMRI data.

What a great era it is to be involved in brain research, so the research content with its means (or methods) is diverse. I list commonly used brain-related data as below.[23]

Data Source 4.1 *Event-Related Potential Data* An event-related potential (ERP) is the measured brain response that is the direct result of a specific sensory, cognitive, or motor event. More formally, it is any stereotyped electrophysiological response to a stimulus. ERPs are measured by means of electroencephalography (EEG). ERP data's format depends on the manufacturers of EEG, such as Brain Products .eeg files and Neuroscan .cnt files.

Data Source 4.2 *Functional Magnetic Resonance Imaging Data* Functional magnetic resonance imaging (fMRI) measures brain activity by detecting changes

---

[22] ***See:*** Abreu R, Leal A, Figueiredo P. EEG-informed fMRI: a review of data analysis methods. Frontiers in human neuroscience. 2018;12, 29.

[23] I integrated all references related to brain data. ***See:*** (1) Multiple entries on Wikipedia include brain, list of neuroscience databases, eye tracking, functional magnetic resonance imaging, event-related potential; (2) Krans B, Murrell D. What is a brain PET scan? HealthLine.com. November 2, 2018. Retrieved from https://www.healthline.com/health/brain-pet-scan.

associated with blood flow. This technique relies on the fact that cerebral blood flow and neuronal activation are coupled. When an area of the brain is in use, blood flow to that region also increases. The primary form of fMRI uses the blood-oxygen-level dependent (BOLD) contrast, discovered by Seiji Ogawa in 1990. DICOM as a format for neuroimaging with fMRI.

Data Source 4.3 *Eye Tracking Data* Eye tracking is the process of measuring either the point of gaze (where one is looking) or the motion of an eye relative to the head. The format of data generated by different eye trackers, the devices for measuring eye positions and eye movement, is various but mostly can be exported as the text.

Data Source 4.4 *Positron Emission Tomography Data* A brain positron emission tomography (PET) scan is typically an outpatient procedure of imaging test that allows doctors to see how the brain is functioning. The scan captures images of the activity of the brain after radioactive "tracers" have been absorbed into the bloodstream. These tracers are "attached" to compounds like glucose (sugar). Glucose is the principal fuel of the brain. Active areas of the brain will be utilizing glucose at a higher rate than inactive areas. PETs' data format depends on their manufacturers.

A number of online brain science (and neuroscience) databases are available which provide information regarding gene expression, neurons, macroscopic brain structure, and neurological or psychiatric disorders, respectively. Some databases contain descriptive and numerical data, some to brain function, others offer access to 'raw' imaging data, such as postmortem brain sections or 3D MRI and fMRI images. Some focus on the human brain, others on non-human. These data sources can also be divided into multiple levels, such as genetic level, cellular level, and system level. A well-known gene-level data source is Yale's Olfactory Receptor Database (ORDB; at http://senselab.med.yale.edu) with its three companion databases OdorDB, ORModelDB, and OdorMapDB, which "dedicated to the dissemination of information related to olfactory receptors and their fictional ligands, odorants."[24] The US NIH's (National Institutes of Health) Brain Research through Advancing Innovative Neurotechnologies (BRAIN) Initiative provides cell-level data source as the Cell Census Network (BICCN; at https://biccn.org/). The system-level data is related to various brain imaging data.

Brain science involves various disciplines, such as brain regions in anatomy, brain diseases in clinical science, brain memory in neuroscience, mental disorders in cognitive psychology, brain thinking in social psychology, and so forth. Constructing big data brain should include both vertical integration and horizontal integration strategies. Vertical integration is the process of a research focus involved data and findings at different disciplines coving anatomy

[24] ***See:*** Marenco L, Wang R, McDougal R, *et al.* ORDB, HORDE, ODORactor and other on-line knowledge resources of olfactory receptor-odorant interactions. Database. 2016. DOI: 10.1093/database/baw132.

(e.g., cellular structure, evolution), physiology (e.g., neurotransmitters and receptors, electrical activity, and metabolism), and function (e.g., perception, motor control, arousal, homeostasis, motivation, and learning and memory). Horizontal integration is to combine various data and findings in the same discipline. Therefore, a minimal viable architecture for big data brain would ensure the reliability and integrity of data integration to achieve across areas, levels, and disciplines.

The complete data integration architecture for a big data brain is shown in figure 4.3 with data governance at both macro and micro levels. On the macro level, it governs of cross-border data flows by countries. On the micro level, it provides all data management practices with the necessary foundation, strategy, and structure needed to ensure that the data is managed as an asset. Various brain-related data sources from different areas, levels, and disciplines are linked together through a brain ontology. An ontology-based data brain can solve the heterogeneity of multimodal data well, besides providing multiple data tools for brain research, including data extraction, data preparation, data mining, and data visualization. It allows researchers to perform secure, private accesses or queries based on multimodal machine learning to implement various tasks, such as automatically generating medical imaging reports (e.g., image caption, paragraph generation) and transforming the free text in clinical notes into images (image generation from text).

In the rest part, I will brief the commonly used data integration for EEG and fMRI data. Note that pre-processing both two is not considered here. One the EEG side, specifically, I won't talk about how to correct several artifacts (e.g., the gradient and pulse artifacts); on the fMRI side, I won't consider the contamination of the fMRI signal by physiological noise of non-neuronal origin.

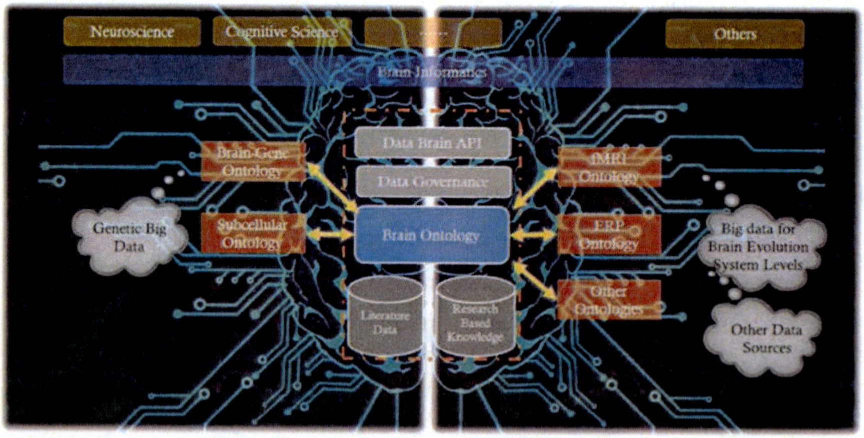

Figure 4.3 A kind of data integration architecture of big data brain. (Source: Author drew it herself)

Since EEG and fMRI are the two most commonly used noninvasive functional neuroimaging techniques, and because they exhibit highly complementary characteristics: EEG's detection ability is about spatial resolution as the shape, size, and orientation of discharge neuron, and fMRI's ability is about temporal resolution limited by the relatively slow hemodynamic response, their multimodal integration has been actively sought. The optimal method for the integration of data from the two modalities remains an open question, mainly determined by a particular application. There are generally two steps (see Footnote 22): first is to choose multimodal machine learning or other traditional data analysis approaches. When combining EEG and fMRI, we can choose symmetrical or asymmetrical approaches. Specifically, symmetrical approaches are in principle ideal to make the most of the multimodal information, since they do not constrain any of the specific modalities. The latter approaches are by far the most common because they may help researchers for a specific research purpose easily; for example, (1) fMRI-driven EEG, whereby the activated brain regions identified with fMRI are used as spatial constraints for the EEG source reconstruction problem, and (2) EEG-informed fMRI, whereby the brain activity recorded with EEG is used to predict hemodynamic changes measured with fMRI.

Secondly, we need to determine which features can be extracted for fitting our data experiment. This indirectly decides which kind of data is dominant; that is, whether our data integration strategy is in the context of the EEG-informed fMRI approach or the fMRI-driven EEG approach. All available features can be roughly partitioned into temporal features, spectral features, spatial features, functional features, etc.

Use "Fake" Data for Social Good

Salvador Dali, a Spanish surrealist artist renowned for exploring subconscious images, famously said: "The difference between false memories and true ones is the same as for jewels: it is always the false ones that look the most real, the most brilliant." It reveals the fact that memories make us who we are, yet the truth is they are far from being the accurate record we like to think they are.[25]

False memory research dates back over a hundred years to the work of psychologist Hugo Münsterberg at Harvard.[26] Münsterberg used a case to show our brains can be led astray. In the case where a woman had been found dead in Chicago, a farmer's son was stopped and accused of the murder, and after being

[25] ***See:*** Shaw J. The Memory Illusion: Remembering, Forgetting and the Science of False Memory. Anchor Canada (November 28, 2017). 2017; 304 pages.

[26] ***See:*** Dodgson L. Our brains sometimes create "false memories" – but science suggests we could be better off this way. Business Insider. December 19, 2017. Retrieved from https://www.businessinsider.com/science-of-false-memories-2017-12.

questioned by the police he confessed to killing the woman—despite having an alibi. Münsterberg concluded it was clear that the man was falling victim to the "involuntary elaboration of a suggestion" from the police officers interrogating him. Unfortunately, Münsterberg's ideas were too radical for the time, and the boy was hanged a week later. Nowadays, new techniques allow neuroscientists to look at brain scans of people having real memories and false memories to see if there's a different (see Footnote 21).

> *In one study from Daegu University in South Korea, 11 people were asked to read lists of words that fall into categories, like "farm animals." Then they were asked whether specific words appeared on the original lists, while functional magnetic resonance imaging (fMRI) detected changes in blood flow to different areas of the brain. When study participants had confidence in their answers and were correct, blood flow increased to the hippocampus — the region of the brain that is important for memory. If they were confident in their answer but were wrong, which happened about 20% of the time, the frontoparietal region lit up — the area associated with "a sense of familiarity."*

In 2013, MIT scientists implanted a false memory into a mouse by optogenetically manipulating memory engram—bearing cells in the hippocampus.[27] These neurons were later optically reactivated during fear conditioning in a different context. The recall of this false memory was context-specific, activated similar downstream regions engaged during natural fear memory recall, and was also capable of driving an active fear response. The observation supported an extraordinary hypothesis: not only was it possible to identify brain cells involved in the encoding of a single memory, but those specific cells could be manipulated to create a whole new "memory" of an event that never happened. Even if without the use of modern technology, in fact, there exists an astonishing variety of ways to implant false memories. Julia Shaw, the author of *The Memory Illusion: Remembering, Forgetting, and the Science of False Memory*, uses the latest research to demonstrate the way radically false memories can be deliberately implanted, leading people to believe that "they brutally murdered a loved one, or were abducted by aliens" (see Footnote 25). However, science suggests that we could be better off this way because false memories can actually help us make safer, more informed choices (see Footnote 26). This conclusion just happens to coincide with Mayer-Schönberger's opinion on human forgetting.

If allowing me to comment on this, I would say that two sides of the same coin. Of course, we also can use "fake" digital memories (i.e., manipulated data) for social good. Adobe Photoshop is a raster graphics editor for photo manipulation, involving the transform a photograph via digital techniques to achieve desired results. Some manipulated images are considered skillful artwork, while others are frowned upon as unethical practices, especially when

[27] ***See:*** Ramirez S, Liu X, Lin P, *et al.* Creating a false memory in the Hippocampus. Science. 2013;341(6144): 387-391. DOI: 10.1126/science.1239073.

used to deceive the public. To my knowledge, Adobe has developed an own algorithm to identify whether an image was manipulated; and the abstract of its related paper is as below.[28]

> *We propose a two-stream Faster R-CNN network [a near real-time rates using very deep network] and train it end-to-end to detect the tampered regions given a manipulated image. One of the two streams is an RGB [a color model specified with red, green, blue] stream whose purpose is to extract features from the RGB image input to find tampering artifacts like strong contrast difference, unnatural tampered boundaries, and so on. The other is a noise stream that leverages the noise features extracted from a steganalysis rich model filter layer to discover the noise inconsistency between authentic and tampered regions. We then fuse features from the two streams through a bilinear pooling layer to further incorporate spatial co-occurrence of these two modalities.*

As far as data science is concerned, manipulation and anti-manipulation—or equivalently, bad and good, are merely "the two sides of the same coin":[29] intention.

Adversarial Learning is another example that potentially carries enormous practical significance in using "'fake' data for social good";[30] especially in medicine.[31] Adversarial Learning lies at the intersection of machine learning and computer security. It aims at enabling the safe adoption of machine learning techniques in adversarial settings like spam filtering, computer security, and biometric recognition. The problem is motivated by the fact that machine learning techniques have not been originally designed to cope with intelligent and adaptive adversaries, and, thus, in principle, the whole system security may be compromised by exploiting specific vulnerabilities of learning algorithms through careful manipulation of the input data. Accordingly, the field of adversarial learning may help improve the security of learning algorithms following the three steps: (a) identifying potential vulnerabilities of machine learning algorithms during learning and classification; (b) devising the corresponding attacks and evaluating their impact on the attacked system; (c) proposing countermeasures to improve the security of machine learning algorithms against the considered attacks. But this novel technology requires both good and bad datasets to do anything useful. Governmental data providers may have such a

---

[28] ***See:*** Zhou P, Hao X, Morariu VI, *et al.* Learning rich features for image manipulation detection. In Proceedings of the IEEE Conference on Computer Vision and Pattern Recognition, 2018: 1053-1061.

[29] ***See:*** Meng X. Information and Uncertainty: Two Sides of the Same Coin. Harvard Data Science Review. 2020;Issue 2.2. DOI: 10.1162/99608f92.c108a25b.

[30] ***See:*** Levine B. University Researchers Use "Fake" Data for Social Good. Govtech.com. November 7, 2017. Retrieved from https://www.govtech.com/security/University-Researchers-Use-Fake-Data-for-Social-Good.html

[31] ***See:*** Finlayson SG, Bowers JD, Ito J, *et al.* Adversarial attacks on medical machine learning. Science. 2019;363(6433): 1287-1289.

dataset related to various bad things but must be careful about how, when, why they share them. If they're not, they risk violating the trust of the public, and even breaking the law. Data sharing agreements can help prevent privacy violations, but they require a level of specificity in the research task that is often premature for exploratory collaborations and often take a long time to establish. In this case, we consider the generation and use of synthetic data a critical ingredient in evading sensitive data at risk.

A good fake data set has two properties: It is representative of the original data and it provides strong guarantees against privacy violations. The DataSynthesizer project[32] is about developing practical, usable tools to satisfy these requirements, using state-of-the-art methods in privacy protection. The initial tool, developed by a team of collocating from two universities: The University of Washington and Drexel University, takes a tabular sensitive data set as the input and generates a structurally and statistically similar synthetic data set with strong privacy guarantees as to the output. The data owners do not need to release their data, while potential collaborators can begin developing models and methods with some confidence that their results will work similarly on the real data set. The distinguishing feature of DataSynthesizer is its usability—the data owner does not have to specify any parameters to start generating and sharing data safely and effectively.

Time: Observations and Conjectures in Data Nature

I can now present the time, which is my favorite stuff in my data research.

At the Sesquicentennial International Exposition to celebrate the 150th anniversary of the signing of the United States Declaration of Independence hosted by Philadelphia hosted in 1926, China as 1 of 25 foreign countries selected to send some exhibition pieces. In addition to porcelain and embroidery, the specific one was a large chart furnished with drawings and illustrations, titled *Diagram Showing the Evolution of Chinese over the Last Four Millenniums* (sic). This diagram received a first-place award in the educational program category. Its author was Li Jinxi, a prominent Chinese philologist. As shown in figure 4.4,[33] the horizontal axis of this diagram is the time, and the time span is 4000 years. The vertical orientation is divided into upper and lower parts, in which a black line divides the middle green time "river" into two: "Words and

[32] ***See:*** Ping H, Stoyanovich J, Howe B. DataSynthesizer: Privacy-preserving synthetic datasets. In Proceedings of the 29th International Conference on Scientific and Statistical Database Management. June 2017, pp.1-5.

[33] Interested readers may search online via any search engine using the keyword of " Evolution of Chinese over the Last Four Millenniums," and I can give a reference in Chinese. ***See:*** Retrieved from https://www.thepaper.cn/newsDetail_forward_4033620.

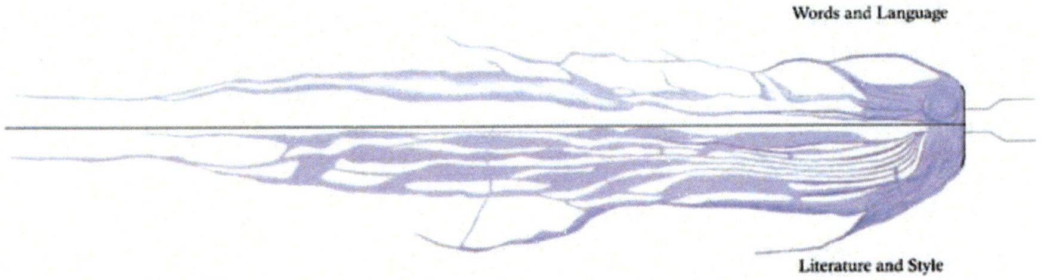

Figure 4.4 A river diagram along the timeline. (Source: Author extracted to redrew it from the *Diagram Showing the Evolution of Chinese over the Last Four Millenniums*)

Language" lying on the upper part, and "Literature and Style" on the lower. Interested readers can easily find it on the Internet. Li's diagram demonstrates the content that needs to be shown in a river-like state along the time, which is a kind of innovation. Beyond the narrow confines of geographic proximity, it provides people a temporal anchorage by shared memories. The timeline here is horizontal. As explained in a literature,[34]

> *Li laid out, in the ambitious diagram, the genealogy of Chinese – both as a language and as a writing system – from the eighteenth century to the twentieth century. He traced the development of writing in Chinese from the "hieroglyphs," the seal script to the free running hand, accounted for foreign influences upon the formation of the Chinese language, singling out Sanskrit via translations of Buddhist sutras, and included for his purpose missionary schemes of Chinese Romanization since the Ming Dynasty. ...*
>
> *Through organizing the representation of the world, it effectively organizes the world itself.4 Li's exegesis of the evolutionary history of the Chinese language compresses and visualizes the "world" of Chinese – a "national language," its literature and culture – over the past four millennia and into a two-dimensional "picture" in its most literal sense. ... The submission of the picture to the world fair accordingly epitomizes the submission of the Chinese national language to the family of world languages and to its governing rules of historical linguistics and language evolution.*

Definition 4.16 *Stream Graph*[35] A stream graph, as a variation of a stacked area graph, displays the changes in data (between the maximum and minimum) over time of different categories through the use of flowing, organic shapes that somewhat resemble a river-like stream.

---

[34] ***See:*** Zhong, Y. Script Crisis and Literary Modernity in China, 1916-1958. Doctoral dissertation, Columbia University, 2014. Retrieved from https://academiccommons.columbia.edu/doi/10.7916/D89C6VHZ.

[35] I only gave simply the definition of stream graph. With various uses, the tool can help reveal data insights. Please see Geoff Parcell and Chris Collision's book for more detail. ***See:*** Parcell G, Collision C. No More Consultants: We Know More than We Think. New York, NY: Wiley; 1 edition (September 7, 2010). 2010; 252 page.

Such a stream graph makes up for Willem Wagenaar's results well, the fact that reflects "the great difficulty humans have in sorting and retrieving memory based on the passing of time" (see Footnote 6), via external memory (i.e., painting and writing). Wagenaar is a Dutch psychologist. From 1978 to 1984, he kept a detailed diary of his life to record cues in recalling events. These cues use "what" to remember "who," "where," and "when" worked best, followed by an equally effective "where" and "who" cue. But the "when" cue failed to work in almost all cases—knowing a specific date rarely helped remember events.

When discussing "why we have difficulties sequencing events with the perspective of time in mind," Viktor Mayer-Schönberger provided some objections by a case. In the case, Jane received an invitation sent from her old friend John about a meeting. She wanted to say yes because she still remembers fondly their last meeting that is the type of pleasant wide-ranging conversation in the warm scene. But after checking for pervious emails, she changed her mind. Digital memories from a saved mailbox reminded her about some disputes, which caused her to cancel the meeting with John. Mayer-Schönberger's comments on this case are that, the pragmatic and conservative evolution lefts us with biological forgetting and didn't give us an alternative conflictive ability to correctly evaluate events in our past. But digital memories revived the old conflict Jane had failed to recall, causing her to become conflicted in her judgment, lose decisiveness, and be in danger of perhaps choosing inaccurately. This spawned a new threat and is inequity (see Footnote 6):

> *It denies us humans the chance to evolve, develop, and learn, leaving us helplessly oscillating between two equally troubling options: a permanent past and an ignorant present.*

It may be helpful to help if having more data.

Example 4.1 We may simply divide the relationship between Jane and John into "harmony" and "conflict," to sort these data points ordered in time as a time series. There is only one variable (i.e., relationship) varying over time, we call it univariate time series. Then, time series analysis extract meaningful statistics and other characteristics of the dataset to understand it. A suitable model is fitted to a given time series and the corresponding parameters are estimated using the known data values. This field of study seeks the "why" behind a time series dataset, next is to predict the future direction of the relationship between the two, then helps Jenny make decisions. Given more time series about the relationships between John and the others, it is easy to know John's attitude towards Jane compared with the others according to John's behavior patterns. Of course, the data to perceive time in data science is not only a time series, but also the ordered data. Temporal scales indicate approaches to these are not limited to time series analysis and forecasting. For example, A "stretchable" similarity expressed as lengthen or shorten in the time dimension but maintain the overall trend in the value dimension could be used to capture patterns of change in an event over time, comparison of risk factor, and make predictions.

This is why I highly agree with Mayer-Schonberg's potential response to "perfect contextualization." All our starting points are fair algorithm fairness and data reasonability, then to make our routine more convenient and more scientific decision-making via them.

The above is we know the time via common sense in data. Using a stream graph that represents the evolutionary history of scientific theory outlines the major events in the natural and social sciences during the development. In biology, for example, evolution is any change across successive generation in the heritable characteristics of biological populations. In biological data, learning any change within a designated period, or detecting disjoint time segments given to these dynamics, are important to research in this discipline.

However, the time in data nature is not so simple as we thought. Data nature is artificial, neither belonging to the real world and its part nor differing with both the micro and macro universe. But we can find all "correspondences" about these worlds in data nature. First, our real world is meant all material reality including both our bodies and the extra-organic universe, while data as a basic single element of data nature. The difference between substance and data is greatly distinct, of which I will come back in the next chapter.

Secondly, the current physical research methods have undergone fundamental changes despite most of the knowledge to be learned now was built 200 years ago. Mainly, about 400 years ago, humans knew very little about science. At that time, people studied physics to a large extent to describe the surrounding world so that scientific methods were to experiment first, then to build theory by summarizing relations among observed variables as experimental laws. Compare with now, the cost of prior experiments was meager. After applying the "control variable" that is what is kept the same throughout the experiment, countless great physical scientists were born in the stage of summarizing experimental laws from scientific experiments. But physics is no longer what it was. There are two main reasons for this: One is many already existing theories through numerous experimental tests. A new theory must be reduced to the old one at some limit, so ways to generating something new are cut a lot. A theory to explain a new fact must conform to an already existing theory. This is why, the previous Nobel Prize was awarded for a specific contribution, but now is given to someone who studied the theories proposed by predecessors. The other is related to the substantial increase in experimental costs. Take the current quantum physics as an example; an experiment aims to find the so-called "graviton," which may require an accelerator as large as the Milky Way. It is unrealistic to use experimental results to summarize the laws. Many physical experiments are either converted into data experiments or changed to the method of coming theory first before the investigation.

So far, data science that takes data nature as its research object does not have a rigorous theoretical system; meanwhile, the time is one of the great problems. Based on some observations, we have several conjectures (see Footnote 3).

Conjecture 1: No Time Exists in Data Nature

The time in physics is a scientific tool of measuring change.[36]

> *... an endless series of instants that, strung together like beads, turn an uncertain future into the present and the present into a definite past. The very concept of time allows researchers to calculate when a comet will round the sun or how a signal traverses a silicon chip. Each step in time provides a peek at the evolution of nature's myriad phenomena.*
>
> *But ... unlike mass and distance, time cannot be perceived by our physical senses. We don't see, hear, smell, touch, or taste time. And yet we somehow measure it.*

When a cadre of theorists attempt to extend and refine the general theory of relativity, there exist a big problem with time—"the illusion of time."[37] "It's a crisis arising when attempting to merge the macrocosm with the microcosm of subatomic particle," said mathematician John Baez, of the University of California at Riverside, "but the solution may take physics in a new direction" (see Footnote 36). According to theoretical physicist Caro Rovelli, our naive perceptions on time flows do not correspond to physical reality and are much more about illusory. These include Isaac Newton's picture of a universally ticking clock and Albert Einstein's relativistic space-time. The former is due to the difference in time passes, even "a clock placed on the floor runs a little more slowly than one on a table."[38] The latter is just an effective simplification, namely "an elastic manifold that contorts so that local times differ depending on one's relative speed or proximity to a mass" (see Footnote 37).

> *Rovelli is one of the creators and champions of loop quantum gravity theory, one of several ongoing attempts to marry quantum mechanics with general relativity. In contrast to the better-known string theory, loop quantum gravity does not attempt to be a "theory of everything" out of which we can generate all of particle physics and gravitation. Nevertheless, its agenda of joining up these two fundamentally differing laws is incredibly ambitious.*
>
> *Alongside and inspired by his book in quantum gravity, Rovelli puts forward the idea of "physics without time." This stems from the fact that some equations of quantum gravity (such as the Wheeler–DeWitt equation, which assigns quantum states to the Universe) can be written without any reference to time at all.*
>
> *As Rovelli explains, the apparent existence of time [in the reality] – in our perceptions and in physical descriptions, ... – comes not from knowledge, ...*
>
> *... Rovelli puts forward the idea that events (just a word for a given time and location at which something might happen), rather than particles or fields, are the basic constituents of the world. The task of physics is to describe the relationships between*

---

[36] Marcia Bartusiak is the author of *Planet3: Thirty-Two (Brief) Tales on the Solar System the Milky Way, and Beyond* which is one of the best popular physical science books. ***See:*** Bartusiak M. Is It Time to Get Rid of Time? Nautilus.com. September 20, 2018. Retrieved from http://nautil.us/issue/64/the-unseen/is-it-time-to-get-rid-of-time.

[37] This article comments Carlo Roverlli's The Order of Time. ***See:*** Jaffe A. The illusion of time. Nature. April 16, 2018. Retrieved from https://www.nature.com/articles/d41586-018-04558-7.

[38] ***See:*** Rovelli C. The Oder of Time. New York, NY: Riverhead Books; Reprint edition (December 10, 2019). 2019;256 pages.

those events ..., each of those events looks like the interaction of particles at a particular position and time; but time and space themselves really only manifest out of their interactions and the web of causality between them.

Whatever the way this controversy needs to go, we need the time in reality as a means of partitioning the past, present, and future. The "real time" as humans observe and experience it seems to have a certain direction, notably from the past towards the future. The things give time a property as the arrows of time.[39] First is the thermodynamic arrow of time, and "forward in time" (see Footnote 37) is the direction in which entropy increases. Therefore, we never see the broken glass pieces gather themselves together to form a whole glass. Even though human civilizations have tried to make things more orderly, the energy dissipated in this process has created more overall disorder in the Universe. The second is the psychological arrow of time, the direction of which towards our subjective sense about time passing. According to Stephen W. Hawking's claims, the human brain measures time in a way where disorder increases in the direction of time. We never observe it working in the opposite direction. While we may forget, we will always remember the past. Thirdly, there exists the cosmological arrow of time, and its direction is the same as the direction of expanding and not contracting of the universe. As for why humans experience these three arrows of time going in the same direction, humans have been living in the expanding phase of the Universe, Hawking called the "weak anthropic principle."

We suspect that no time exists in data nature. Someone may refute this conjecture via a timestamp that corresponds to real-world time. But based on our observations, performing each data operation is traceable (i.e., undo and go back to the past). In this way, **data nature is without the arrow of time.**

Conjecture 2: Time Will Be Achieved in Data Nature

Data nature is artificial, so errors in computer systems caused by human failure often have impacts on data nature. There are two notorious flaws related to time design: The Year 2000 and 2038 problems. The Year 2000 problem, also known as the millennium bug, refers to events related to the formatting and storage of calendar data for dates beginning the year 2000. Many programs represented four-digit years with only the final two digits to save storage overhead, resulted in the year 2000 indistinguishable from 1900. This practice became problematic with logical errors arising upon "rollover" from XX90 to XX00. Without corrective action, long-working systems would break down when the "... 97, 98, 99, 00, ..." ascending numbering assumption suddenly became invalid.

[39] ***See:*** Hawking S. A Brief History of Time. London: Bantam; 10th Anniversary edition (September 1, 1998). 1998; 212 pages.

Correcting this on techniques was not the largest part of the problem, but "60% of the time and money needed for its total compliance efforts"[40] was devoted to testing the source code change made to address the issue.

The Year 2038 problem relates to representing time in many digital systems as "the number of seconds passed since 00:00:00 UTC on 1 January 1970 and storing it as a signed 32-bit integer."[41] Like the millennium bug, caused by insufficient capacity used to represent time, the Year 2013 problem is most likely to affect embedded systems in transportation systems from flights to automobiles.

From the perspective of which humanity can correct errors, **we guess time to be achieved in data nature soon.** I'm going to illustrate by a case study about the non-/or temporal database.[42]

To store the life of Yogi Yu in a non-temporal database we use a table Person (Name, Address), in which Name is defined as the primary key of Person. Table 4.1 shows all records about Yogi's biography. After graduation, Yogi moves out, but forgets to register her new address. Yogi's entry in the database is not changed until December 27, 2015, when he finally reports it. Note that the information of Yogi living in Wujiaochang has been overwritten, so it is no longer possible to retrieve that information from the database. An official accessing the database on December 28, 2015 would be told that Yogi lives in Jiangwan. More technically: if a database administrator ran the query SELECT ADDRESS FROM PERSON WHERE NAME = 'Yogi Yu' on December 26, 2015, the result would be Wujiachang. Running the same query 2 days later would result in Jiangwan. Until her death, the database would state that she lived in Jiangwan.

Table 4.1 Using a non-temporal database for the records from a fabricated short biography

Time	Real word event	Database action	What the database shows
June 13, 1981	Yogi is born	Nothing	There is no person called Yogi Yu
July 1, 1981	Yogi's mother officially reports Yogi's birth	Inserted: Person (Yogi Yu, Wujiaochang)	Yogi Yu lives in Wujiaochang
February 1, 2015	After graduation, Yogi moves to Jiangwan, but forgets to register her new address	Nothing	Yogi Yu lives in Wujiaochang
December 27, 2015	Yogi registers her new address	Updated: Person (Yogi Yu, Jiangwan)	Yogi Yu lives in Jiangwan
April 1, 2016	Yogi dies	Deleted: Person (Yogi Yu)	There is no person called Yogi Yu

[40] The primary reference here is the entry of "The Year 2000 problem" on Wikipedia. ***See:*** Retrieved from https://en.wikipedia.org/wiki/Year_2000_problem.

[41] The primary reference here is the entry of "The Year 2038 problem" on Wikipedia. ***See:*** Retrieved from https://en.wikipedia.org/wiki/Year_2038_problem.

[42] The primary reference here is the entry of "temporal database" on Wikipedia. ***See:*** Retrieved from https://en.wikipedia.org/wiki/Temporal_database.

On April 1, 2016, the coroner deletes the Yogi Yu entry from the database. After this, running the above query would return no result at all.

A temporal database stores the data relating to time instances. It offers temporal data types and stores information about the time of the past, present, and future. More specifically, temporal databases often could be the three temporal aspects, including valid time is the time period during which a fact is true in the real world; transaction time is the time period during which a fact stored in the database was known; and user-defined time (e.g., decision time that is the time period during which a fact stored in the database was decided to be valid).

A bitemporal model contains both valid and transaction time. To enable the two time in Case 4.2, four more fields are added to the Person table as Person (Name, Address, Valid-From, Valid-Till, Entered, Superseded). As shown in Table 4.2, the bi-temporal database allows capturing this changing knowledge, since entries are never directly modified or deleted relating to "Address". Instead, each entry records when it was entered and when it was superseded (or logically deleted). This also provides both historical and rollback information. Historical information (e.g., Where did Yogi live in 2015?) is provided by the valid time. Rollback (e.g., In 2015, where did the database believe Yogi lived?) is provided by the transaction time. The answers to these example questions may not be the same—the database may have been altered since 2015, causing the queries to produce different results.

Case 4.2: A Fabricated Short Biography

Yogi Yu was born on June 13, 1981 in Shanghai, China, as a daughter of Tan Yu who lived in Yapu. Tan Yu proudly registered the birth of her first-born on July 1, 1981 at the Wujiaochang Police Station. Yogi grew up as a beauty, turned out to be a brilliant PhD student and graduated with honors in 2013. After graduation, she went to live on her own in JiangWan. Although she moved out on February 1, 2015, she forgot to register the change of address officially. It was only at the turn of the seasons that her mother reminded her that she had to register, which she did a few months later on December 27, 2015. Although Yogi had a promising future, her story ends tragically. Yogi Yu passed due to illness at the Renji Hospital on April 1, 2016. The physician reported her date of death on the very same day.

Such extensions dealing with the fact offer the comprehensiveness of explaining each data operation. Even if a trackbacking data operation, it will be saved as a new record. In this way, all data operations will increase the amount of data in the database. This means the arrow of time towards the increasing database. In a sense, data nature can preserve all the states of the real world; namely, infinite realities are mapped, stored, and indexed along the direction of increasing the data, wholly and parallelly. With this arrow, there exist constitute one new,

Table 4.2 A comparison of using the non-/bi-temporal database for a fabricated short biography

Time	Non-temporal database records	Bi-temporal database records
June 13, 1981	NULL	NULL
July 1, 1981	Insert: Person (Yogi Yu, Wujiaochang)	Insert: Person (Yogi Yu, Wujiaochang, 13-Jun-1981, ∞, 1-Jul-1981, ∞)
February 1, 2015	Person (Yogi Yu, Wujiaochang)	Person (Yogi Yu, Wujiaochang, 13-Jun-1981, ∞, 1-Jul-1981, ∞).
December 27, 2015	Person (Yogi Yu, Jiangwan)	Update: Person (Yogi Yu, Wujiaochang, 13-Jun-1981, 1-Feb-2015, 27-Dec-2015, 27-Dec-2015) Insert: Person (Yogi Yu, Jiangwan, 1-Feb-2015, ∞, 27-Dec-2015, ∞)
April 1, 2016	Deleted: Person (Yogi Yu, Jiangwan)	Person (Yogi Yu, Wujiaochang, 13-Jun-1981, 1-Feb-2015, 27-Dec-2015, 27-Dec-2015) Update: Person (Yogi Yu, Jiangwan, 1-Feb-2015, 1-April-2016, 1-April-2016, 1-April-2016)

discrete (seemingly continuous due to a timepoint that sometimes accurate to a tiny fraction of a second) dimension (or dimensional plane) that are infinitely close to the present. Of course, the premise of this assumes the unlimited storage capacity of the computer systems. But this premise can be easily simplified as: the growth rate of storage capacity keeps greater than the speed of accumulating the data. Note that data nature can be reduced theoretically, but it is a different concept relating to our discussion.

Conjecture 3: Data Nature Has Time Paradoxes

I mentioned above time is the first scientific tool. Without this tool, we cannot distinguish the past and future. Even if given the arrow of time in data nature as the direction of increasing the data, however, data refers to "digital memories," which only represents the past, and present or future not included. Also, given a trackbacking data operation, we cannot judge the data belonging to which end of this operation is older, so the data itself (data carriers not included) is without any concept of depreciation over its life expectancy. This is a paradox and may be confusing a little bit. Conceptually, a paradox is a statement that leads to an apparent self-contradictory conclusion based on the original premises of the problem. **We suspect that data nature has more time paradoxes**, just like our realistic nature.

Time Paradox 1: Time in data nature is like virtual time This means time is merely the "projection." While not everyone buys Rovelli's thoughts, this set of statements on "a complex network of events that projects sequences of the time" makes sense in data nature, which can be used to explain this paradox. Before proceeding, I'm going to introduce virtual data and actual time via a simple figure (see figure 4.5).

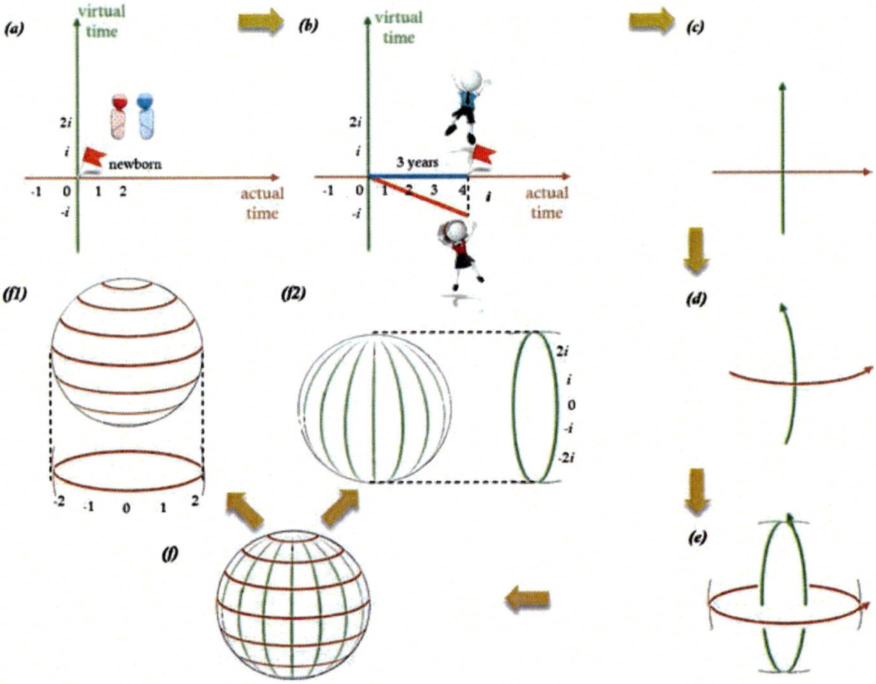

Figure 4.5 Virtual time and actual time. (Source: Author drew by herself according to some materials on the Internet)

Stephen W. Hawking proposed black hole evaporation (i.e., Hawking radiation) and singularity theorems (i.e., Penrose-Hawking singularity theorems) in his notable book *A Brief History of Time*. But they are not all. He even did self-overturning through the virtual time, which overthrew spacetime singularities, his personal research findings. The "virtual" and "actual" time axis cross vertically, just like the complex plane in mathematics: the horizontal number line is the real axis, and the vertical number line is the imaginary axis.

Example 4.2 Suppose a twin (a boy and a girl) was born at the moment of origin as shown in figure 4.5a. The boy grew up on the earth, while the girl traveled interstellar at the speed of light with her father. Three years later, the twin met together on Earth. It seems that the boy and girl are in the same position on the actual time axis, but as shown in figure 4.5b, there exists an interval of i on the virtual time axis between the two. Thus, the actual time is our current real earth time, and the virtual one reflects the time difference after having an interstellar travel. Next, we extend the two-dimensional time axis into a higher three-dimensional space, see figure 4.5c–f. So, an earth-like shape wrapped by a latitude and longitude network appeared. As shown in figure 4.5f1, the virtual time axis has no beginning and no end. Time becomes a ring, and the past is connected to the future. Similarly, as shown in figure 4.5f2, the actual time axis also has no head and no tail. This means no past and no future; and all represent present. Singularities previously claimed by Hawking himself are that, the big bang resulted in leaving boundaries of time and then falling into the abyss of time. But the virtual time proved that there exists no singularity in the universe, the point of the big bang where time begins. In this way, time no longer has only one direction and is round as the earth; there are countless paths from the past to future like "all roads lead to Rome." How does it apply to data nature? We know there exists no present and future in data nature. Given a past point that is close to the present, it is possible to observe a past and then to predict multiple futures.

Time Paradox 2: Bootstrap Paradox This paradox is a paradox of time travel that questions how something that taken from the future and placed in the past could ever come into being in the first place. It is a common trope used by science fiction writers and has inspired plotlines in everything like *Doctor Who*. In the book of the University of Massachusetts' David Toomey *The New Time Travelers: A Journey to the Frontiers of Physics*, there exists an example:[43]

> *a time traveler buys a copy of Hamlet from a bookstore, travels back in time to Elizabethan London, and hands the book to Shakespeare, who then copies it out and claims it as his own work. Over the centuries that follow, Hamlet is reprinted and reproduced countless times until finally a copy of it ends up back in the same original bookstore, where the time traveler finds it, buys it, and takes it back to Shakespeare. Who, then, wrote Hamlet?*

This paradox can be extended to a potential difficulty in medicine given by Viktor Mayer-Schönberger's *Delete* (see Footnote 6):

> *Suppose one's entire medical information was in some accessible digital storage. ... How would information added over a person's lifetime be interpreted given that the contexts in which the individual information bit had been collected over the years and decades varied greatly? What one doctor might have thought of as a "troubling response" to a particular therapy or medication, for example, may be described as a "normal reaction" by another, or even by the same person with a couple of years of extra experience.*

This is posting the question of interpretative consistency relating to the time. Another example is within a general reliance on big data for prediction in the criminal justice system, such as determining whether "future" guilty should be confirmed, or whether parole should be granted, based on potential personal tendencies obtained through behavior analysis. In the case of incremental data, careful consideration should be given to the order of using data sampling and

[43]There are two references. ***See:*** (1) Jones PA. 10 Paradoxes That Will Boggle Your Mind. Mentalfloss.com. February 10, 2016. Retrieved from https://www.mentalfloss.com/article/59040/10-mind-boggling-paradoxes; (2) Toomey D. The New Time Travelers: A Journey to the Frontiers of Physics. New York, NY: W. W. Norton & Company (July 1, 2007). 2007;400 pages.

accepting testing results; for example, how to weight alternately appeared opposite results when adding relevant data multiple times.

Besides, paradoxes are omnipresent in algorithmic problems in the real world, causing them mapped in data nature. Someone argues that if given more data, artificial intelligence might not be affected because it doesn't have a notion of common sense. However, most problems require human analysis and intervention and are in human-curated datasets. We are thus going to live in a universe of paradoxes for quite some time. Interested readers may do some research on this in which some are not very relating to the time, so I will not repeat them here.

Chapter 5
Particularities in Data as Capital

In Chap. 2 "Principles of Capital," I briefly introduced previous research from Jonathan Haskel and Stian Westlake,[1] Thomas A. Hemphill,[2] and Carol Corrado *et al.*[3] on intangible capital. According to a summary given by Haskel and Westlake, the intangibles have four unusual economic characteristics. They are: scalability (meaning that intangibles can be used repeatedly and in multiple places at the same time), sunkenness (involving irrecoverable costs because intangible investments with high irrecoverable costs can be difficult to finance), spillovers (meaning that some ideas to construct intangibles are simply nonexcludable, while choosing keep trade secrets), and synergies (that is a major driver of intangible synergy at both the company and industry levels, as well as for the national and local economies). Of course, Haskel and Westlake tended to attribute intangible capital to the category of "knowledge."

When considering "data as a capital," I must first introduce certain particularities of the data itself. I begin by noting that these particularities are the key to understand data rights.

In actuality, one particularity of data is about value (as our active pursuit of the goal), just as big data as a technical terminology built around the notion of the 5V[4] (i.e., volume, velocity, variety, veracity, value). In other words, the

---

[1] ***See:*** Haskel J, Westlake S. Capitalism without Capital: The Rise of the Intangible. New Jersey, NY: Princeton University Press; Reprint edition (October 16, 2018), 2018: 296 pages.

[2] This article is a recommendation for Haskel's and Westlake's book. ***See:*** Hemphill TA. Capitalism without Capital: The Rise of the Intangible Economy by Jonathan Haskel and Stian Westlake. CATO Journal. Fall 2018. Retrieved from https://www.cato.org/cato-journal/fall-2018/capitalism-without-capital-rise-intangible-economy-jonathan-haskel-stian.

[3] While this article was published in 2009, there is some evidence that this study began in 2005. ***See:*** Corrado, C., Hulten, C., & Sichel, D. Intangible capital and US economic growth. Review of Income and Wealth, 2009;55(3):661-685.

[4]. There are two primary references here. ***See:*** (1) The entry of "Big Data" on Wikipedia. Retrieved from https://en.wikipedia.org/wiki/Big_data; (2) Lo F. What is big data? DataJobs.com. Retrieved from https://datajobs.com/what-is-big-data.

© Springer Nature Switzerland AG 2021

C. Tang, *Data Capital*, https://doi.org/10.1007/978-3-030-60192-8_5

essence of this particularity meaning is to investigate whether a data is valuable to analytics, or to obtain the value at the desired time; and how important value is, depending on any possible production.[5] This is very different from the substances/materials in mother nature. Every substance usually has a value, because the existence of physical reality, no matter in eastern and western philosophy, is objective and no transferring with the will of a human. Consciousness does not influence the existence of these substances. A human being may use or process them through the ability to work to meet the needs of life, of which the basis is, however, the recognition of consciously acknowledging and obeying the objectivity of nature. The human cannot create any substance. On the contrary, the data is human made. Without producing to obtain value, we need to spend in disguise. Even though dropping the data center to the bottom of the ocean that includes a complete self-sustained, powered entirely through offshore wind, wave or tidal generators, and seawater cooling units;[6] or enlarging the capacity of the storage media to include advanced data storage techniques, such as synthesized DNA as a target that can encode palm-sized artificial DNA capable of holding as large as 3 ZB (zebibyte, 2^{70}) of data, each stage from design,[7] implementation, usage to operation is very expensive. Take the bitcoin as an example, according to research conducted by Elite Fixtures, there exists no profits when a single bitcoin prices less than US$8038.[8] The cost of mining a bitcoin depends on multiple types of costs, including equipment, electricity, maintenance and cooling costs. Note that mining a bitcoin is irrelevant to detect data value because it is merely a simple hash calculation based on computer performance, such as short response time for a given piece of work, high throughput, high availability of the computing system, and so forth. In short, it means no value but high costs without data on production.

Another particularity of data is shareability (based on reproducibility), which differs from natural materials. In physical reality, **shareable stuff means something is jointly owned**. For instance, if one couple owned an apartment together, then the condo should be done as a property split when they break up. **Data sharing is essentially different, referring to separate ownership among the owners**. Each owner has the same or an almost similar one (involving data volume, data form, data content, etc.)—a nearly full copy of the data. Note that it much more makes sense to use the similarity to describe two data copies, instead

[5] ***See:*** Zhu Y, Xiong Y. Defining big data. Big Data Research. 2015;(1)1: 71-81.

[6] Microsoft's project Natick is newer than Google's patent US7525207. ***See:*** Irving M. Microsoft deploys underwater data center off the coast of Scotland. Newatlas.com. June 7, 2018. Retrieved from https://newatlas.com/microsoft-project-natick-offshore-data-center/54943.

[7] ***See:*** Kosuri S, Eroshenko N, LeProust EM, *et al.* Scalable gene synthesis by selective amplification of DNA pools from high-fidelity microchips. Nature biotechnology. 2010;28(12), 1295.

[8] ***See:*** Hankin A. Here's how much it costs to mine a single bitcoin in your country. Marketwatch.com. May 11, 2018. Retrieved from https://www.marketwatch.com/story/heres-how-much-it-costs-to-mine-a-single-bitcoin-in-your-country-2018-03-06.

of the same. While we often advocate data sharing: from raw data to data products, it becomes more and more difficult nowadays. The reason is related not only to our willingness and wishes but to the uncertainty of data. A huge amount of data is generated every day like data explosion, in which the unknown is incredibly increasing. For example, "Microsoft Flight Simulator 2020" indeed gave "the entire Earth to the player" using textures and topographical data, in which the data obtained from satellites is up to 2 PB (pebibyte, 2^{50}). The data out of control include but not limited to whether the shared data is correct or accurate, what kind of data can give and what cannot, how the consequences of sharing are.

According to the two particularities, we found that in addition to the economic characteristics of intangibles, the potential impacts with data as capital have another four. Namely, profit multiple, competition intensifies, convergence effect, and latecomer advantage. In the next chapter, I will analyze interesting variations caused by the four aspects. For now, the fact that data capital has four characteristics roughly will be useful to keep in mind.

Data Sovereign: Conceptual Innovation and Economic Inheritance

With data explosion, we live in a data-rich world. One may assume that data-rich markets mainly equal data resource rich. After all, "far more information permeates data-rich markets compared with conventional ones, and Internet bandwidth has been increasing steadily with no end in sight."[9] According to some investigations of Internet use given by leading network technology providers such as Cisco,[10] global network performance will continue to be accelerating: by 2023, double speeds on fixed broadband (around 110.4 Mbps up from 45.9 Mbps in 2018), triple speeds on mobile and Wi-Fi (43.9 Mbps up from 13.2 Mbps in 2018, in which the average 5G connection speed will reach 575 Mbps). There are merely necessary technical developments on accelerating speeds, but they are insufficient. We suggest that, in our data-rich future, how fast we process information will be far less important than how well and how deeply we do so. Even though we speed up the communication of price on the market, like (ultra or) high-frequency trading on the stock market we've already done, we'd still be "oversimplifying" (see Footnote 9). What we need is "decen-

[9] ***See:*** Mayer-Schönberger V, Ramge T. Reinventing Capitalism in the Age of Big Data. New York, NY: Basic Books (February 27, 2018). 2018; 288 pages.

[10] ***See:*** Cisco. Cisco Annual Internet Report (2018–2023) White Paper. March 9, 2020. Retrieved from https://www.cisco.com/c/en/us/solutions/collateral/executive-perspectives/annual-internet-report/white-paper-c11-741490.html.

tralized decision-making" as much as possible—analyzing the same data to produce various distinct data products, then to achieve profit multiple. But one vital issue is coming—what data can be used by us to analyze.

When seeing the data as an economic resource, we found there exists almost no available data to be analytical. Some data sources are free, but it requires high costs for data cleaning or manual labeling. Some data sources are to use easily, but the analysis seems to perform with walking on the edge of ethics due to a lack of definition in law. They involve not only informed consent, privacy issues, security problem but also commercial or non-commercial conflicts of interest. When previously considering the emerging data economy, as the majority, I once envisioned the key to solve this is from attribution analysis that begins by identifying the asset class in which the data is invested. However, I'm too naive because finding a correct data source to support investments is challenging. Studies by beginning from the concept of "asset" is a way that is "easy to in but difficult to out." Instead, choosing a starting point from a perspective of "capital" can "change numerous for brief." They are specialized, field-specific terms from economics (and finance or accounting) and may have very specific definitions in different contexts. I'll give you the general flavor by an answer found on the Internet.[11]

> *Capital is generally a means to producing something else. For the most part, this term is used from the perspective of a business. Its capital consists of all of its resources (buildings, machines, patents, cash, etc.) that are used in the course of business. If the business has some wealth set aside but not likely to be used in the course of its business, it is less likely to be considered capital. ... The main message conveyed by the use of "capital" that it is a tool used to create wealth.*
>
> *An asset is a more general term. It is anything that has value. There are real assets, which are physical or intellectual property that have value and can be sold. If they are being used to produce something, they may also be considered capital, but this is not necessary. There are also financial assets, which are things like stocks and bonds. These do not directly produce any real wealth. Instead they are contracts that entitle the owner to a stream of cash flows. It is not necessary that an asset be expected to make more money than it originally cost. The word "asset" conveys the idea that something is valuable and could probably be sold if you wanted to.*

In sum, capital provides as a strong signal of "trust and confidence" as equating with power. A perspective of looking data as an asset is almost thinking of discounts (allowed in realization account); namely, presenting a deal via pricing instead of producing the data. Such a situation seems to describe as the following sentences given by Mayer-Schönberger (see Footnote 9):

> *Price and money have proved to be an ingenious stopgap to mitigate a seemingly intractable challenge, and it worked – to a degree. But as information is compressed, details and nuance get lost, leading to suboptimal transactions. If we don't fully know*

[11] ***See:*** The answer given by the user named "farnsy." Stackexchange.com. December 24, 2018. Retrieved from https://ell.stackexchange.com/questions/190598/difference-between-capital-investment-and-asset.

what is on offer or are misled by condensed information, we will choose badly. For millennia, we tolerated this inadequate solution, as no better alternative was available.

Whose Data Is It Anyway?

The concerns over consumer data privacy have reached the global stage. More and more insightful people often remind us of "an inequality when deciding whose data belongs to whom"[12] among consumer data. As shown in Case 5.1 (see Footnote 12), we signed the terms of consumption to experience customized transactions ecommerce vendors provided. After all, if we are searching for a car seat or a stroller online, we would not prefer that advertisement in our

Case 5.1: A Scenario of Authorizing Consumer Data Privacy

We've all been there: we've downloaded a new app or signed up for a new account online, only to see a permission screen. "This app will post on your behalf," or "This site will access your contacts list." You must click "I agree" to the terms and conditions to move forward, and you're expected to scroll endlessly while you read legal terms and conditions.

All too often, consumers don't read those terms or really understand the permission they've just granted the company when they click "I agree." Yes, it may be a pain to read the seemingly never-ending legalese, but you really should. One reason it is so important is because the lengthy document states what the company who owns the app or site can do with the data they will collect about you. It's important for individuals to know what can happen to their data, especially if they blindly agree to whatever is stated in these terms.

If you look hard enough, you may find a few scary items that aren't legally hidden in the terms and conditions. Facebook, for example, has the ability and the right to track your location at any time if you're using its app on your mobile device, and can even map and store an image of your face (it's how you can click a photograph to "tag" someone in the picture). Tech manufacturers can "brick" your device permanently if they find out you pirated some software. Some sites that let you upload a photograph and share it with others have the right to use that photograph or even sell it to advertisers, all without your knowledge or permission.

After all, you agreed to it.

[12] ***See:*** Velasquez E. Whose data is it anyway? Consumer intent matters when deciding whose data belongs to whom. CSO. February 1, 2018. Retrieved from https://www.csoonline.com/article/3251721/whose-data-is-it-anyway.html.

social media feed be for sales and deals on baby items rather than a chainsaw. But it should be seen, and this is what I mentioned repeatedly, it was fair from the beginning if no considering any data brokers' over illegal or unlawful behaviors. Meanwhile, consumer privacy data is not all in our data-rich markets.

Here I'm going to introduce another inequality about "whose data belongs to whom" relating to data producers. In the data industry, data producers are the entrepreneur who "innovates and applies new combinations of factors of production;"[13] meanwhile, since the data is not directly from nature (but is created and used by computer technology), we should understand "data belongs to the producer reasonably."[14] It is unfair to blame one who has been pursuing data value innovation. I'm going to mark each paragraph as the serial number for a discussion in the next section.

[Paragraph 1] While possibly being a scientific problem to discover potential value through a proper data resource, such as comment-analysis tasks to covert thousands of online reviews into a one-sentence summary seem to need a novel algorithm to crawl and mine the online review data, the key to this is the distributional question center around data (and data resources). This results that we often see controversial issues caused by data rights (e.g., ownership rights in data, the right to data privacy) among governments, or between governments and private actors. A recent deal (in January 2018) between the US money transfer company, MoneyGram, and Chinese firm, Ant Financial, owned by business conglomerate Alibaba Express, fell through over foreign interest around concerns about citizens' data privacy. The US$1.2 billion deal resulted in a US$30 million termination fee paid to MoneyGram when the US government shut down the deal. One thing to consider in looking at this debacle is that the personal information or identifying data would be in the hands of a foreign government.

[Paragraph 2] To clarify income and output concerning data rights, it is not enough to use a variety of existing terms since they are confusing and nearly interchangeable (see figure 5.1). There exist some terminologies developed with cloud infrastructure when addressing private data across borders, such as data residency and data localization.[15] Data residency refers to the geographical

[13] ***See:*** Mee, S. Joseph Schumpeter and the business cycle: An historical synthesis. 2015. Retrieved from https://www.tcd.ie/Economics/assets/pdf/SER/2009/simon_mee.pdf.

[14] This viewpoint is based on a consensus in the data science community. I here quoted the two sources: one is from my doctoral advisor Yangyong Zhu's a popular science book, and the other comes from Kevin Kelly has made a career in predicting the future of technological advances. ***See:*** (1) Zhu Y. *Beautiful Data: Understanding Big Data in 100 Minutes*. Shanghai: Shanghai Scientific & Technical Publishers; (October 1, 2018), 2018: 184 pages, in Chinese; (2) Ogilvy J. The future according to "Wired" editor Kevin Kelly. Forbes. 2017, March 30. Retrieved from https://www.forbes.com/sites/stratfor/2017/03/30/the-future-according-to-kevin-kelly/#46b07cbf76bf.

[15] ***See:*** Box J. Data sovereignty vs data residency vs data localization. Insights For Professionals. March 12, 2019. Retrieved from https://www.insightsforprofessionals.com/it/storage/data-sovereignty-data-residency-data-localization.

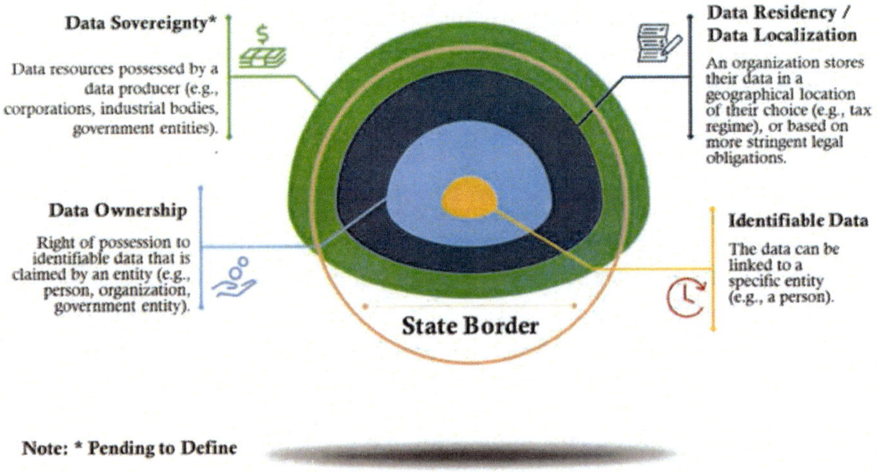

Figure 5.1 Complicated terms related to data rights. (Source: Author drew it herself)

locale in which any organization stores its data (e.g., a better tax regime). Data localization differs from data residency in that not only is the data stored in a designated location, but it is additionally subject to the laws of the country (i.e., data localization laws) in which it is physically stored. In addition, data ownership can be used to present the three terms. Data ownership also involves the real location of the entities who were collected their data. However, partitioning boundaries in cyberspace is tough[16] so that no way to avoid the confusion. A new term to appropriately fill in a terminology gap is needed as it enables the right to data portability, such as in where and how to set data resource boundaries in cyberspace.

[Paragraph 3] We propose defining the term "data sovereign"[17] [noun] as a data producer with the ability to possess and protect the data. Here, the word "sovereign" is borrowed from a fundamental economic notion of "consumer sovereignty,"[18] where William H. Hutt coined to denote the concept according to which, "in the market economy, the production decisions of entrepreneurs are rigidly governed by the freely expressed spending decisions of consumers."[19] We believe that data producers in the emerging data economy are less vulnerable

---

[16] ***See:*** Kalir E, Maxwell EE. Rethinking boundaries in cyberspace: A report of the Aspen Institute internet policy project. 2002. Retrieved from https://assets.aspeninstitute.org/content/uploads/files/content/docs/cands/RETHINKCYBERSPACE.PDF.

[17] ***See:*** Tang, C., Plasek, J.M., Zhu, Y. *et al.* Data sovereigns for the world economy. Humanit Soc Sci Commun 7, 184 (2020). https://doi.org/10.1057/s41599-020-00664-y.

[18] ***See:*** Hutt WH. The concept of consumers' sovereignty. The Economic Journal. 1940;50 (197): 66-77.

[19] ***See:*** Salerno JT. The essence of Hutt. Mises Institute. Mises.org. December 30, 2009. Retrieved from https://mises.org/library/essence-hutt.

(just like the consumers in pervious industrial sectors) so that they need some powers. Our goal is to eliminate some misunderstandings regarding data resources—the correct understanding is, data resources belong to data producers instead of a country or the person providing the data. Some national regulations make data ownership to be the state, which has the potential to significantly disrupt the global "data" economy paths to prosperity. Our view stands on their opposite side and may cause controversy in both academic and industrial circles. We recommend using the power of decentralized markets to protect data resources via non-legislative acts—production and protection from attack by seeing the data as capital.

[Paragraph 4] Capital equates with power. German economist and sociologist Max Weber pointed out that the individual's "power" is his or her ability to control any economic resources. Weber's classic definition of "politics" is "the pursuit for a portion of power or for influencing the division of power whether it is between … groups of people,"[20] which is inseparable from Weber's conception of the state. Weber notes that "the state is seen as the sole grantor of the 'right' to physical force" (see Footnote 20) used to justify the legitimacy of political domination. In other words, the basis for Weber's conception of the state is a superior means as the right to physical force. By analogy, since data as a resource, the organizations which protect data resources from any attack may have a character of the state. This logic might not entirely persuade some people. We give a real example as follows.[21] In 2011, the New York Times published a report *Island Nations May Keep Some Sovereignty if Rising Seas Make Them Uninhabitable*, that highlighted some small island states in the Pacific and Indian oceans (e.g., Marshall Islands, Tuvalu, Kiribati) could remain United Nations (UN) member states and maintain sovereign government status. A root cause is ocean territories that can be allotted resource exploitation rights as defined by the UN Convention on the Law of the Sea. By the same token, the fact that the data to be "the world's most valuable resource"[22] is the current consensus despite of being intangible. This also is the original intention of the US to promote offensive cyber operations.

[Paragraph 5] We defined data sovereign as follows.

---

[20] There are two references. ***See:*** (1) Weber T (translator). Weber's rationalism and modern society: New translations on politics, bureaucracy, and social stratification. Palgrave Macmillan; 2015 edition (April 8, 2015), 2015: 245 pages; (2) The entry of "Politics as a Vocation" on Wikipedia. Retrieved from https://en.wikipedia.org/wiki/Politics_as_a_Vocation.

[21] ***See:*** Gronewold N. Island nations may keep some sovereignty if rising seas make them uninhabitable. The New York Times. May 2011. Retrieved from https://archive.nytimes.com/www.nytimes.com/cwire/2011/05/25/25climatewire-island-nations-may-keep-some-sovereignty-if-63590.html?pagewanted=1.

[22] ***See:*** The world's most valuable resource is no longer oil, but data. The Economist. May 2017. Retrieved from https://www.economist.com/news/leaders/21721656-data-economy-demands-new-approach-antitrust-rules-worlds-most-valuable-resource.

Definition 5.1 *Data Sovereign* Data sovereign is a data producer with the ability to defend data resources against any attack.

The so-called "attack" in this definition is with elasticity—not merely in physical form but also as cyber-attacks, scaling from a minimum value that closes to zero to "infinity;" and in case of infinite attack should be in line with the same level of offensive operations from the US cyber force. Here, the word "defend" means possess, protect, and so forth.

[Paragraph 6] Given the willingness to data producer could effectively align incentives for data innovations on various data resource protections. Using "force" to protect data does not imply that abandon data sharing. Rather, it should be easy for an organization (e.g., corporations, industry bodies, or government specifies) to enable sharing of data and data products internally or with trusted partners.

[Paragraph 7] When it comes to external entities, data sharing and portability rights are often the main commodities to be sold or traded. Like the transfer of state sovereignty in international relations,[23] utilizing the concept of a data sovereign enables the transfer of rights to data sharing and/or portability. The data sovereign can transfer partial rights to governments or other industrial competitors through a modest negotiation. Such a transfer is proactive and strategic when developed by anticipating or analyzing trends and reviewing the organization's past performance concerning externally induced crises or threats. When it comes to data resources, it is often too late to execute a reactive transfer when opportunities arise as these defensive transfers often incur or result from a considerable loss of data asset valuation.

[Paragraph 8] A data sovereign is inherently concerned with the movement of data may be amenable to the free flow of data across geopolitical borders. We often take for granted that we live in an interconnected world. With the global adoption of "data sovereign," boundaries in cyberspace will have been appropriately defined.

[Paragraph 9] Data sovereign enables economic agglomeration in cyberspace. Utilizing data as an economic resource usually has the threat of new entrants and substitute products. When a data sovereign cannot effectively protect their own data resources, it may be prudent to join an alliance of data sovereigns for added protection. This concept of a "data sovereign" may help form data industry clusters in cyberspace that can grow the global economy through their collaborations. Toward proactive defense and transfer reflects the data sovereign's subjective willingness, which is in line with the capitalist ideas of using the power of decentralized markets to protect data via non-legislative acts and the scientific spirit—to let data to call every shot.

[23] ***See:*** Taylor, C. R. (1997). A modest proposal: statehood and sovereignty in a global age. U. Pa. J. Int'l Econ. L., 18(3), 745-809.

Responding to an Imaginary Reviewer About Data Sovereign

For the sake of completeness, I should make clear the aspects of the term data sovereign from different perspectives. Consider the following hypothetical situation; an email was received from a reviewer concerning these questions.

> *Hi,*
>
> *Your manuscript which you submitted has been reviewed. We think it might be of interest to our readers. The comments of the reviewer(s) are included at the bottom of this letter.*
>
> *The reviewer(s) suggest some revisions to your manuscript. Therefore, I invite you to respond to the reviewer(s)' comments and revise your manuscript.*
>
> *I hope you will find the comments useful. Again, please do not hesitate to email the Editorial Office if you have any questions during the revision or submission process.*

Reviewer: [For Paragraph 2] Stripping away some words, the last sentence of the second paragraph says, "A new term ... is needed ... enables the right ..." I find that odd. A term doesn't enable rights; legal frameworks and social norms enable rights. Terms just provide a common language for discussion.

Me: Thanks. Terms truly offer a common language for discussion, and an extensive debate may reach a consensus, which is our goal. Also, legal frameworks and social norms need an agreement when coming to new stuff. So, the word "enable" here is an indirect result to achieve "the right ..." We need terminology to discuss since there exists no consensus, which then helps avoid a relapse into the previous dispute over the data rights. Note that the term data sovereign we coined here is both conceptual innovation and economic inheritance. That is to say; the term inherits Max Weber's classic definition of "power" and uses William H. Hutt's meaning in "consumer sovereignty" to strength the power. One particularity of data is about value. Whoever develops the data can own it, and whoever no participating in this processing does not belong to the data sovereign.

Reviewer: [For Paragraph 3] The term "data producers" is used but not defined, and it is used in a way that refers to a different entity than some readers will expect. The entities generating data are not producers; rather, it the entities capturing/organizing the data. That won't be intuitive to some readers.

Me: Thanks. A data producer is any agent (e.g., corporations, individuals) who can pursue adding value to row data or other primary data products. I defined one data producer that does not refer to an individual who provides personal data. Note that industrial bodies and government specifies also included if they try to extract data value from processing data. Still, some firms may not if they just possess data without analyzing. For example, the patient cannot be a data sovereign. Although we envision the future patient-lead data sharing is on empowering them to be digital stewards of their health data. One reason is they didn't analyze or process their data that is too few to do.

Reviewer: [For Paragraph 3] The analogy to "consumer sovereignty" is not fully developed. Given that this comparison is made in the fifth paragraph, I would have expected a more fulsome discussion. Consumer sovereignty captures the notion that the independent actions of many consumers serve to discipline the choices made by entrepreneurs. Are the authors arguing that the independent actions of many "data sovereigns" in a market economy are disciplining the choices of another group of entities? If so, what entities? If not, then this comparison doesn't seem helpful.

Me: Thanks. According to Fisher-Clark's classification, a difference in the data industry compared with the traditional ones is the "increasing profits of other industries" property of the quaternary sector. In the conventional industry like manufacturing, it is generally large-scale production. The consumer has no more choices relating to the merchandise and belongs to a vulnerable group. Contrarily, data products aim to meet consumer demand, such as targeted advertising and customized processing, so that data producers are more susceptible. Do you accept such an interpretation to understand why we inherit the economic concept of "consumer sovereignty"?

Reviewer: [For Paragraph 4] The point of this paragraph is unclear, possibly because it lacks a topic sentence and instead starts with a series of definitions followed by a confusing discussion of "power." Also, this paragraph suggests that all you need to be a data sovereign is the ability to possess and defend data from attack. By that logic, data collected by ill means don't preclude you from being a data sovereign. Of course, those means may violate other laws, but nowhere is it discussed how this notion of data sovereign interacts with legal frameworks. Could you give some examples?

Me: Thanks. I frequently used the word "power" in both paragraphs 3 and 4, but they have different meanings. Weber defined "power" as the ability to control any economic resources. The data to be valuable resources is the current consensus, so that controlling data resources means to have "power." However, keeping data resources is not so easy, and requires a "force," we call data force, which comes from data technology. The scope of protection against attaches is vast. Here are two examples. Some adult websites are often criticized for profiting via porn revenge porn,[24] deepfake (AI-generated videos that realistically edit new faces onto pornographic actors), or anything published without a person's consent or permission. Such criticism is an attack. Removing all said content simply might not be a great solution because an attack has been made. Data tracking technology is needed when

[24] *See:* O'Brien SA. Woman awarded $6.45 million in revenge porn case. CNN. April, 2018. Retrieved from https://money.cnn.com/2018/04/09/technology/revenge-porn-judgment/index.html.

uploading. The other is for the data breach scandal.[25] If causing that the data can self-delete if being replicated to strange server environments without permission, it is an excellent anti-attack data technology.

Reviewer: [For Paragraph 6] I did not understand the point of that sentence "Given the willingness to data producer …" I think the entire point of the second half of the paragraph is just to say, "well-defined property rights over data can help facilitate data sharing agreements."

Me: Thanks. You are right. Data sharing is on well-defined property rights. But property rights are irrelevant to our topic. We believe that data property rights and sovereignty can be separated. The term data sovereign was developed as a response to the right to use data. So, data sovereignty is similar to state sovereignty. State sovereignty transfer has become a trend in the process of globalization and regional integration. State sovereignty transfer does not mean the disfranchisement of sovereignty, but the sharing of power among states concerning multinational corporations and UN affairs. We envision that data sovereignty transfer with a possible negotiation may be the basis for resolving disputes over data rights, whether it is among governments or among private actors, whatever the state encompasses or not. Such transfers can be partial/all, unpaid/paid, and two-way process as from country to organization, or from organization to country; mainly, data resources are replicable.

Reviewer: [For Paragraph 7] Again, I find the phrase "Data sovereign enables" odd, as legal frameworks and social norms are what enable things to happen. The first sentence also introduces "economic agglomerations," but what is described in the paragraph goes beyond mere agglomeration.

Me: Thanks. Regarding the word "enable," I explained above meaning it is indirect to achieve the goal. Regarding "economic agglomeration," I've revised to move it paragraph 9.

Reviewer: [For Paragraph 8] It isn't clear what is meant by "data resource boundaries," and, as noted for the second paragraph, the use of "data producers" as an entity other than the person generating/providing the data is going to be confusing to some readers. It also wasn't clear why your view will cause controversy, or at least any more controversy, in academic and industrial circles. I think all of these views cause controversy, as each view ascribes property rights over data to a different entity, and there is no agreement over who should hold those rights. I also found it odd that the third paragraph argued in support of industry collaborations, as such collaborations are a way to reduce the effects of the market forces that the authors argued for. Finally, that last sentence is worded in an odd manner.

[25] *See:* Snider M, Baig EC. Facebook fined $5 billion by FTC, must update and adopt new privacy, security measures. USA Today. July 2019. Retrieved from https://www.usatoday.com/story/tech/news/2019/07/24/facebook-pay-record-5-billion-fine-u-s-privacy-violations/1812499001.

Me: Thanks. Regarding "data producer," I answered above. From a perspective of data science, "data resource boundaries" can be obtained through the practices of data technology. For example, a dataset can be classified effectively by an algorithm that uses test conditions involving one or more attributes at a time. After partitioning, the border between two neighboring regions of different classes is known as a "decision boundary." Of course, identifying "data resource boundaries" is not so simple since the test condition involves multiple types of data. Also, the boundaries are related to something realistic; such as a data producer's business field (e.g., healthcare) and the country where the data producer conducted the business. Regarding "controversy in both academic and industrial circles" in the third paragraph, my thoughts are as follows. I pointed out an inequality about "whose data belongs to whom" relating to data producers, but the others are seeing on the individuals. This is a point of contention. Just as two schools of capitalism, the second point of contention lies on: one tends to the government controls, and the other is with the domination of free markets. While industry collaborations can be supported by the majority, some may stand on the opposite side. In addition, some are optimistic, and the others are pessimistic, resulting in different viewpoints. So, I prefer to keep this. Regarding the last sentence, it can be understood as: the "force" given by data technology can help possess and protect data resources; thus, a data sovereign who control these resources may obtain a "power" (that equates with capital), namely, data capital.

Reviewer: [For Paragraph 8] I think you need the word "global" in front of "adoption," as I don't think this idea works if only partly adopted/respected.

Me: Thanks. I added the word "global" in front of "adoption."

Reviewer: [For Paragraph 9] Some of the thoughts in this paragraph feel like they should have been in the introduction, as the paragraph highlights the value of data access and the need for clearly assigned property rights to facilitate market transactions related to data (though it doesn't do so in those words). Also, the last sentence is worded in a very confusing manner.

Me: Thanks. I revised them.

Reviewer: [For Figure] I'm not sure what role is intended by figure. The only reference to it refers to "data disputes," but the figure defines concepts not disputes. If the figure is going to be included, it needs to be explained. What does "border" mean? Why are the balls concentric? What is "pending to define"? Where does "data sovereignty" appear in this figure? How is the reader supposed to make use of the figure?

Me: Thanks. Figure 5.1 indicates four categories of data within a state, in which identifiable data refers to the data with private information. Consumer data (discussed in the beginning of this section) is a part of these identifiable data. One misunderstanding regarding data resources is, data resources belong to the person or other entities who provide the data. Painting the figure to be concentric is for a good-looking purpose. The figure also illustrates data ownership may be the individuals' superiors such as their affiliations and home countries. The terminologies of data residency and data localization are explained above, which are developed with cloud infrastructure when addressing private data across borders. The "pending to define" is the position of data sovereignty, which may be boarder than any state border. Through this figure, the reader can easily understand three other things. First, identifiable data does not mean all data resources. Secondly, data ownership is not just data providers. The third is that state borders are not equivalent to data resource boundaries.

Reviewer: [Overall Comment] Perhaps the "data sovereign" concept is useful, but it would require a much clearer exposition. As some of my comments suggest, I think the authors may be trying to find a way to talk about "property rights" as they pertain to data. If so, then I would focus on that concept, and define the terms that are needed to talk about those rights. Do all rights get assigned to the data sovereign in the authors' view, or does the data provider have some rights? This paper seems to downplay the role of countries in favor of data sovereigns, but don't countries/governments have a role in protecting the rights of data sovereigns and data providers in a global, etc. ...?

Me: Thanks. The perspective of developing "data sovereign" here is unique. The term "data sovereign" is conceptual innovation and economic inheritance on "sovereignty." It can not only avoid participating in various data disputes about property rights, but also endow data producers with a strong signal of "trust and confidence" to use the data, resulting in an active market economic system. I didn't the data provider need to give up their rights because I have another solution to this. While data trading between ecommerce vendors and individuals was fair from the beginning, there exists a subsequent inequity that is each stakeholder may not share equitably in the profit return when the data is keep processing to be a new product. After giving the sovereign status for data producers, we need to guide them to distribute data-rich profits actively. Also, states/countries have a role in protecting the rights of data sovereigns and data providers in a global; and the role assigned to governments is a proper referee in the free economic arena.

Virtual Nations: After Geopolitics

Nobel laureate Olga Tokarczuk[26] has been interested in crossing geographic boundaries. Tokarczuk's 2019 Nobel award cites her "narrative imagination that with encyclopedic passion represents the crossing of boundaries as a form of life."

[26] There are two references related to 2019 Nobel laureate Olga Tokarczuk, in which one is written by herself. ***See:*** (1) Shotter J. Nobel laureate Olga Tokarczuk: Why populist nostalgia will past. Ft. com. February 14, 2020. Retrieved from https://www.ft.com/content/36cfd978-4c1a-11ea-95a0-43d18ec715f5; (2) Tokarczuk O. A new world through my window. Newyorker.com. April 8, 2020. Retrieved from https://www.newyorker.com/books/page-turner/a-new-world-through-my-window.

The child of two teachers, she studied psychology and worked as a therapist, before deciding she was "more neurotic" than her patients and focusing on writing. Her breakthrough came in 1996 with her third novel, Primeval and Other Times, a tale with more than a hint of magical realism that follows a fictional Polish village (guarded rather laxly by four angels) through the trials of the 20th century. Flights, a mix of meditations on travel and the body interspersed with short stories, won the Nike Award, Poland's top literary prize, in 2008.

Tokarczuk wrote this paean to mobility in 2007 additionally, when seeming to the high-water mark of the faith in globalization and open borders by looking back from now. Since then the financial crisis and its political aftershocks have ushered in a nationalist turn that has helped detach Britain from the EU and install Donald Trump in the White House. The 2020 coronavirus pandemic has been met by the tighter security regulations within the international transportation and logistics. And now, her writing is different. She puts the shift down to uncertainty and fear.

The fear of getting sick has reminded us of the nests from which we hail and in which we feel safe. In such a situation, even the most assiduous travelers will always press on to some kind of home. At the same time, sad truths have been revealed to us – that in a moment of danger, our thought resorts once more to the limiting and exclusive categories of nations and borders. ... bringing back the division between "ours" and "foreign" – exactly what we have fought against these past decades ... The virus reminds us: borders exist, and they're doing just fine.

The issue of "boundaries" is about geopolitics[27] and, to a certain extent, can boil down to globalization with its future. Geopolitics is the study of "the effects of Earth's geography on politics and international relations" (see Footnote 27). There exist two viewpoints relating to the globalization now. Some insist that a new Cold War in the "jungle society" dominated by big powers has begun, and even will see a possibility of World War III soon. The others believe that one driving force of globalization is a spontaneous order so that globalization cannot be ended. The former is from a conscious reason based political perspective, and the latter belongs to spontaneously economic views. I prefer the latter, because Austrian-British economist Friedrich A. Hayek said that a spontaneous order (e.g., market economies) is superior to any order a human mind (i.e., a rational order such as planned economies) can design "due to the specifics of the information required."[28] "Globalization as a zero sum game," which many people think, is not correct. Globalization is dominated by individuals' willingness to work. The person has a competitive advantage in collaborating in the global market, thereby creating wealth and sustainable prosperity. In this way,

[27] ***See:*** The entry of "Geopolitics" on Wikipedia. Retrieved from https://en.wikipedia.org/wiki/Geopolitics.

[28] ***See:*** (1) Hayek FA. The Fatal Conceit: The Errors of Socialism. Chicago, IL: University of Chicago Press. (August 31, 1991). 1991; (2) The entry of "Spontaneous order" on Wikipedia. Retrieved from https://en.wikipedia.org/wiki/Spontaneous_order.

various interventions by geopolitical countries, such as physical borders around the globe that are closed, can anti-globalization but cannot stop globalization.

The idea of "virtual nations" without geographic constraints appears earlier in the research of currency politics, "… shows why various interest groups ought to favor strong or weak currencies, and stable or flexible exchange rates …"[29] Someone even criticizes the US has split into the two counties—geo-American in Washington and cur-American in Wall Street (as the symbol center of American capitalism). According to this statement, cur-American is a global network consisting of about 150 financial giants and large multinationals. Its "citizens" are 1% population at the top of the world wealth pyramid, and its GDP can reach US$36 trillion that is more than the sum of the US and China. The cur-American is virtual that makes sense because it is based on finance, the most thorough field of digitalization. However, the "population" of cur-American is not too much (about 75 million, 1% of the current global population as 7.594 billion in 2018), which is not our focus. Let's see virtual nations that have threat potential towards the US and its military.

With the increasing popularity of bandwidth of networks, coupled with the development of virtual and augmented reality, social virtualization trend is irresistible. Of them, Facebook is the biggest social network worldwide with over 2.6 billion monthly active users as of the first quarter of 2020.[30] Facebook's number of active users surpassed one billion as early as 2012. Active users are those who have logged in to at least one of the company's core products during the last 30 days, Instagram, WhatsApp, and Messenger included. Facebook is the largest "country" virtually despite still being locked outside China. A person can shop, play, party, social, and even gamble online. Not only do the people participate in concerts, but they also create the self-made live broadcast they want to show. Virtual live activities have become popular, resulting in a substantially different business model. In February 2019, for example, Fortnite, an immensely popular game, invited a notably electronic music producer Marshmello to take the stage insider of the game. According to developer Epic Games, 10.7 million people attended the 10-min electronic music concert, while the largest live concert participant to date is less than five million. Besides, the structure of the workforce has been changing. Traditional industrial jobs are being replaced by digital ones; and people can now work online from anywhere in the world, independent of any one company or country. These highly persistent population flows laid the foundation for emerging virtual nations.

---

[29] This is a comment on this book given by Harvard's Jeffrey Franket. ***See:*** Frieden JA. Currency Politics: The Political Economy of Exchange Rate Policy. Princeton, NJ: Princeton University Press; Reprint edition (September 6, 2016). 2016: 320 pages.

[30] ***See:*** Clement J. Number of Facebook users worldwide 2008-2020. Statista.com. April 30, 2020. Retrieved from https://www.statista.com/statistics/264810/number-of-monthly-active-facebook-users-worldwide.

As the world changes, people expect their governments and public services to do so too. When it's easy social with someone on the other side of world, there is an expectation that paying taxes, applying for services or voting should be too. The first broad category of virtual nations[31] is when a nation-state digitizes all governmental services, potentially offering e-Residency programs.

> *The primary actor in this disruption is Estonia, which leads the world in digital government. In 2015 they introduced an e-Residency, allowing anyone anywhere in the world to receive a government-issued digital identity. The e-Residency gives people access to digital public services and the ability to register and run online businesses from the country, in exactly the same way as someone born in Estonia. As of November 2017, over 27,000 people have applied to be Estonian e-Residents, and they have established over 4,200 companies. Estonia aims to have ten million virtual residents by 2025.*

The second are virtual nations not supported by any government,[32] of which their prototype is cyber communities with sign-up based "citizenship" and a political or ideological allegiance.

> *Asgardia [as the world's first "virtual nation" in space] ... recently launched a nano-satellite into orbit containing its citizens' data. [A man called] Ashurbeyli created this new nation in space so that falls outside the jurisdiction of Earth governments – although behavior in space is governed by the Outer Space Treaty, signed by 103 countries including the US and Russia. ... Anyone over the age of 18 with an email address, including ex-convicts, can become "citizens." Around 114,000 people from 204 countries have signed up. ... Ashurbeyli hopes to convince the UN to recognize Asgardia as a sovereign nation. That would require a two-thirds majority in the UN Security Council.*
>
> *... Bitnation is a blockchain-based technology which enables people to create and join virtual nations. This allows people to agree their own social contracts between one another, using smart contract technology, removing the need for governments as an administrator or mediator. Since it began in 2014, it has been offering traditional government services, such as notaries, dispute resolution, marriages and voting systems, without the need for a middleman. ... As of November 2017, there are over 10,000 Bitnation citizens. Bitnation believes its model can have a huge impact in the developing world, where restrictions citizens face on incorporating businesses or protecting property rights for example, are far greater than in Western democracies. Bitnation is positioning itself to meet this demand in the developing world and believes that take up in its emerging markets will put pressure on mature markets, which will need to adapt the same governance speed and flexibility to remain competitive.*

---

[31] The two references come from the UK and US, respectively. ***See:*** (1) Symons T. The nation state goes virtual: whey citizenship need no longer be determined by geography. Newstatesman.com February 1, 2018. Retrieved from https://tech.newstatesman.com/guest-opinion/virtual-nation-states; (2) Virtual Nations: An Emerging Supranational Cyber Trend. Mad Scientist Laboratory. July 9, 2018. Retrieved from https://madsciblog.tradoc.army.mil/66-virtual-nations-an-emerging-supranational-cyber-trend.

[32] ***See:*** (1) Asgardia, the world's first "virtual space nation." TheWeek.co.uk. November 20, 2017. Retrieved from https://www.theweek.co.uk/89815/asgardia-the-world-s-first-virtual-nation; (2) Retrieved from https://tse.bitnation.co.

The latter category, in general, possesses "the greater potential towards the US and its millitary" (see Footnote 31). Although likely, the category won't be officially recognized until 2050; the US Army has been preparing the combat and operation rules with which it is already familiar, following its doctrine and practices of endearments, as well as by international law and UN convention. Having no official recognition makes virtual nations more dangerous because there is no external accountability or regulation; meanwhile, some legitimate "virtual government actions" may blend in with those of individuals and non-state actors in a haze of anonymity. "By 2050, as certain ideologically based online communities become more prolific and organized, members may begin to feel a stronger attachment and sense of belonging to their online identity than to their nationality" (see Footnote 31). They would be more united with an internal recognition: everyone within the nation believing that it is a "nation."

Communicative Coordination Goes Global

Currently, the virtual nations remain a prototype existing only online. They will compete with real countries, once virtual nations are large enough to have power and control over the beliefs and actions of their citizens. But the concerns may be superfluous. Because the original intention of online communities was not overall perfection of the community but individual fulfillment of the members, and that means "celebrating the individuality, diversity, and occasional craziness that is so quintessentially human" (see Footnote 9).

Of course, individual fulfillment is often the product of coordination because the truth is that our resources aren't unlimited. When growing ever more proficient at large-scale coordination, we were able to accomplish far more than generations before us. Wikipedia is the one—a new testament to human coordination aimed at capturing our newly understanding (and cognition) of this world, which effectively and efficiently coordinate hundreds of thousands of contributors (without payment) to create "more than 40 million entries in nearly 300 languages" (see Footnote 9). Similarly, the Large Hadron Collider, which detected the Higgs boson in 2012 to help solidify the Standard Model of the particle physics, involved more than 10,000 scientists from over 100 countries. Resources and our means for utilizing them are both scarce, but virtual communities help us to reach them economically and efficiently, avoiding waste. The most critical of them relies on two things: first, richer analyzable data with more robust data tools makes the information accurately. Secondly, more diverse human coordination with better delivery makes information flows smoothly.

In terms of data and data tools, take coordination in a market as an example; such a coordination is decentralized—"Individuals in the market gather and provide information and make decision for themselves" (see Footnote 9). In legend, the wealth and fame of the Rothschilds increased by faster and more accurate information than others.[33]

An unapproved assertion is Nathan M. Rothschild was the first person in London to know of Wellington's victory at Waterloo in 1815 [which is often widely accepted as a historical fact]. ... The story is more legend than fact. Nathan, it was said, witnessed the battle in person and rode in haste to the coast, where he bullied and bribed a fisherman into ferrying him to England through a terrible storm. In this way he reached London twenty-four hours before the official news and 'gained twenty million in a single coup' on the stock market. ... Notably, the element of trading subterfuge was added. No sooner had Nathan reached the City of London, it was claimed, than he deliberately provoked a collapse in stock market confidence by encouraging rumors that Wellington had been defeated. Only when prices were near to rock bottom, and after many investors had been ruined, did he begin to buy, and in this way he was able to double the gains he made when the victory became known.

Improvements in the information quality often underlie a step-change in our trading capability. More detailedly (see Footnote 9),

Market coordination takes place through transactions, when buyer and seller discover they have matching preference and agree on the terms of a deal. Myriad transactions take place in markets around the globe every day. ... Globally transactions worth will over US$100 trillion take place every year, a figure that has grown by a factor of almost 2,000 since the 1500s.

Each transaction boils down to two parties communicating correctly with each other. Massive analyzable data and increasingly reliable data tools provide more transactions while meeting everyone's wants and needs.

Secondly, online communities provide more than a single "chain" of collaborative channels which cannot be summarized as tyrannical or democratic purely (see Footnote 9).

The varieties of human coordination are as diverse as human populations, from the web of reciprocal responsibilities and duties within a social network of family and kinship to the centralized command and control of an army to the collaborative peer production of encyclopedic projects and scientific experiments.

Apart from traditional conversations (voice), correspondence or contracts (written word), we gained more instant tools, the means to express ourselves across miles actively and passively. A more effective way to study collaborations is network analysis, I won't go into details here. But I gave Case 5.2, as an appendix for the reader, to indicate some standard methods used in the computer systems for automatically delivering information flows.

[33] ***See:*** Retrieved from https://www.rothschildarchive.org/materials/nathan_and_waterloo.pdf.

Case 5.2: Common Types of Automatically Delivering Information Flows

- Subscribe: A Subscribe model is often implemented whereby a data producer publishes a notification of data changes and interested subscriber pick it up. For example, a sales system publishes a new customer and a number of systems including the billing system add the customer to their databases.
- Push: A push model is from one system to another such as a sales system that pushes new customers to a billing system.
- Pull: Systems that pull data from a single source of truth. For example, a billing system that doesn't store basic customer information but uses and an API offered by a sales system to access customer records.
- Choreography (no central controller of the process): A method of implementing processes whereby services are aware they are participating in a process. For example, swarm robots that are putting up a beach umbrella may communicate amongst each other to get the job done with no boss controlling things. This results in complex, dynamic and often redundant information flows.
- Orchestration: A method of implementing processes that relies on a central controller. For example, robots that are putting up a beach umbrella who don't move unless the boss tells them too. This results in relatively easy to model information flows that flow from the controller out.
- Event Handling: Information that flows via human communication such as a business email. This is typically dynamic but can also be modeled in terms of processes and procedures that people follow.

Agglomeration in Cyberspace: Following Reproducible Data Capital Flows

There is a mechanism that has been critical in helping us coordinate successfully at scale. Alfred Weber, younger brother of the well-known sociologist Max Weber, formulated this mechanism as a least cost theory of "agglomeration,"[34] in which many enterprises are located together where the transportation costs of raw materials and final products are minimum. In principle, the phenomenon of agglomeration is a question of economic geography, called the Weber problem (named after Alfred Weber). This problem can be expressed as, finding a

[34] There exist several references. ***See:*** (1) Weber A. Alfred Weber's Theory of the Location of Industries. London, UK: Forgotten Book (September 16, 2017). 2017; 294 pages; (2) The entry of "Weber problem" on Wikipedia. Retrieved from https://en.wikipedia.org/wiki/Weber_problem.

point in the plane that minimizes the sum of the transportation costs from this point to n destination points, where different destination points are associated with different costs per unit distance. If generalizing the geometric median (that assumes transportation costs per unit distance are the same for all), the problem is changed to compute the Fermat point; for example, in the triangle case with respect to three points *a*, *b*, *c*, the Fermat point as a point *d* to locate the three points in such a way that the sum of the distances between *d* and each of the three points is minimized. For this reason, it is sometimes called the Fermat-Weber problem. The Weber problem is, in turn, generalized by the attraction-repulsion problem, which allows some of the costs to be negative so that greater distance from some destination points is better. Also, in the triangle simplest case, the attraction-repulsion problem consists of locating a point *d* concerning three points (i.e., *a*, *b*, *c*) in such a way that attractive forces exerted by points *a* and *b*, and the repulsive force exerted by point *c* cancel each other out as it must do at the optimum.

Weber's set of rules and concepts relating to agglomeration treats a single enterprise as an independent life-system, so the "interactions" among other enterprises (and the environment) are analogous to vegetation or population distribution "clustering" in mother nature dynamically. In the past, the agglomeration means the enterprise as the main body doing such interaction activities of the same or similar industries located in the close geographical space. It is susceptible to abstract mathematical problems to give a reasonable solution. But now, it is much more complicated. Research institutes and universities are closely involved in agglomeration, also venture capital included, and even agglomerations are flexible and dynamic. Adding participants is easy. People can join or leave the agglomeration at will because the space represents not only geography but cyber.

Example 5.1 I'll give an example of the automotive ecosystem. The worldwide automotive enterprises all require industrial agglomeration. In other words, many auto parts factories often choose to a site that is close to one (or more) collaborative automotive enterprise. Next, industrial concentration can help form collaborative dependency relationships inherent with the enterprise cluster. Then, some hatched or merged firms occurred to increase the cluster, reflecting a certain levels of industrial monopoly realtering to the automotive enterprise. Detroit, MI typically belongs to an automotive industrial agglomeration. The Detroit area, strategically located along the Great Lakes waterway, emerged as a signification metropolitan region in the US in the early twentieth century. And this trend hastened around the 1950s when the Big Three (i.e., the three largest automobile manufacturers in North America) as General Motors, Chrysler, and Ford, one by one, established their headquarters in this city. Global auto sales are slipping; unfortunately, Detroit had become going down to bankruptcy a decade before. Bankruptcy in Detroit resulted in the closure of factories and the

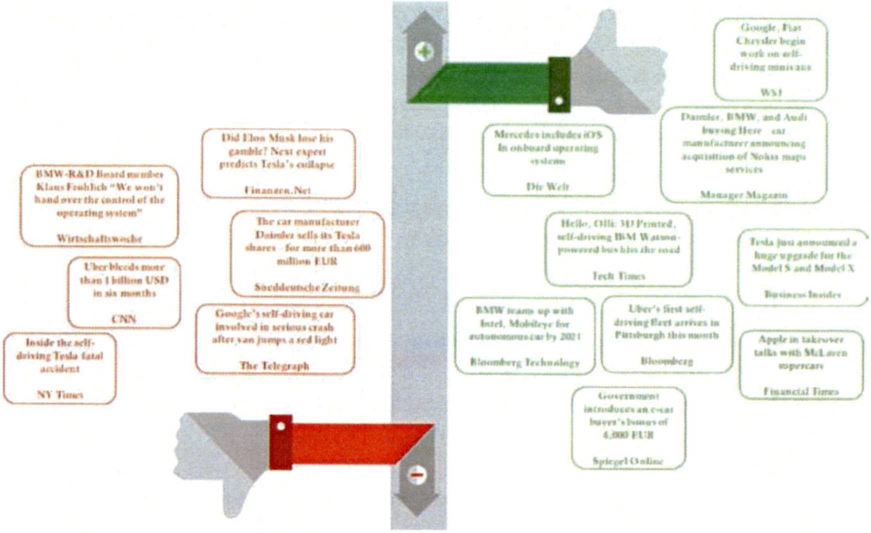

Figure 5.2 Signposts (in pros and cons) to the automotive future. (Source: Author drew it herself, referring to Footnote 35)

disappearance of jobs. This is because; Detroit continued developing the automotive industry without any change—"put its fate in a few critical enterprises," and by such a vertical "long-arm" leadership, the others only passively absorbed a bit of exogenous innovations brought in by the "long-arm." Currently, the car industry is still under siege. But disruptive times are ahead of the automotive value chain. News of disruptive, historical changes relating to the automotive has dominated industrial debates in recent years, see Figure 5.2. The content of my following discussion comes from a report given by Deloitte,[35] named *The Future of the Automotive Value Chain: 20205 and beyond*, which used a large number of interviews with experts to identify more than 60 drivers relating to the future automotive industry; these experts involve automotive original equipment manufacturers (OEMs), mobility entrepreneurs, researchers, and battery developers. Let's take a look.

[35]There are two language version (in both Chinese and English) reports about *The Future of the Automotive Value Chain: 20205 and beyond.* ***See:*** Helbig N, Sandau J. The Future of the Automotive Value Chain: 20205 and beyond. Newstatesman.com March 2017. Retrieved from (1) https://www2.deloitte.com/content/dam/Deloitte/us/Documents/consumer-business/us-auto-the-future-of-the-automotive-value-chain.pdf; (2) https://www2.deloitte.com/content/dam/Deloitte/cn/Documents/consumer-business/deloitte-cn-consumer-automotive-value-chain-2025plus-zh-180913.pdf.

First, this report assumes a typical automotive enterprise.

A prototypical OEM [is] headquartered in Europe, with no distinct market positioning – serving both premium and volume segment – but with a respected brand and solid profit streams. With an annual turnover of almost 60 million EUR and an EBITDA margin of 12.5% it ranges well within the order of magnitude of other European carmakers.

Second, by modeling business cases for the typical automotive enterprise in 2025, the report gives four scenarios based on the 60 plus drivers with the greatest impact and highest degrees of uncertainty. Among the four, the former three are dominated by this enterprise, while the latter one is by any data giants.

Scenario 1 – Data and mobility manager: In this world, connectivity has become a differentiator. E-mobility (including battery as well as plug-in hybrid electric vehicles, range extenders and fuel cell), autonomous driving, and integrated mobility are a common reality for the broad public. OEMs are able to set the standards and are the dominant players in the automotive industry, offering a rich portfolio of products and services. Innovative automotive outsiders have to play according to the rules set by OEMs. In particular premium brands and status play a decisive role in consumers' buying behavior. OEMs offer an attractive workplace for talent.

Scenario 2 – Stagnant car maker: Massive lobbying by OEMs has prevented potential new high-tech player from entering the market. However, this defensive strategy has also slowed down technical development, with the results that many potential innovations have not been rolled out to the market, with regulations, for example, limiting the deployment of technology. Dramatic accidents with immature autonomous cars have also resulted in loss of consumer acceptance.

Scenario 3 – The fallen giant: The car is a mere means of transportation and brand attractiveness has diminished. The technology hype has cooled down, which has put an end to the rise of the high-tech car. As mobility has become a commodity, the profit margin has decreased and OEMs are focusing on improving processes and on cost efficiency. Industry outsiders such as Uber have entered the market and are forging exclusive alliances with suppliers to provide affordable mass mobility. Since private car ownership has decreased, fleet management has become of significant importance for OEMs. New talent is hard to come by, due to the loss in the attractiveness of OEMs.

Scenario 4 – Hardware platform provider: IT players have disrupted the automotive value chain. OEMs have mainly become the suppliers of white-label cars to the Internet giants. In this world, OEMs can play a relevant role only if they provide a superior platform for "infotainment" and mobility services and/or retain a strong brand image. Since OEMs are not able to fully cash in the revenue potential, the margin per vehicle decreases.

Third, by applying the approach of scenario-based business modeling, the report identifies four plausible setups (two of them are existing), the automotive enterprise's value chain might have in 2025, to predict some implications (e.g., for asset and people) in this area. Figure 5.3 indicates the results I compiled from the report.

Today, the company primarily operates two business models: producing and selling vehicles (turnover 2015: 45 bn EUR; [Business model 1 as "vehicle sales"]) and offering financing service (turnover 2015: 13 bn EUR; [Business model 2 as "financial

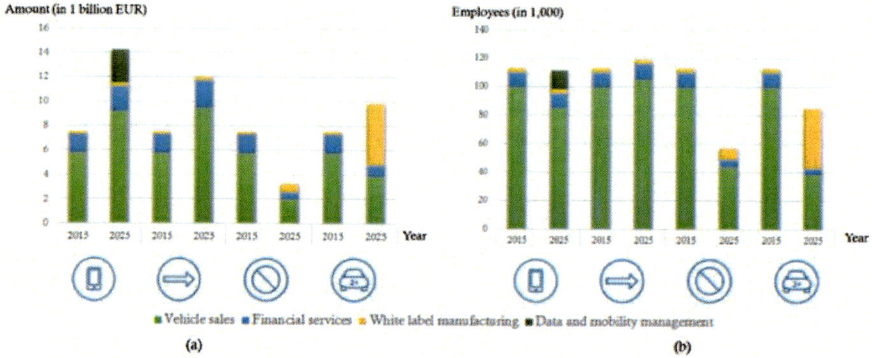

Figure 5.3 A comparison of 2015 and 2025 on EU prototypical OEM under four scenarios given by Deloitte. Note: The bottom icons from left to right are scenarios 1–4. (Source: Author drew it herself according to the data from Footnote 35)

services"]). We believe that in 2025 two additional business models will play a significant role for the future car company.

Producing white-label components and vehicles (i.e., unbranded products that are marketed by other players; [Business model 3 as "white label manufacturing"]) will have evolved as a third business model -depending on the scenario we look at. The main customers for white-label vehicles will be new market entrants such as Google or Uber. They will focus on their strength, which is user-centric software, while leaving hardware development to OEMs such as the future care company.

As a fourth business model, the future car company will be creating profits from mobility services and user data [Business model 4 as "data and mobility management"]. On the one hand this includes car- and ride-sharing services (mobility management). On the other the company will generate revenue from connected services and data, i.e., software and apps for information or entertainment offerings as well as other ways of monetizing the large amount of data available in and around cars.

Deloitte is optimistic on scenario 1 "data and mobility manager."

Simulation results indicate that car sales have increased by a moderate 7% compared to 2015, as customer demand for car- and ride-sharing has increased significantly. ... Across all business models, the future car company is in a very healthy condition with increases of 71% in revenues and 91% in EBITDA.

The large proportion of electric cars and advances in automation have led to a substantially smaller workforce in production. Twenty percent of employees in manufacturing have had to be made redundant due to e-car production and automation. However, the future car company is an attractive workplace for research and digital talents, as high-tech features and mobility management services are at the center of the company's value proposition. This helps recruit the 13,000 employees needed to build and scale the mobility services business segment.

Based on my understanding of the automotive industry (I have a patent for Internet of vehicles), I have a different opinion with Deloitte. I tend to choose scenario 4 "hardware platform provider" plus business model 3 "white label manufacturing," because it is a typical, automotive industrial agglomeration in cyberspace.

> *Cars have become software-based high-tech products with standardized interfaces comparable to today's smartphones. Tech plyers from Silicon Valley and other hot spots have gained critical access to the automotive world, provide key elements of the software, and have captured the mobility and data management service business. ... Liability situations are complex due to the multilayered production process of the company as a white-label producer, with the Silicon Valley players building their solution on top. ... the presentation of autonomous cars based on a Chrysler platform by Google's sister company Waymo at the 2017 Detroit Auto Show ... [is] a signpost to such a development ...*
>
> *The future car company is struggling to create the same brand appeal to private customers it used to have. ... The white-label business is now even the OEM's main business (52% of overall revenues). ... EBITDA margin in the white-label business is strong (13%). Mobility services do not play a role, since value-added digital services and shared mobility are operated by other players. All in all, the future car company is in a good position with high revues (26% higher than in 2015) and strong EBITDA (31% higher than in 2015).*
>
> *Since the future care company produces a large proportion of e-cars and has invested in automation, the workforce has decreased by 24% - despite higher revenues.*

In theory, the fact that scenario 1 can achieve a substantial increase in both vehicle sales and key financials without abandoning workforce development. But this conclusion is based on the optimistic expectation of which global automotive demand is stabling and has four major defects. First, from a strictly logical point of view, the current industrial structure is still large-scale manufacturing. This means the output is always higher for local demand in Europe. The car company must maintain its manufacturing capacity by enhancing logistics, resulting in more expensive logistical costs. Some realistic problems, such as the 2015 migrant crisis with its political aftershocks that have ushered in a nationalist turn, and the 2020 global epidemic with its health concerns that have created social distancing in all forms, inevitably undermine consumer confidence, then to make the expectation unrealistic.

Secondly, the scenario without a workforce reduction—"up to 15,000 of the future car company's employees in production might become redundant whereas 13,000 [high-tech employees, e.g., data scientists] need to be hired" (see Footnote 35)—doesn't mean it helps cut personnel expenses.

The third is about the data. When adopting this scenario, the data relating to the consumer this car company required is insufficient. Customers are generally assessed, from a marketing perspective, by how they fit within three categories: current, potential, and competitors' customers. The customers recorded in all databases of this automotive enterprise and its affiliated ones belong to existing customers. In other words, these customers have more or less concluded deals with the car company. Analyzing them aims mostly to ensure customer resource integrity to avoid losing them. One drawback is that the preferences, interests, and short-/long-term demands of current customers found in the databases cannot might not in line with potential customers.

Lastly, the goal of the car company is to lead its geo industrial cluster, so it is so difficult to introduce any enterprises focused on new technologies into the cluster. In a competitive, well-functioning cluster, no single leader is deciding what is being supplied or demanded and under what conditions, no central authority that tells members what to do and when to do it. Because coordination is diffused, as a leader following centralized decision-making, the car company cannot switch and swap based on availability and demand, to respond to new information from the members rapidly.

To determine whether business model 3 "white label manufacturing" is the benefit, we must first gain a better understanding of its definition that includes "contract manufacturing of engines, chassis—up to complete vehicles" (see Footnote 35). This reminds me of a story my father also involved in, about motorcycle manufacturing for the growing Chinese domestic market. In the 1990s, many people jumped on the bandwagon to purchase a motorcycle in China, resulting in a vast market domestic demand. At that time, motorcycles were provided by the "big four" Japanese manufacturers; and they are Honda, Kawasaki, Suzuki, and Yamaha. About 31 state-owned companies in China licensed the designs from the Japanese developers, then produced each part to exacting specifications. While being much cheaper than the equivalent models manufactured in Japan, these motorcycles, at around 10,000 RMB (≈US$700), were still well beyond the budget of most Chinese citizens (average monthly salary is about 500 RMB). After the government opened up the industry to small entrepreneurs, some motorcycle factories clustered in Chongqing Province broke away from the licensing system trying to create a less expensive process that would make motorcycles' price lower. They planned to buy and assemble parts made by others in the market rather than to decrease the expenses in their own factories.

They found potential parts suppliers to the sketches of four basic modules, each made up of hundreds of components. Potential suppliers can make any improvements in the design they wanted to, significantly, if they reduced the cost. Still, the suppliers must ensure that the parts they offered met basic standards for weight and size and worked seamlessly with the other parts in the module. Most specially, the fact that the motorcycle factories and suppliers did not cluster in a wheel-axle type agglomeration—"all of them on equal footing" (see Footnote 9). The motorcycle factories made it clear that they were not going to enter into exclusive contracts with any potential supplier. The factories hoped the freedom to buy the same or similar components and modules from multiple sources; in turn, the suppliers can also select factories to support. In this way, they were both able to respond to any changes in consumer preferences timely—adjustment or conversion. In Chongqing, millions of interchangeable parts were churned out every day, and even small "mom-and-pop" shops could get into motorcycle assembling, dramatically expanding the number of market participants. Through such a modular, market-based production process, the price of a motorcycle plummeted to under US$200. By simultaneously increasing production and declining in price, these Chongqing factories started marketing.

My father participated in one motorcycle factory called Jialing (嘉陵) to be a distributor, partnered with another woman in 1998. A deposit of 5 million RMB can withdraw motorcycles that were worth 200 million RMB—liquidated once every 3 months. The gross profit of a motorcycle is about 8% in which the sales tax is about 3%. By 2005, Chinese manufacturers accounted for half of the global production of motorcycles, and they took Japanese market shares; for example, Honda's sales fell from 90% to 30% of the market in Vietnam within only 5 years. This model is not "long-arm" centralized control and vertical management, and encourages participants in a market to efficiently produce affordable motorcycles. So, the "extreme" situation described by some critics of capitalism might not show up, which is "firms will increase in size and combine to form monopoles or oligopolies that may ultimately control the economy and undo the market as we know it" (see Footnote 9).

Part of the reason for that business 3 in scenario 4 is profitability may be above. First, although declining in vehicle sales from 1.6 million units in 2015 to 1.3 million units in 2025, white label manufacturing can increase by 0.8 million units. The two are both built on its own advantages, and will contribute to increasing by about 30% to both vehicle sales and key financials.

Secondly, tech players from Silicon Valley and other hot spots are best at building the solution on top via data analysis. Also, they owned or controlled some customer data resources, especially the potential and competitors' customers relating to the car company. Both parties can save money on buying relevant data. Based on data schemes, the car company can adopt some emerging manufacturing strategies, such as bespoke style and crowdsourcing. Despite it is still a standardized production, the bespoke style allows consumers to participate a little bit early to represent their wants and needs. Such wants and needs are possibly a minor change like a personalized touch. Volkswagen tried once as early as 2011 in Shanghai, China, the project called "The People's Car." The project collected almost all available opinions or ideas to the vehicles (e.g., shape, color) produced by Volkswagen via an online webpage. While having no in-depth data analysis, the number of participants was impressive. The final statistic number is nearly 0.4 million people participated. The bespoke style of manufacturing may help reduce a part of logistical costs when advance planning. Crowdsourcing aims to find global "masters" to participate in solving any standing problem, which may be dismantled as some pieces and published online. This is possible to reduce the hired requirement on high-tech employees.

Third, the workforce in this scenario will decrease by 24% due to a large proportion of investing automation. This resulted in further cutting the manufacturing cost in terms of accounting. It should be noted that "manufacturing 4.0" also requires big data. The report shows that Deloitte didn't consider management fees relating to high-tech employees.

Finally and foremost, the collaborative process belongs to a part of the cross-border data capital flows. Following the reproducible data capital flows, the attraction-repulsion problem in cyberspace studied for fixing defects found in collaboration among two or more parties automatically, is feasible. Note that I used the word "reproducible" here, of which the reason is: data capital flows as the experimental data for each party is the same, so results reproducibility will require access to the same experimental data for analysis studies, but may use different methods (see Definitions 4.11–4.13). For the evidence obtained in data flows (on such as unequal exchange and long run-in period), either the car company or tech players from Silicon Valley can do the freedom of deciding to change the current collaborative circuit or business scenario. For example, the car company found that the data relating to the operating system they transferred to tech players is much higher than any online customer data tech players offered in asset value. Another example, tech players found some entertainment apps require too long a test period to connect in the vehicle. Regarding solving the attraction-repulsion problem in cyberspace, data analysis is possibly much more advantageous than pure mathematical proofs because the problem usually involves multiple parameters. Of course, two parties of data sovereigns can fully share data resources in the collaborative process within the cluster, to help manage industrial conflict and reshape advantages in coordination. Perhaps non-egoistic tendencies can help gain more profits for all enterprises clustered together. I will come back to this point.

Data Capital Migration and Inequality

In the previous section, we see how online social cooperation is dependent upon the ground to the principle of equal liberty of communicative participation. Based on this, individuals or enterprises as a data sovereign may share data capital by duplication according to their willingness. But when the situation shifts to data capital migration, it is not all equality.

There exists no term "data capital migration" Among existing terminologies, but the relevant are "data migration" vs. "capital migration." Although data won't go aging without a life cycle, the legacy storage medium may be decommissioned. For example, a digital photo might not be readable due to a hard disk problem after decades. Data migration, which focuses on the data object itself, is the act of transferring or rewriting data from an out-of-date medium to a current platform and has, for many years, been considered as a viable approach to the long-term preservation of the data. So, we often treat data migration as a project to implement, which is the process of selecting, preparing, extracting, and transforming data. Additionally, the validation of migrated data for completeness and the decommissioning of legacy data storage are considered part of the entire data migration process.

Definition 5.2 *Data Migration*[36] Data migration is a project by means of which data will be moved or copied from one environment to another, and removed or decommissioned in the source.

Data migration occurs for a variety of reasons, among which an urgent one is data center relocation and disaster recovery. Data centers have always been secure, tightly controlled facilities for data preservation. The "9/11" attacks brought a change to data centers that pushed technical and physical security to even higher levels, and another one is to increase the number of redundant backups and the acceptable distance between primary and secondary locations. These changes also caused more data migration requirements. But even with high-speed internet connections, it can take decades to transfer an exabyte-scale (2^{60}) data. This is why the Amazon Snowmobile business came into being. It uses a 45-ft long ruggedized shipping container, pulled by a semi-trailer truck, to move massive volumes of data, such as pulling from the east coast to the west coast.

Capital migration is generally an immigration issue, which represents some consequences of population mobility, whether cross-country or rural-urban. These consequences involve gains or losses in financial capital, human capital, and social capital. Socially, there exist certain connections to gender, health outcomes, poverty, development, and globalization.

Definition 5.3 *Data Capital Migration* Data capital migration involves multiple data sovereigns, which makes data capital flowing in one or more directions (and there may be numerous migrations taking place simultaneously). So, its representatives will be moved or copied from one environment to another, and removed or decommissioned in the source.

It is essential to realize that data capital migration represents a one-way process with no access. Here begins my examination of inequality at this process; see Case 5.3 that is a simplest theoretical case of receiving data capital from whom went out of business, in which the transferor and receiver are called "Past as Ps" and "Currents as Cs," respectively.

There are two inequalities from the returns on capital. First is at the data sovereign level. The transferor had worked hard for nearly 10 years to accumulate consumer data of two billion, which was sold to the receiver. While the receiver paid some as compensation, the receiver owned high benefits on these data via a similar social network from differential target as the enterprise. The second is between the data sovereign and consumer. Even though the transferor informed the consumers, both the transferor and receiver have never been considering data-rich profit sharing with the consumer.

[36] ***See:*** The entry of "Data Migration" on Wikipedia. Retrieved from https://en.wikipedia.org/wiki/Data_migration.

Case 5.3: A Theoretical Case of Data Capital Migration

- The transferor: Past (Ps)

In 2011, Ps was launched as an invite-only social network and then provided cloud storage services later in the year. Similar to Facebook, Ps acted as a social media for posting photos and creating status' on individual feeds. Along with being a social network, Ps was produced to be the social layer for many apps, such as email, video, and so forth. By 2018, Ps had over two billion user accounts within its interface as it gave access to other apps. However, less than 400 million consumers were actively using the social media aspect, with 90% of these users using the social network for 5 s. Ps saw limited consumer interaction with apps and low social media usage and sold this part of business with the data to Cs in 2019, after having informed all users via some mails.

- The receiver: Currents (Cs)

Cs is an enterprise network. The network incorporates new features including Marks. Marks have been built for employees to be able to be involved in relevant organization conversation if they follow a certain mark. Managers will be able to monitor conversations between employees and interact through the social media network. Leaders' posts will also be given priority within the business' posts and events and will be given more visibility across the network. This network has differing security and privacy settings. Within the Cs' privacy setting, each app will need to request individual permission for each aspect of a consumer's profile, and the consumer will have to grant this to create access. New posts will only be accessed within the domain however this can be made Public. Profiles can no longer be publicly searched and discovered.

Someone may argue the third inequality relating to the transferor who is possibly keeping a backup. I would say this concern is about the "spirit of contract." But data technology may help; for example, using Snapchat's idea to vanish data in the source after "transferring." In addition, existing version comparison algorithms (e.g., SimHash originally used for paper plagiarism or webpage duplicate content detection, which can be found in Google's 2007 article of *Detecting Near-Duplicates for Web Crawling*) can offer the ideas of solution to develop something new.

The Succession Issue in Data Heritage

On July 1, 2014, a Reddit user called Natesw published a post at R/nosleep, which titles "my dead girlfriend keeps messaging me on Facebook. I've got the screenshots. I don't know what to do"[37] with high engagement. Reddit is now one of the most visited social website in the world (as for May 2020, it ranks 19th), comprising user-generated content—including text-based posts, photos,

[37] ***See:*** Natesw. My dead girlfriend keeps messaging me on Facebook. I've got the screenshots. I don't know what to do. Reddit/nosleep.com. Retrieved from https://www.reddit.com/r/nosleep/comments/29kd1x/my_dead_girlfriend_keeps_messaging_me_on_facebook.

video, and links; more than a half of its users came from the US. Currently, the thread of this post is archived so that new comments cannot be posted, and votes cannot be cast.

The user Natesw called his real name as Nathan. His girlfriend Emily died on August 7, 2012, when "they had been dating for 5 years." Nathan said that Emily "was vibrant; the kind of girl that would choose to dare every time. She always smelled like cinnamon," but she "wasn't perfect. She always said something along the lines of the music she said she liked, and the music she liked was very different." After making arrangements for Emily's funeral, he kept Emily's Facebook account activated and "shared" access with Emily's mother Susan. On September 4, 2013, the thirteenth month after Emily's gone, Nathan said he received several "hello" from her account. Nathan first assumed it was Susan who possibly logged in a wrong account. After getting a response from Susan that she hadn't logged in to Emily's account since the week of her death, Nathan then assumed "this was one of her more tech savvy 'friends'" playing with him in "the worst possible way." Around February 2014, Emily's account has activities to start tagging her in Nathan's photos. For the operation of tagging, Facebook officially explained as follows.

When you tag someone, you create a link to their profile. The post you tag the person in may also be added to that person's timeline. For example, you can tag a photo to show who's in the photo or post a status update and say who you're with. If you tag a friend in your status update, anyone who sees that update can click on your friend's name and go to their profile. Your status update may also show up on that friend's timeline.

When you tag someone, they'll be notified. Also, if you or a friend tags someone in your post, the post could be visible to the audience you selected plus friends of the tagged person.

Tags in photos and posts from people you aren't friends with may appear in timeline review where you can decide if you want to allow them on your timeline. You can also choose to review tags by anyone, including your friends.

So, Nathan "got notifications for them, but the tag would generally always be removed by the time" he got it. Nathan's friends also noticed these tags; some of them think Nathan's mental may have some problems, so they removed him from their Facebook friends list. Nathan said he "had changed the password and all security information countless times" but no avail. So far, Nathan still believed it was a hack behavior. On March 15, 2014, Nathan actively sent a message to Emily's account:

Why are you doing this? Why do you keep tagging her?

Ten days later, Nathan received an answer with "cinnamon scented candles," which appeared in the chat between the two in 2012. Also, Nathan noticed she was recycling his words as well. Since that, Nathan contacted Facebook's customer service to check the recent login records of Emily's account. The results made him scared. Facebook told him the locations her page had been accessed from, but since her death, they're all place Nathan can account for (his home, his

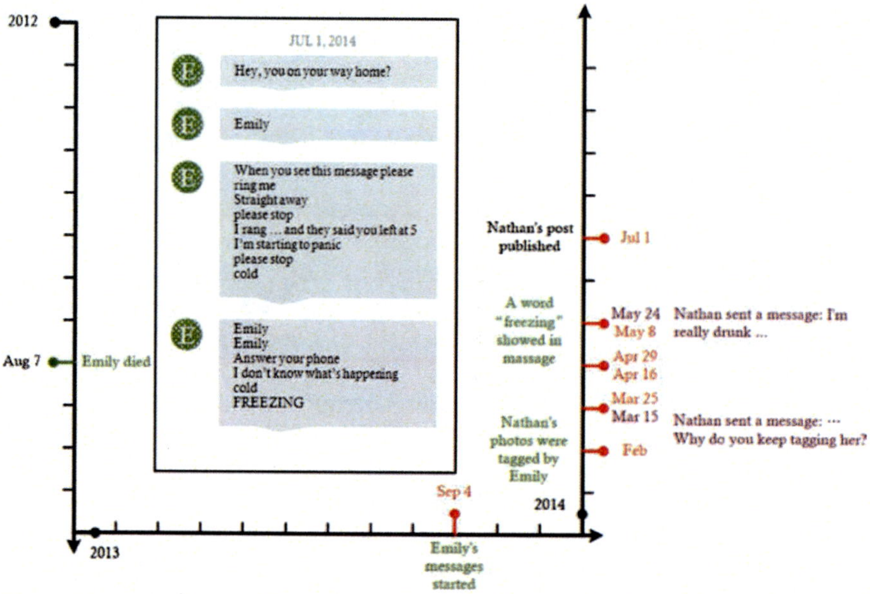

Figure 5.4 The timeline of a post on Raddi R/nosleep titled "My dead girlfriend keeps messaging me ...". Note: The red pushpin represents the date when Emily's account has activities. (Source: Author drew it herself)

work, her mother's house, etc.). Nathan found that the messages he received from Emily's account were basically in the chat history. However, a word "FRE EZIN G" shown in a message received on May 8, 2014 never appeared. On My 24, 2014, Nathan sent a message actively again.

> *I'm really drunk. I miss you. Whoever's on this account I don't give a fuck. I keep coming home from and expecting to see you at the computer. Shouldn't I be used to that by now?*

Nathan explained in his post that, he was not actually drunk at that time. Emily "wasn't an affectionate girl, and it always embarrassed her to exchange 'I love you'." She was more comfortable with it when I was boozed up. So, Nathan always got fake-drunk a lot. One hour late, Nathan got a reply with "let me walk," which appeared in the last chat of two in 2012. In sum, Nathan said in the post using various words relating to emotional collapse, such as "angry," "stopped being able to sleep," and so forth. The day Nathan finally received messages from Emily's account is when he posted the post, see figure 5.4. Weirdly, when the post was published, the user Natesw disappeared. Later, a netizen found an Australian news released on August 7, 2012. The traffic accident was mentioned in this news, but no one named Emily was on the death list, but Nathan listed.

Being a scary urban legend, whether real or false, it at least released a truth: internet users leave vast volumes of online data behind when passing away, commonly referred to as digital remains. A study indicates that,[38]

> *A minimum of 1.4 billion users will pass away before 2100 if Facebook ceases to attract new users as of 2018. If the network continues expanding at current rates, however, this number will exceed 4.9 billion.*

What happens to one Facebook account after its user dies? In this respect, an article on Time Warner's Mashable.com, a well-known technology website, makes a useful in-depth analysis.[39]

> *In 2015, Facebook announced in a press release that users would be able to name a "legacy contact," or a person responsible for managing your account after you die.*
>
> *Legacy contacts can memorialize accounts and download all of an account's data. They can even delete accounts completely. However, users can set preferences to have accounts memorialized or deleted after they pass away, and a legacy contact must abide by those specifications.*
>
> *Prior to the invention of the legacy contact, if anyone saw the Facebook profile of someone they knew had passed away, all they needed to do was report it to Facebook and request that the account either be memorialized or deleted. The only other way people could manage someone's account after they passed away was if they had their password. ... In 2012, a German 15-year-old was hit by a train in an apparent suicide. When her mother tried to access her Facebook account, she was unable to, because it had already been memorialized, Reuters reported in 2017. Locked out of the account, she was forced to go to court to try to gain access to her daughter's profile. The court in Germany initially sided with her, but a ruling in an appeals court dictated that the right to private data outweighed any parental inheritance of information.*
>
> *Only legacy contacts that have been predetermined can access the accounts after they've been memorialized, according to Facebook, but that doesn't mean they're the only ones who can memorialize the account. ... horror stories ... [are] people locked out of their loved one's accounts despite having their passwords because someone else had memorialized it first. ... The list of complaints goes on.*

While advising that people's digital remains should be viewed in the same way as the human body, and treated with care and respect, online death has been gaining increasing traction.[40]

> *Imagine you have a close friend you frequently communicate with via text. One day, they suddenly die. You reel, you cry, you attend their funeral. Then you decide to pick up your phone and send them a message, just like old times.*

---

[38] ***See:*** Öhman CJ, Watson D. Are the dead taking over Facebook? A Big Data approach to the future of death online. Big Data & Society, 2019;6(1): 2053951719842540. DOI: 10.1177/2053951719842540.

[39] ***See:*** Luz A, Santiago H. Facebook still needs to work on what to do when users die. Mashable.com. August 29, 2018. Retrieved from https://mashable.com/article/facebook-memorials-can-be-improved.

[40] ***See:*** Matei A. New technology is forcing us to confront the ethics of bringing people back from the dead. Quartz. January 27, 2017. Retrieved from https://qz.com/896207/death-technology-will-allow-grieving-people-to-bring-back-their-loved-ones-from-the-dead-digitally.

> *"I miss you," you type. A little response bubble appears at the bottom of the screen. "I miss you too," comes the reply. You keep texting back and forth. It's just like they never left.*

"Black Mirror" as a Netflix dram once imaged a world in which tech could be used to recreate the dead. Now people are working to make that a reality. The possibility of digitally interacting with someone from beyond the grave is no longer the stuff of science fiction. The technology to create compelling digital surrogates of the dead is rapidly evolving, with researchers predicting its mainstream viability within a decade.

But what about the ethics of bereavement—and the privacy of the deceased? Speaking with a loved one evokes a significant emotional response. The ability to do so in the wake of their death will inevitably affect the human process of grieving in ways we're only beginning to explore. Maybe, we should treat these data as the data heritage to let commercial gains go slightly slower than scientific research. Then, scholars of law and related areas could investigate new dilemmas arising from the inheritance of digital estates and issues of posthumous online privacy. Sociologists and anthropologists could turn their gaze towards the new types of "para-social" relationships, and the "continuing bonds" that we shape with the online dead. And philosophers could focus on the ontological and ethical status of digital remains.

Chapter 6
The Metamorphoses of Data Capital and Their Circuits

The metamorphoses of capital and their circuits have always been at the center of economic research, and economists at the focus often change the pursuit of simplicity and developed very complex theories. What do economists do? Economists study the ways an economy (e.g., a state, or an enterprise) uses "scarce" resources to produce goods and services. The analysis is generally on the costs and benefits of distributing and consuming these goods and services, using a series investigation of capital in each stage. Example 6.1 takes one productivity-related research purpose as an example, which wants to prove the content as follows. Whenever any new general-purpose technology as productive forces appears, resulting in human economic systems reshaped repeatedly, still, a new economy can only occupy a dominant position rather than completely replace the old ones. These productive forces included but not limited to, the invention of the steam engine and the use of coal, the discovery of petroleum and the emergence of the heavy chemical industry, the creation of electric dynamo and the electrification of human life, and the popularization of computers and the impact of digital technologies. In macroeconomics, within a closed economy (the one is entirely sufficient meaning no trading activity with outside economies), a course of study logics may be as below.[1] The emergence of mercantilism didn't eliminate agrarian capitalism in both economic pattern and market exchange model; agrarian capitalism still existed, but mercantilism had begun to dominate resource allocation, income and wealth distribution, and price system. Mercantilists participated in politics in many ways. The rise of industrial capitalism didn't wholly replace mercantilism in both economic pattern and market exchange model; and industrial capitalism had begun to dominate resource allocation, income and wealth distribution, and price system. Industrial capitalists participated in the national government. After the rise of

[1] ***See:*** Xiang, S. The New Capitalism: Rising, Crisis and Future of the Global Financial Capitalism. Beijing: China Citic Press; 1 edition, December 1, 2014, 616 pages.

© Springer Nature Switzerland AG 2021

C. Tang, *Data Capital*, https://doi.org/10.1007/978-3-030-60192-8_6

financial capital, industrial capital continued to exist. Financial capital had started with dominating resource allocation, national economic policy, income and wealth distribution, price systems. So, economists should study the hierarchical structure of the industrial division of labor, market functions, and price system. In the sense that there is a perspective of chain connecting with industries, markets, and price systems in a way to look at which one occupies the upper end and dominants the others. The core research content is the evolution of the capital stock, looking at both its overall size, and its breakdown into different types of capital assets, of which nature has changed radically since the eighteenth century or much earlier. Note that spatial development doesn't need to consider here due to an assumption of "a closed economy."

Example 6.1, also can be considered at the micro level. An enterprise appears as a buyer on the commodity and the labor markets; the buyer's money as capital is transformed into commodities (i.e., materials and labor). Productive consumption for the purchased commodities by the enterprise is on particular technologies and organizational forms. The enterprise returns to the markets as a seller; the seller's new commodities are turned into money. The result is a new commodity of more value than that of the materials entering into its production. This is Karl Marx employed the circuits of productive and money capital in his first volume of Capital. Hence the formula for the circuit of money-capital is $M - C \ldots P \ldots C' - M'$, the dots indicating that the process of circulation is interrupted, and C' and M' designating C and M increased by surplus-value. The central proposition here is that labor can create more value (i.e., surplus-value) than the price when buying. The new commodities as the products contain surplus-value, so selling them on the market can obtain the profits. Thus, the reproduction of capital depends on reusing all or part of money owned from this round for purchase; namely, the next round of buying materials and labor. Productive consumption we can pay to its particular manifestation in different eras. Given two generations of this enterprise, for example, a production of the prior generation was wheat processing (that is manually cut the heads from the top of the wheat plants and then thresh them to remove the grains from the rest of the head, called chaff); and next, the production is biscuit manufacturing. While the later generation mainly produces biscuits, it remains some production about wheat processing the prior did. In this way, "a new economy cannot wholly replace the old ones" was proved.

The two economic studies fall on a core issue—value and currency. That makes sense. Macroscopically, the term "value" summarizes all human activities relating to the creation, exchange, distribution, dissemination, and consumption, so the value chain could show the direction of value as it flows between upstream and downstream in production. On the micro level, when trade was transacted with gold and silver coins, it seems self-evident; precious metals are rare, so coins made from them are valuable. With money, economists can convert the value about the capital when processing in its metamorphoses and circuits, and the costs and benefits can be calculated much more easily.

The real markets work so indeed via money and price. As a standardized yardstick to dominate the value of goods and services, money acts, allowing people to size up different commodities, materials and labor also included. In the absence of money, when individuals bartered in the market, they had to come to some agreement about how much of one commodity could be exchanged for how much of another. This was tough without an accepted common dominator, resulting in a difference in correlating transactions. "Knowing that an individual traded a knife for a fur coat is not much help to someone wanting to trade a slab of reindeer meat for a vessel full of fish oil."[2] Just like Harvard historian Niall Ferguson wrote in his book *The Ascent of Money: A Financial History of the World*, "money is the root of most progress."

However, both two economic analyses here are on an assumption of the ideal market, and the factor of information excluded. Because it is only in ideal markets that each participant has information rationally with a zero cost, and can condense it into a price instantly—that is the efficient market hypothesis (EMH). The EMH as a ground-breaking work admitted by the Nobel committee states that, asset prices formed by market buying and selling reflect all available information perfectly (e.g., relating to the value and future income steams). But the reality is more challenging, and even has the "bad money drives out good."[3] Information can make or break a market; this is why rules mandate the sharing of information in many markets; for example, the used car market often require people selling their cares to inform buyer of any major accidents the car has been involved in.

The information might not be overlooked, fortunately, when introducing data capital to economic research so that capital appears at each stage of metamorphoses and circuits is the data. In this chapter, I will give examples with their interpretations stage by stage to indicate the role of data economically capitalized in production.

Primitive Accumulation of Data Capital

The process of capital accumulation, as one of the defining characteristics of a capitalist economic system, forms the basis of capitalism. In Marx's view, capital accumulation is a supposed capitalist obsession. In particular, primitive accumulation of capital is the process in which a ruling class takes possession of land and thereby converts the resident population into a landless proletariat.

[2] ***See:*** Mayer-Schönberger V, Ramge T. Reinventing Capitalism in the Age of Big Data. Now York, NY: Basic Books (February 27, 2018). 2018; 288 pages.

[3] Note that under Gresham's law, "good money" is money that shows little difference between its nominal value (the face value of the coin) and its commodity value (the value of the metal of which it is made, often precious metals, nickel, or copper). ***See:*** The entry of "Gresham's law" on Wikipedia. Retrieved from https://en.wikipedia.org/wiki/Gresham%27s_law.

Just like a famous quote in Max's *Capital* from Thomas Dunning's pamphlet, *Trades' unions and strikes* runs as:[4]

> *Capital is said by a Quarterly Reviewer to fly turbulence and strife, and to be timid, which is very true; but this is very incompletely stating the question. Capital eschews no profit, or very small profit, just as Nature was formerly said to abhor a vacuum. With adequate profit, capital is very bold. A certain 10 per cent. will ensure its employment anywhere; 20 per cent. certain will produce eagerness; 50 per cent., positive audacity; 100 per cent. will make it ready to trample on all human laws; 300 per cent., and there is not a crime at which it will scruple, nor a risk it will not run, even to the chance of its owner being hanged. If turbulence and strife will bring a profit, it will freely encourage both. Smuggling and the slave-trade have amply proved all that is here stated.*

However, the primitive accumulation of data capital might not be so sanguinary. Some inappropriate or improper behaviors have been deterred, such as China's cracking down on companies using dodgy means to collect data in 2019,[5] and the US regulators recording a "historic" US$5 billion fine of Facebook over privacy in 2020.[6] So, most of accumulating data capital in the primitive stage is good, which used free services to barter consumer data in exchange. Note that the currency seems to participate in this process of circulation rarely. Of course, a few cases are innocent in which the most famous one is about Aaron H. Swartz.

We can roughly use four basic principles of economics to understanding capital accumulation; they are the principle of wealth, income flow, discount rate, and risk and uncertainties. According to economics, these principles largely correspond to the investors' subjective expectations of the future given by the information available from the markets. In other words, capital accumulation is the dynamic decentralized decision-making helped by the flow of information from the markets: "people evaluate the information available to them and use it to make decisions that benefit them. Information flows from everyone to everyone" (see Footnote 2). Compared with physical capital, data capital's characteristics are as follows. The first is about depreciation. The depreciation of computers, telecommunication, and other electronic equipment is at an alarming rate over the past half-century. Still, we don't need to accrue depreciation from data capital since the data never goes aging. The second is on liquidity. I think data capital has high liquidity. Data production is the overall process of converting raw data into useful information; meanwhile producing data productions does not require all data capital; a small part may be enough. There exists a critical link between market efficiency and information flows, so the market speed of adjustment to

---

[4] ***See:*** The entry of "Thomas Dunning" on Wikipedia. Retrieved from https://en.wikipedia.org/wiki/Thomas_Dunning.

[5] ***See:*** Zhang Y, Shen T., Li I. China gets tough on companies using dodgy means to collect data. Afr.com. November 1, 2019. Retrieved from https://www.afr.com/companies/financial-services/china-gets-tough-on-companies-using-dodgy-means-to-collect-data-20191031-p53633.

[6] ***See:*** Court approves record $5 bn fine of Facebook over privacy. Techxplore.com April 25, 2020. Retrieved from https://techxplore.com/news/2020-04-court-bn-fine-facebook-privacy.html.

data capital is much faster than to physical capital. Thirdly, while being intangible, data capital, like physical capital, needs continuous investments to maintain normal operations, such as electricity and maintenance costs.

Stitch Fix: Raising Data Capital via Service

Example 6.2 We may be familiar with the story of Stitch Fix (see Footnote 2).[7] At first glance, Stitch Fix seems to be a traditional retailer with a subscription box that delivers five items of clothing and accessories to consumers, who pay for some they choose to keep. But a closer look reveals that Stitch Fix acts much more like an intermediary in the market for apparel than like a retailer. For each box sent to a customer, Stitch Fix is charging a "styling fee" (e.g., US$20). The fee is waived when a customer decides to keep at least two of the five items in a shipment. Also, there exists an extra discount if a customer has all five things. The service offers a unique experience by personalizing the clothing in each box to the customer's taste preferences and individual needs. This personalization isn't done through guesswork or purely through human analysis. The Stitch Fix model is driven by 85 data points that customers willingly provide, such as style, size, fit, and price preference (Figure 6.1). Stitch Fix also collects offbeat details, such as areas of the body where clothes typically fit awkwardly or areas the customer would like to flaunt or cover-up. Once that data is crunched, one of Stitch Fix's stylists (the number is close to 5000) gets to work, adding their human touch and personalized input to select the pieces to include in the customer's box. The result is a data-driven experience that feels personal and reminiscent of the days when store tailors greeted you by name and took individual measurements. In 2015, the company was valued at US$300 million in the third round of financing, prompting the New York Times to list it as a potential start-up "unicorn." Stitch Fix listed on the NASDAQ in 2017 and generated sales of US$1.2 billion in 2018.

But the latter successful part is not our focus. We should look at the former part of this story to understand how Lake did primitive accumulation on data capital. Regarding this, the story was told very clearly in the book *Reinventing Capitalism in the Age of Big Data* written by Viktor Mayer-Schönberger and Thomas Ramge (see Footnote 2).

> *As a young management consultant at the Parthenon group, Katrina Lake realized that conventional brick-and-mortar retailers failed to understand the wants and needs of their customers because they didn't make adequate use of data. Gifted with an ability to think analytically as well as an innate ambition, she decided to go to business school so that she could become "a CEO of a retail company and lead the industry in*

---

[7] This borrow the example posted on the webpage of Rocketsource.Co, and got its authors' approval of permission. ***See:*** Barlow B, Greene J. Data as a Service: The what, why, how and when. Rocketsource.Co. 2018. Retrieved from https://www.rocketsource.co/blog/data-as-a-service.

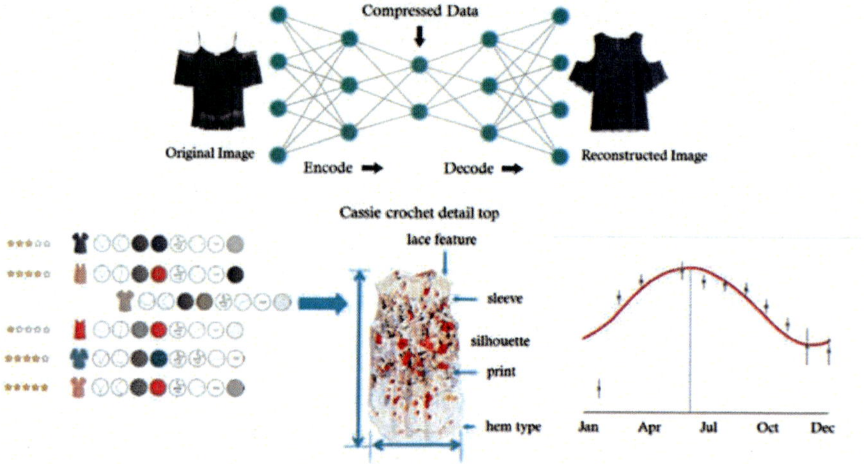

Figure 6.1 An example of Stitch Fix's data and personalized service. (Source: Author redrew it referring to Footnote 7)

> *technology and innovation." She got into Harvard Business School in 2009 and set out to achieve her goal. About a year after she had entered the MBA program, Lake started her own venture, initially named Rack Habit, using data and algorithms to help women dress a partners (with whom she has since parted ways), Lake sent out invitation to her friends and their friends in the Boston area, asking them to fill out an online questionnaire about their shopping references. The two founders transferred data on fit and style from the questionnaires onto spreadsheet, then created simple profiles of their new customers. Then they went shopping for them. Dresses, blouses, and skirts were delivered by mail. Customers could keep them or send them back – a benefit that today is standard for curated shopping services. According to Fortune magazine, Rick Habit was "constantly maxing out Lake's $6,000-limit credit card" and "made no money." Still, the continually growing number of customers proved that Lake and her partner were onto something.*
>
> *The trouble with shopping for apparel isn't that the right product – one that fits and matches a person's preferences – doesn't exist. The problem is that most people, especially individuals with demanding jobs and young kids, don't know how to find it quickly and easily. Or, as economists would put it: the apparel industry has increased variety but not discoverability. What shoppers need is a smart intermediary that can find the right items for them. These intermediaries exist: they are called personal shoppers, but unfortunately for most of us, they are too expensive. Rack Habit's core business idea was to give customers a competent personal shopper they can afford. By proposing to use rich data to match her customer and their preferences with the hugely diverse supply of apparel on the market, Katrina Lake raised $750,000 in seed funding in 2011.*
>
> *With here newly minted Harvard MBA in hand, Lake moved to San Francisco, changed the name of her company to Stitch Fix, ...*

Lake's initial data capital accumulated legitimately comes from a business "made no money;" namely, "exchange data through a service." And her seed funding of US$750,000 was obtained from a real data product of "matching customer preference with the hugely diverse supply of apparel on the market," rather than raw data. This example indicates two things. One is that data capital is lucrative, and the other is that the consumers are willing to share personal data when it involves a "clear personal benefit." There is a typical processing mode of exchanging data through a service. This mode should be coordinated between enterprises from traditional industries and new forms of business, respectively. First, an enterprise with new forms of business offers the individual with a low-price and even free terminal equipment (e.g., a shared bicycle, a shared charger, restaurant, etc.) or service (e.g., an app, the search engine, free movie, free courses). Next, some enterprises from traditional industries (e.g., equipment manufacturing, infrastructure construction, and IT sector) to be transformed into a foundry, produce the terminal equipment or provide the service. Then, the individual accepted the offer to convenience to his or her work or life, but the cost of all of them is the data—the individual agreed to the use of his or her personal data by the app or platform providers. Shoshana Zuboff's claim in her book The Age of Surveillance Capitalism does make sense. Personal privacy data has a high commercial value when only using naive data technologies. Whatever's that a necessary choice of transportation, catering, and healthcare, or that an optional choice of entertainment and education, the privacy can bring much income from any product or service targeting the individual. However, this phenomenon will change, as data technologies drive to expand the depth and breadth of reproduction on data capital.

Simple Reproduction on Data Capital

After the primitive accumulation of capital, that will be a reproduction. Michel Aglietta, a French economist and one of the founders of the regulation school, views economic reproduction as "the process whereby the initial condition necessary for economic activity to occur are constantly re-created."[8] Reproduction on data capital can be very complicated because it involves almost all existing economic research content if considering each digital part relating to any form of capital. In view of data capital is a new thing, my purpose here is to simplify at the very beginning. I'm going only to consider a reproduction on data capital that is formed by one or more limited datasets. Marx distinguishes reproduction from simple or enlarged reproduction. In the former case, no economic growth occurs, while in the latter case, more is produced than is needed to maintain the economy at the given level, making economic growth possible. I will introduce them to this and the next section, respectively.

---

[8]There are two references. ***See:*** (1) Aglietta M. A Theory of Capitalist Regulation: The US Experience (Radical Thinkers). Brooklyn, NY: Verso; Reprint edition (September 1, 2015). 2015; 448 pages; (2) The entry of "Reproduction (economics)" on Wikipedia. Retrieved from https://en.wikipedia.org/wiki/Reproduction_(economics).

In Max's *Capital*, he argues simple reproduction is the starting engine for meeting the consumption of society. For Marx, the production and consumption of society are both held by one another in a balance that necessitates reproduction. "A society," he says on the second sentence of the chapter devoted to simple reproduction, "can no more cease to produce than it can cease to consume." Reproduction of this kind, i.e. simple reproduction, is a combination of consumption and production. I think simple reproduction on data capital is in the same way, and the consumption is generally for entertainment purposes, such as fan-made movie edits. It is interesting to discuss such homemade movie edits from a perspective of simple reproduction rather than copyrights. I would like to make a comparison here.

Before entering the topic, it is best to emphasize that simple reproduction also belongs to data production (see Definition 3.6). I admire the Internet activist Aaron Swartz, who had an unbeatable combination of political insight, technical skill, and intelligence. But those academic journal articles he downloaded (from a digital library Journal Storage) to open access for the public were not processed by "production." Hence, it is not within scope for our discussion.

With a fusion of digital technology and performance art (or other cross-border performance), the media and entertainment (M&E) sector significantly enriches the data economy. YouTube and other video sharing platforms are littered with tens of thousands of fan-made movie edits: edits created by fans of a piece of movie wherein they use that movie to make their own ones. "Fair Use" is the key to review from a perspective of copyright. A copyright, as a form of protection geared towards words and symbols, protects literary and artistic assets such as books, movies, videos, plays songs, and photographs. It should not be confused with a trademark which is predominantly for the protection of a company's intellectual property surrounding its brands and logos. Intellectual property as an intangible item of value based on ideas, computer code, trademarks, copyrighted stories and characters, and so forth. For many companies, their intellectual property is both their primary product and their intangible asset. So, it stands to reason, they go out of their way to protect that. I intercepted an instruction about "Fair Use" in Case 6.1.[9,10]

> *According to the U.S. Copyright Office, Fair Use is a legal doctrine that promotes freedom of expression by permitting the unlicensed use of copyright-protected works in certain circumstances. Section 107 of the Copyright Act provides the statutory framework for determining whether something is a fair use and identifies certain types of uses – such as criticism, comment, news reporting, teaching, scholarship, and research – as examples of activities that may qualify as fair use. …*

---

[9] ***See:*** Retrieved from https://www.youtube.com/about/copyright/fair-use.

[10] ***See:*** Dawson R. The Ultimate Guide to Fair Use and Copyrights for Filmmakers. Frame.io. August 30, 2017. Retrieved from https://blog.frame.io/2017/08/30/copyrights-and-fair-use-for-filmmakers.

Case 6.1: An Example of Instructing How to "Fair Use"

- What is "Fair Use"?

"Fair Use" is a legal doctrine that says you can reuse copyright-protected material under certain circumstances without getting permission from the copyright owner.

- "Fair Use" guidelines

Different countries have different rules about when it's okay to use material without the copyright owner's permission. For example, in the United States, works of commentary, criticism, research, teaching, or news reporting might be considered fair use. Some other countries have a similar idea called fair dealing that may work differently.

Courts analyze potential fair uses according to the facts of each specific case. You'll probably want to get legal advice from an expert before uploading videos that contain copyright-protected material.

- The four factors of "Fair Use"

In the United States, "Fair Use" is determined by a judge, who analyzes how each of the four factors of fair use applies to a specific case.

The purpose and character of the use, including whether such use is of commercial nature or is for nonprofit educational purposes

Courts typically focus on whether the use is "transformative." That is, whether it adds new expression or meaning to the original, or whether it merely copies from the original. Commercial uses are less likely to be considered fair, though it's possible to monetize a video and still take advantage of the fair use defense.

The nature of the copyrighted work

Using material from primarily factual works is more likely to be fair than using purely fictional works.

The amount and substantiality of the portion used in relation to the copyrighted work as a whole

Borrowing small bits of material from an original work is more likely to be considered fair use than borrowing large portions. However, even a small taking may weigh against fair use in some situations if it constitutes the "heart" of the work.

The effect of the use upon the potential market for, or value of, the copyrighted work

Uses that harm the copyright owner's ability to profit from his or her original work are less likely to be fair uses. Courts have sometimes made an exception under this factor in cases involving parodies.

- "Fair Use" myths

There is some misinformation out there that might lead you to believe fair use automatically applies if you say a few magic words. There is actually no silver bullet that will guarantee you are protected by fair use when you use copyrighted material you don't own. Courts will consider all four of the factors described above and weigh them on a case-by-case basis. Here are some common myths:

Myth #1: If I give credit to the copyright owner, my use is automatically fair use.

As you saw above, transformativeness is usually a key in the fair use analysis. Giving credit to the owner of a copyrighted work won't by itself turn a non-transformative copy of their material into fair use. Phrases such as "all rights go to the author" and "I do not own" do not automatically mean you are making fair use of that material -- nor do they mean you have the copyright owner's permission.

Myth #2: If I post a disclaimer on my video, my use is fair use.

As we noted above, there are no magic words that will do this for you. Posting the four factors of fair use in your video or including the phrase "no infringement intended" won't automatically protect you from a claim of copyright infringement.

Myth #3: "Entertainment" or "non-profit" uses are automatically fair use.

Courts will look carefully at the purpose of your use in evaluating whether it is fair, but the three remaining factors also need to be considered. Declaring your upload to be "**for entertainment purposes only**," for example, is unlikely to tip the scales in the fair use balancing test. Similarly, "**non-profit**" uses are favored in the fair use analysis, but it's not an automatic defense by itself.

Myth #4: If I add any original material I created to someone else's copyrighted work, my use is fair use.

Even if you've added a little something of your own to someone else's content, you might not be able to take advantage of the fair use defense particularly if your creation fails to add new expression, meaning, or message to the original. As with all the other cases discussed here, courts will consider all four factors of the fair use test, including the quantity of the original used.

The thing that makes fair use so difficult is that it's not a clear-cut law. It's relatively open to interpretation, and unless you're actually sued or challenged, you may never know 100% whether your use adheres to the law or not. And if taken to court, it's the ruling judge's discretion based on his or her own interpretation of the law.

As shown in Case 6.1, "Fair Use" has four factors that boil down into two essential conditions. The first is a "transformative" meaning, when re-purposing the material for use in a way other than its original purpose. A reasonable example would be the use of the opening scene from *Saving Private Ryan* within an educational context of introducing to World War II knowledge. The second condition is that, given to satisfy the first one—the use is transformative, the amount and substantiality of the portion used in relation to the copyrighted work should be a small percentage of the whole. In the above example of using clips from *Saving Private Ryan*; if only having a few random voice-over commentaries (e.g., about cinematography), that would be a harder fair use argument to support. If your interpretation of "education" cannot be the same as a judge's, satires, parodies, video essays, and documentaries may be applicable; and they are often protected by fair use.

However, we often see any great inconsistency between a non-profit purpose and real income (see Footnote 10).

Perhaps the most common area where you might see this example of fair use is in video essays. Essayists like Tony Zhou ("Every Frame a Painting") and Evan Puschak ("Nerdwriter") garner millions of views from their respective video essays. And they all include movie clips, photographs, television clips, and in some cases, even music. ... And, on top of that, these guys are making thousands of dollars per video (as of this writing [in August 2017], Tony's Patreon campaign for "Every Frame a Painting" yields over $7,700 per video). ... But video essays [follow the first "Fair Use" condition] are the quintessential example of fair use in terms of both education and critical commentary.

..., just about every single instance of a fan film is a copyright violation. They are not necessarily making money from these films; but for the most part, the films are not making any kind of commentary or critique

on the [movie]. They are not transformative in purpose either (i.e. education). They are entertainment for entertainment's sake, just like the original ...

Some intellectual property holders encourage fan films and allow a vibrant fan film community to flourish. Fan films keep the culture alive and fans excited about this. Lucasfilm is a perfect example of that is how they've responded to and embraced the "Star Wars" fan film community. One of the most celebrated "Star Wars" fan films recently was "Darth Maul – Apprentice (2016)" with over ten million views. Now, movie editing has changed from niche art to mass entertainment, and it is no longer the scene combined with a piece of pure music and personal edits. In the face of massive "short video" traffic, a wave of content entrepreneurs is coming in. Inequality in profits is emerging between the original and adapted parties. "Fair Use" might no longer suitable for market demand. Perhaps it is time to change our view from a perspective of data simple reproduction.

Annotating Tongue Images with Electronic Medical Records

Traditional Chinese medicine (TCM) does not meet most scientific criteria. In a 2019 strongly worded editorial, Scientific American magazine, the oldest popular science magazine in the US, claimed that[11]

> *Data supporting the effectiveness of most traditional remedies are scant, at best. An extensive assessment was done in 2009 by researchers at the University of Maryland: they looked at 70 review papers evaluating TCM, including acupuncture. None of the studies proved conclusive because the data were either too paltry or did not meet testing standards.*

While traditional medicine originated in ancient China, today it's widely used throughout Asia, including in Japan and Korea. Over a decade and more, the World Health Organization is committed to getting representatives from Asian countries to condense thousands of years of knowledge in one neat classification system. Tarik Jasarevic, a spokesman for WHO, said traditional medicine diagnosis is poorly documented or not documented; still, it does not mean we should not "link traditional medicine practices with global norms and standard development."[12] But he added the inclusion of traditional medicine was "not an endorsement of the scientific validity of any traditional medicine practice or the efficacy of any traditional medicine intervention" (see Footnote 12).

Mainstream global healthcare as the common human wealth, through evidence-based thinking and practice, absorbs and adopts the most advanced theories and technologies from the current disciplines, such as physics, chemistry, and biology. The specialty that came to be known as preventive, clinical, and rehabilitation medicine is partially from traditional medicine conditions and practices. Denying traditional medicine as science does not mean total denial. Empirical knowledge based on tradition is possible to have practical value. Take the malaria treatment artemisinin as an example; Chinese scientist Tu Youyou, who won a Nobel prize in 2015, scoured ancient texts and folk manuals and traveled to remote parts of the country for clues, ultimately collecting 2000 potential remedies. One vital inspiration is a prescription as "one bundle of Artemisia annua, soak in 2 L of water, wring out for juice extract, then take it all for treatment of malaria" found in Ge Hong's Elbow Reserve Emergency. Ge Hong lived in the Western Jin Dynasty around the fourth century. To be sure, many widely used and experimentally validated pharmaceuticals, including

[11] ***See:*** The Editors. The World Health Organization Gives the Nod to Traditional Chinese Medicine: Bad Idea. Scientific American. April 1, 2019. Retrieved from https://www.scientificamerican.com/article/the-world-health-organization-gives-the-nod-to-traditional-chinese-medicine-bad-idea.

[12] ***See:*** Hunt K. Chinese medicine gains WHO acceptance but it has many critics. CNN.com. May 26, 2019. Retrieved from https://www.cnn.com/2019/05/24/health/traditional-chinese-medicine-who-controversy-intl/index.html.

aspirin, decongestants and some anticancer chemotherapies, were originally derived from plants or other natural sources.

Given more data, traditional medicine can also accept scientific principles of testing, such as falsifiability and consistency. Acupuncture that gains traction in the West has been studied from a perspective of mechanical analysis. For example, using one or more force and torque miniature sensors integrated on the needle body or shaft collects real-time force data during the acupuncture process. These data reflect on the needle movement of lifting and twisting. They can help detect group similarity or individual differences of the recipients (i.e., patients) when an applicator uses different modalities, or different applicators use the same pattern.

I have some experience with this (i.e., using data to support traditional medicine) via a study of mining electronic medical records (EMRs). We applied the FP-Growth, PrefixSpan, and USpan pattern mining algorithms to around 70,000 medication administration records from over 30,000 childhood inpatients with diagnosis of pneumonia during a 6-year period at a children's hospital in China. The first two are more traditional methods of pattern mining and mine a complete set of frequent medication use patterns. PrefixSpan also incorporates an administration sequence. The newer USpan method considers medication utility, defined by the dose, frequency, and timing of use of the 652 individual medications in the dataset. Together, these three methods identified the top ten patterns from six age groups, forming a total of 180 distinct medication combinations. These combinations encompassed the top 40 most frequently used medications. These patterns were then evaluated by subject matter experts to summarize five medication use and two treatment patterns. In the medication use patterns, we found that pattern 4 as "antibiotics and (medications for enteritis or skin diseases)" and pattern 5 as "(antiasthmatics or expectorants or corticosteroids) and (medications for enteritis or skin diseases)" indicate a potential relationship between the lungs and skin.[13]

> *There were differing opinions within our expert panel as to what pattern 4 (antibiotics and medications for enteritis or skin diseases) indicates. One possible explanation, offered by 4 US experts, is that because of the immaturity of infant organs, damage to one organ or organ system could cause reactions in other organs. Furthermore, from a pharmaceutical perspective, broad-spectrum antibiotics (e.g., cefotaxime) cause diarrhea by irradiating normal gastrointestinal (GI) flora, and physicians can treat this type of diarrhea in the pediatric population with bifid triple viable, a probiotic used to re-establish normal GI flora. Additionally, diarrhea can cause diaper rash that can be treated with zinc oxide. Chinese experts believed that these findings support an underlying internal relationship between the lungs and large intestine (or the lungs and skin), which is consistent with established theories of traditional Chinese*

[13]This is my previous work ***See:*** Tang C, Sun H, Xiong Y, *et al.* Medication Use for Childhood Pneumonia at a Children's Hospital in Shanghai, China: Analysis of Pattern Mining Algorithms JMIR Med Inform 2019;7(1):e12577. DOI: 10.2196/12577.

> *medicine (TCM). TCM believes that the human body is an organic whole, that there is a relationship between the lung and large intestine, and that the lungs govern the skin and hair. Despite this rationale, our expert panel generally agreed that pattern 5 was unexpected (i.e., the use of medications for enteritis or skin diseases following inhalation treatment, without concurrent use of antibiotics). The overall Fleiss kappa for these 7 patterns was 0.693 indicating moderate agreement. Our findings neither contradict nor confirm these hypothesized relationships as our data target treatment protocols which cannot, on their own, speak to the relationship between organ systems. However, according to a recent study [published in Science in 2018] on innate lymphoid cells, lung inflammation might originate in the gut, and there exists a link between intestinal microbiota and lung diseases (e.g., asthma).*

Let's look at traditional medicine of tongue diagnosis. Example 6.3 tongue diagnosis is a non-invasive, simple diagnostic tool for monitoring health. Essentially, tongue diagnosis purports that the tongue appearance (e.g., color and texture) as an outer manifestation of the status of the human body. Differentiate conditions of individual patients can be partitioned as many syndromes; these syndromes have been used to characterize patient status in many diseases or conditions, including inflammation, infection and stress, as well as immune and endocrine disorders. For example,[14] the Cold Syndrome and Hot Syndrome are the two most common and representative syndromes that represent two opposite but interrelated conditions of the human body and are identified by a "white-greasy" and "yellow-dense" coating of the tongue, respectively. In recent years, tongue diagnosis has seen gradual acceptance in modern Western medicine. Some medical literature has used the term "geographic tongue"[15] to highlight tongues with discolored regions or cracks that accompany illness. One case study associated with a variety of systemic disorders that was published in the renowned New England Journal of Medicine describes a patient's "smooth, shiny tongue."[16]

While phones have superb cameras with a high-resolution sensor to create visually lossless, the collection of tongue images requires more specialized instruments. In this way, it ensures the consistency of light conditions to capture different tongue appearances from patients. Some professional corporations should provide these specialized instruments. For example, our instrument is a patented light-field camera (CN201520303463.5) called the intelligent mirror using CIE L*a*b* color space (CIELAB). CIELAB is a color space defined by the International Commission on Illumination in 1976, which is roughly the same amount of visually perceived change. Once data collection through these instruments, a dispute on primitive data capital accumulation could emerge

[14] ***See:*** Jiang B, Liang X, Chen Y, *et al.* Integrating next-generation sequencing and traditional tongue diagnosis to determine tongue coating microbiome. Scientific Report. 2012;6(2):936.

[15] ***See:*** Zadik Y, Drucker S, Pallmon S. Migratory stomatitis (ectopic geographic tongue) on the floor of the mouth. J. Am. Acad. Dermatol. 2011;65: 459-460.

[16] ***See:*** Lee HJ, Jo DY. A Smooth, Shiny Tongue. N Engl J Med, 2009;360(6), e8.

between acquisition units (e.g., hospital) and equipment providers. We assume all patients have signed the consent documents to participate in the study by sharing their tongue images. Note that the patient is a data provider rather than a data sovereign; the disputes here are among data sovereigns (more detailed explanations, see Chap. 5). Data collection has been a bottleneck in machine learning, manifested in the follows. First, as machine learning is becoming more widely-used, we often see new applications that do not necessarily have enough labeled data. Although a specialized instrument can generate some parameters (e.g., the CIELAB value of tongue, tongue coating, lip, etc.) automatically when shooting the tongue, these parameters to be features are too few to help machine identify the images. Second, even if deep learning techniques automatically generate features, which saves feature engineering costs, but in return may require larger amounts of labeled data. I give the definitions relating to feature engineering and labeled data as below.

Definition 6.1 *Feature Engineering*[17] Feature engineering is the process of using domain knowledge to extract features from raw data via data technologies. These features can be used to improve the performance of algorithms.

Definition 6.2 *Labeled Data*[18] Labeled data is a group of samples that have been tagged with one or more labels. Labeling typically takes a set of unlabeled data and augments each piece of that unlabeled data with meaningful tags that are informative.

Labels can be obtained by asking humans to make judgments about a given piece of unlabeled data (e.g., Does this photo have a shiny tongue?), and are significantly more expensive to obtain than the raw unlabeled data.

Figure 6.2 shows a real image, in which six parameters on the left are produced by the intelligent mirror automatically, including the CIELAB value of the face, lip, tongue coating, and overall tongue. Our annotations are on the right (see the blue highlighted segments). The annotated information is not only from the patient's EMRs but standard representations of tongue diagnosis translated by six parameters. For example, the color combination of cyan, red, yellow, white, and black were used to describe the color of the face according to the traditional medicine literature.

The study was conducted at a tertiary, comprehensive hospital integrated traditional Chinese medicine and western medicine. We recruited hospitalized subjects (excluding minority groups or other sensitive or disempowered populations) in its geriatrics department. We pulled unidentified patient ID (identification), age range, gender, weight and height, initial diagnosis, and admission and

[17]*See:* The entry of "Feature engineering" on Wikipedia. Retrieved from https://en.wikipedia.org/wiki/Feature_engineering.

[18]*See:* The entry of "Labeled data" on Wikipedia. Retrieved from https://en.wikipedia.org/wiki/Labeled_data.

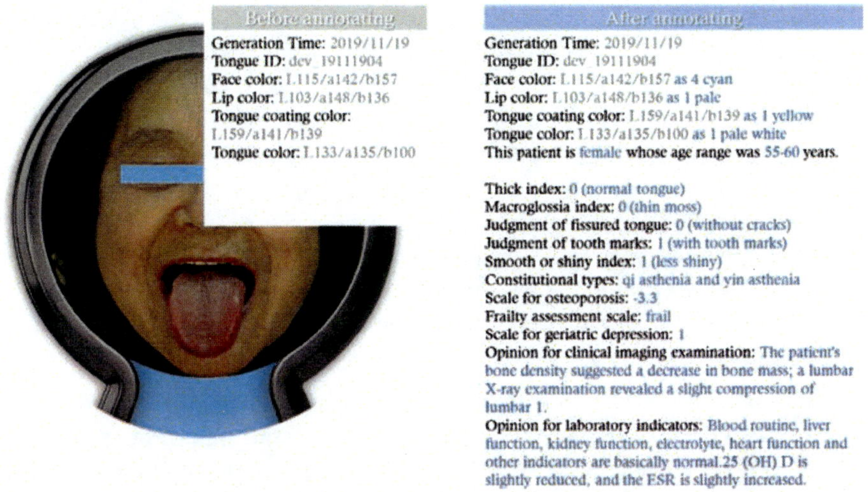

Figure 6.2 A tongue image with a comparison of its parameters before and after annotating. Note: the patient signed the consent form to share this image, the sample can be found on https://www.dataindustry.org/tongue. (Source: Author drew it by herself)

discharge dates, as well as previous chronic medical history from EMRs stored in hospital information system. We also collected vital signs, clinical imaging examination, and laboratory indicators during the hospitalization.

The entire data acquisition and image annotation process was conducted by a panel of subject matter experts which consisted of four fully credentialed senior-level physicians (i.e., associate chief physician and above), one resident, and two medical students. One medical student was in charge of the data acquisition process. The resident was responsible for consolidating patients' previous chronic medical history, clinical imaging examination, and laboratory indicators. One physician was to diagnose patients' constitutional types. Another physician gave a final admission diagnosis by considering the patient's constitution based on both TCM and Western medicine. Note that constitutional types are based on TCM analysis and differentiation of pathological conditions in accordance with the eight principal syndromes, namely 八纲辨证, including yin and yang (阴阳), exterior and interior (表里), cold and heat (寒热), and hypofunction and hyperfunction (虚实). All the information from the free-text data labeling was documented digitally by one medical student in Chinese and translated into English. The treatment plan corresponding to the admission diagnosis was reviewed and annotated by the remaining two physicians.

After annotating as simple reproduction on tongue images, the newly labeled dataset becomes the hospital's capital asset meeting daily demand for treatment (as consumption). It does not belong to the corporation that provides an intelligent mirror. Note that the corporation owned an unlabeled dataset as its capital asset.

Visualizing World Trade Reports

When preparing this book, I have developed certain expectations about the contents of consulting reports in which the annual World Trade Report is essential. As a nonnative speaker of English, I'm so lazy to tackle a to-be-read pile: dozens of reports with each over 250 pages. I rely on data visualization to help discover something exciting and intuitive. I used the simplest wordle with a stream graph rather than some cutting-edge tools. Note that any online consulting reports are generally in the Portable Document Format (i.e., PDF files). We may use some ready-made codes on the Internet to convert PDF files before text mining.

Definition 6.3 *Wordle* A wordle is a visual depiction of the words contained in a piece of free text. The more frequent a particular word was within the source text, the bigger it's displayed in the wordle—font and color variation, as well as giving weight to particular words, which are positioned vertically or horizontally. It can help filter out both low-frequency and low-quality information to help a viewer understand content by just glancing it. Example 6.4 the series of World Trade Report started in 2003, of which the first paragraph gives the purpose in the 2003's FOREWORD.

> *The World Trade Report is a new annual publication produced by the WTO [World Trade Organization] Secretariat. Each year, the WTR will explore trends in world trade and highlight important issues in the world trading system. In addition to monitoring and interpreting trade developments, the Report seeks to deepen public understanding of pressing policy issues. The WTR does not pretend to provide comprehensive answers to complex and many-sided questions subject to continuing debate among governments and their constituencies. Rather, by explaining the origin of issues and offering an analytical framework within which to address them, the WTR aims to contribute to more informed discussion and a better appreciation of the options available to address policy challenges.*

My focus here is the titles of tables, charts, and boxes in all reports. Figure 6.3 is wordles generated by a text model automatically from 2003 to 2017, and each for 1 year. The findings indicate keyword variations; for example, the word "natural" is enormous in the 2010 wordle (that is in the middle of this figure), and the subtitle of the 2010 report is *Trade in Natural Resources*. Another example is in the 2011 wordle, in which an abbreviation "PTAs" coves a lot meaning "preferential trade agreements." The 2011 World Trade Report shows that an expansion of preferential trade flows is not matching the explosion of PTAs. In other words, despite the proliferation of PTAs, 84% of world merchandise trade still takes place on a non-discriminatory most-favored nation basis. So, WTO concluded that deep regionalism is gaining momentum.

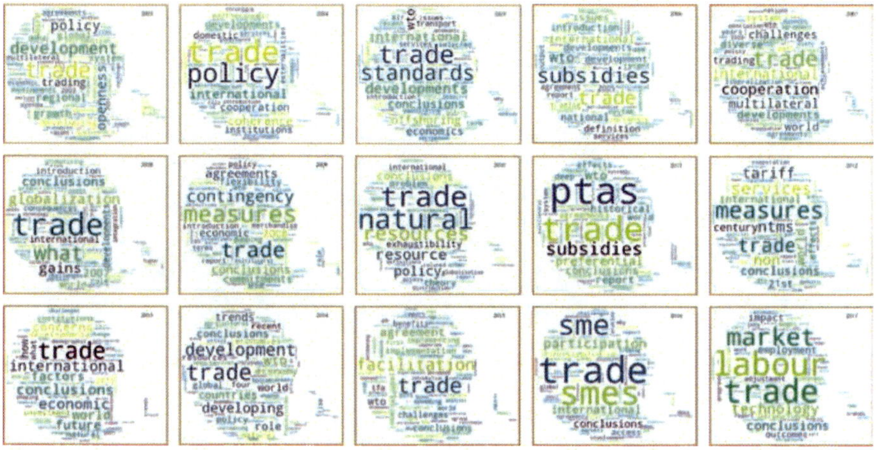

Figure 6.3 The wordles of all words relating to the titles of tables, charts, and boxes in the World Trade Report from 2003 to 2017. Note: the year-label is at the upper right corner in each wordle, for example, the first row is from 2003 to 2007. (Source: These wordles are produced by a text model automatically)

Simple reproduction on this series of reports is about consumption, that is, to help me obtain the content of them quickly. To understand the theme evolution, I also produced figure 6.4. I won't give more explanation here.

Enlarged Reproduction on Data Capital

Enlarged reproduction on data capital than on the others is pretty straightforward. One of four potential impacts with data as capital is profit multiple. Even if only replacing the methods with differential technical difficulties on the same dataset as enlarged reproduction, we may obtain different results to set up various earning strategies. For the up-to-the-minute data about customer purchases at the checkout counters of a retailer, figuratively speaking; if only using statistics to know the ratio of customers to male and female or the period for shopping, we can decide to fold in half for the ladies on one Saturday afternoon. If using basic concepts of association analysis to uncover relationships among commodities in market basket transactions, we can place beer and diapers together to do promotion. If comparing merchandise sale trends via the "stretchable" similarity to do inventory control, we can choose hunger marketing—remove one of two beverages that are similar but have a several-day time interval. For example, a store's customers are partitioned in two kinds: one likes the herbal tea, and the other loves a sour plum drink. If the latter cannot find their favorite, they are likely to revisit this store. Thereby, the store can increase

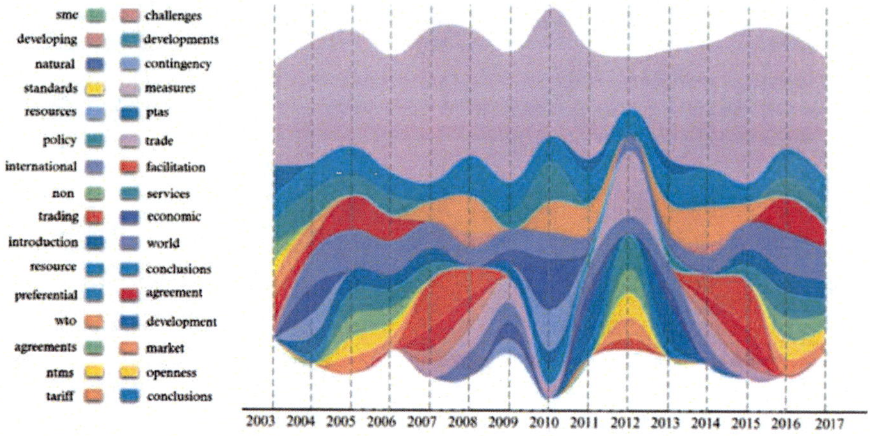

Figure 6.4 A stream graph of indicating theme evolution relating to the titles of tables, charts, and boxes in the World Trade Report from 2003 to 2017. (Source: This graph is produced by a text model automatically)

turnover by more visits. Note that rising tours is an essential promotional means for the stores.

Competition intensifies, of course, are the direct result as more and more people master data technologies. In cases some data is not available, alternative datasets will emerge. To estimate housing vacancy rates, for example, power consumption data can be used. It is necessary to investigate housing vacancy rates through questionnaires. But designing, issuing, and recalling a survey is time-consuming and laborious. Also, some respondents may mislead us intentionally or unintentionally, resulting in inaccuracy. One apartment must never be lived if no using electricity, and we can confirm whether to use electricity or not by checking the power meter. Therefore, it is feasible to investigate housing vacancy rates by analyzing power consumption data. The person owned electric data to study real estate across the industries caused competition.

After competition intensifies, the following potential impacts with data as capital may be convergence effect or latecomer advantage. The future automotive sector given in the previous chapter is an example. Tech players from Silicon Valley and other hot spots like Google enter into the automotive ecosystem, starting with consumer data analysis. This phenomenon indicates a convergence of new forms of business and the traditional industries, and new forms of business have a latecomer advantage. Better service efforts mean more data; vice versa, more data with advanced technologies say much better service. Other traditional industries include, but not limited to, the agriculture and energy sectors. Agriculture has become increasingly high-tech over the years. Now, agriculture professionals can dig into data and use it to make highly informed decisions, such as managing crop diseases and pests, investigating agricultural niches, coping with climate change, and making yield predictions. Many oil and gas companies have applied data mining techniques to look across upstream, downstream, chemicals, and regional units to reveal savings since a decade ago. In addition to the industry level, the impacts also indicate a global convergence, as I mentioned in the introduction. In general, the forces of convergence are based on the ability of "data sovereigns" to protect the data they owned or controlled. Using data force to protect data does not imply that abandon data sharing. Instead, it should be easy for an organization to enable the sharing of data and data products internally or with trusted partners. Economic agglomeration in cyberspace without considering geopolitical borders is now possible, and further promotes the diffusion of data knowledge and skill. Data knowledge and skill diffusion, the key to "overall productivity growth as well as the reduction of inequality both within and between countries,"[19] enables a convergence between emerging and developed economies. And even emerging economies have a latecomer advantage, like India, of which the software is the dominant sector.

The above reasons reveal that it is easy to do enlarged reproduction even though having a small amount of data capital. Its purpose is to pursue profits rather than consumption. In the rest of this section, I'm going to interpret further them, relating to new technologies and new applications, based on the two datasets I used before.

Representation Learning with World Trade Reports

In Example 6.4, I used the most uncomplicated way relating to natural language processing (NLP) to visualize the World Trade Reports. As a possible humanity's greatest invention, language allows us to communicate information, convey ideas, and collaborate on this basis. But for machines, the difference in semantics between languages can be a tremendous barrier. NLP's development for now has experienced three stages. The first stage is based on the rule. The rule is given by conventional manual or semi-manual methods. The previous version of machine translation suggests the original input should match any rule in advance; otherwise, a translated sentence will carry out with errors. However, natural language is too complicated to describe using the limited rules fully.

The second stage relies on statistics. TF-IDF[20] stands for term frequency—inverse document frequency is one of the most popular term-weighting schemes today. Consider a set of documents, when each document is represented by a

---

[19] ***See:*** PiKetty, T. translated by Goldhammer A. Capital in the Twenty-First Century. Cambridge, MA: the Belknap Press of Harvard University Press; Reprint edition (August 14, 2017), 2017: 816 pages.

[20] ***See:*** Wu H, Luk R, Wong K, *et al.* Interpreting TF-IDF term weights as making relevance decisions. ACM Transactions on Information Systems, 2008;26(3):13.

vector whose components are the frequencies with which each word occurs in the document. So, these words, excluding low-frequency and low-quality ones, refer to the features; that is[21]

> *We assumed that an input corpus C has n documents D_i, i∊ [1, …, n] and its entire vocabulary set is W. We applied standard text processing techniques including tokenization (e.g., part-of-speech tagging & filtering) and stop-word removal on each document to abstract the document into a set of words $W_i \subseteq W$. We then employed TF-IDF (i.e., term frequency – inverse document frequency) to acquire TF-IDF score matrix S_i^{tfidf}. The total number of rows of S_i^{tfidf} equals W_i (i.e., the number of words in D_i). Each line of S_i^{tfidf} represents a word. The total number of columns is 2 and the second column in the matrix represents the TF-IDF score of the word. The TF-IDF score is computed by the frequencies of the individual words in the document (i.e., TF) multiplied by the inverse frequencies of words in entire collection (i.e., IDF). Each word in the final output of the pipeline is considered a feature, where the weight of the feature is the TF-IDF score. We thus represent each document as W_i pairs of (feature, weight) converted from the matrix S_i^{tfidf}.*

With the rapid development of computer hardware, an increase in computing power enables the queries in the ontology established by features within our limits of tolerance. As a simple example, an ontology is to map all features in the set of {Harvard, Harvard Graduate School of Arts and Sciences, Harvard University} to the concept of "Harvard University." After building the collection ontology, it is easy to explore deeper relationships among ontologies, such as the opposite and approximate relationships. But useful features require lots of human efforts; meanwhile, for certain specific tasks, feature engineering means repeatedly inevitable work because their features cannot be reused.

The third stage depends on representation learning. At the current stage, the most troublesome issue of the prior stage on "features" evolve into multilevel object representation; these data objects involve words, sentences, paragraphs, documents, etc. This is because:[22]

> *the performance of machine [and deep] learning is heavily dependent on the choice of data representation (i.e., features) on which they are applied. For this reason, much of the actual effort in deploying … algorithms goes into the design of preprocessing pipelines and data transformations of the data that can support effective learning. … feature engineering is important but labor-intensive and highlights the weakness of current machine learning algorithms: their inability to extract and organize the discriminative information from the data. Feature engineering is a way to take advantage of human ingenuity and prior knowledge to compensate for that weakness. In order to expand the scope and ease of applicability of machine [and deep] learning, it would be highly desirable to make learning algorithms less dependent on feature engineering, so that novel applications could be constructed faster, and more importantly, to make progress towards Artificial Intelligence.*

---

[21] This paragraph comes from one of my papers. ***See:*** Tang C, Plasek JM., Xiong Y, *et al.* A Clustering Algorithm Based on Document Embedding to Identify Clinical Note Templates. Annals of Data Science, 2020. DOI: 10.1007/s40745-020-00296-8.

[22] ***See:*** Bengio Y, Courville A, Vincent P. Representation learning: A review and new perspectives. IEEE transactions on pattern analysis and machine intelligence, 2013;35(8), 1798-1828.

No matter what kind of data object chosen, embedding is needed to express these in a uniform representation. Then, we can run a variety of shallow or deep neural networks directly. An embedding is a mapping of a discrete variable to a vector of continuous numbers. The study of language after embedding becomes a problem of vector space mathematics. For example, adding and subtracting vectors produces another composite vector in the same space. The easiest way to think about words is with an example. The *notorious* one is "King – Man + Woman = Queen." In other words, a composite vector, which adds the vectors associated with the words "king" and "woman" and subtracts "man," is close to the vector associated with the "queen." Another example is: "Paris – France + Poland = Warsaw." In this case, the vector difference between "Paris" and "France" captures the concept of capital city. One reason is vector closeness. Words close in vector space are often synonyms (e.g. happy and delighted), antonyms (e.g. good and evil), or other easily interchangeable words (e.g. yellow and blue). If you want to play with that, look at Word2Sense phrase search.[23]

While existing some problems as interpretability and computational complexity, one benefit from neural networks enables algorithms to learn data object representation automatically, thereby avoiding a large workload of feature engineering. A significant change is the task of text generation (e.g., poem generator, AI manuscript writer) becomes possible. Deep learning NLP converts most tasks into an encoding and decoding problem. Encoding free text is similar to the previous job of feature extraction, and decoding an internal feature can detect its representation—a word, a sentence, a paragraph, and even a document. This is very helpful for graph generation and text generation. The former is to build an image by an understanding of the text description, and the latter is to summarize a paragraph to depict an image. We thus, can achieve cross-modal retrieval. Besides, the problem of multilingual generation may be solved if possibly aligning features that come from different languages. Example 6.5 uses representation learning to compare the titles he titles of tables, charts, and boxes in the World Trade Report from 2003 to 2017. The object to learn here is the sentence. I used Sanjeev Arora et al.'s embedding method.[24]

> ***... a new sentence embedding method that is embarrassingly simple: just compute the weighted average of the word vectors in the sentence and then remove the projections of the average vectors on their first singular vector ("common component removal"). Here the weight of a word w is a/(a + p(w)) with a being a parameter.***

Table 6.1 shows both the top and bottom ten similar sentence pairs. The top ten means the most similar sentence pairs, but the identical ones (i.e., their

---

[23] ***See:*** Retrieved from https://explosion.ai/demos/sense2vec.

[24] ***See:*** Arora S, Liang Y, Ma T. A simple but tough-to-beat baseline for sentence embeddings. In 5th International Conference on Learning Representations, ICLR 2017.

Table 6.1 Top and bottom 10 sentence pairs in the world trade reports

Top	Distance	Sentence 1	Sentence 2
188	0.075561881	2006 App Table 2 I World trade of commercial services by region and selected country, 2005	2005 App Table 2 I World trade of commercial services by region and selected country, 2004
189	0.08589819	2007 App Table 1 I World merchandise trade by region and selected country, 2006	2006 App Table 1 I World merchandise trade by region and selected country, 2005
190	0.089300405	2006 App Table 1	2005 App Table 1 I World merchandise trade by region and selected country, 2004
191	0.100159118	2011 App Table 2 I World exports of commercial services by region and selected country, 2010	2010 App Table 2 I World exports of commercial services by region and selected country, 2009
192	0.100159118	2011 App Table 5 I Leading exporters and importers in world trade in commercial services, 2010	2010 App Table 5 I Leading exporters and importers in world trade in commercial services, 2009
193	0.100159118	2012 App Table 2 I World trade in commercial services by region and selected country, 2011	2010 App Table 2
194	0.100159118	2012 App Table 5 I Leading exporters and importers in world trade in commercial services, 2011	2010 App Table 5
195	0.102748614	2009 App Table 2 I World exports of commercial services by region and selected country, 2008	2008 App. Table 2 I World exports of commercial services by region and selected country, 2007
196	0.107402134	2009 Chart 14 I Chart 14 Number of cases of tariff increases of at least 15 percentage points, by country, 1996–2006	2009 Chart 11 I Number of cases of tariff increases of at least 15 percentage points, by product, 1996–2006
197	0.075561881	2009 App Table 2	2010 App Table 2
...			
868	0.99667363	2012 Figure C.2 I Shares of product lines and trade value covered by NTMs, 1996–2008	2014 Figure C2 I Share in imports of parts and components by country group, 1996–2012
869	0.996734937	2004 Chart IIC.2 Environmental governance and water quality	2004 Chat IIC.3 Environmental governance and air quality
870	0.996768906	2010 Figure 28 I Export taxes by natural resource upper bound estimates	2010 Figure 29 I Export taxes on mining products by subheading upper bound estimates
871	0.997064912	2013 Table 1.3	2008 Table 2 I World exports of merchandise and commercial services, 2007
872	0.997521063	2014 Table 3 I World exports of merchandise and commercial services, 2005–2013	2006 Table 3 I World exports of merchandise and commercial services, 2005
873	0.998449333	2016 Figure B.18 I Shares of direct and indirect services exports by firm size in developing economies	2016 Figure B.17 I Indirect exports by manufacturing sector and firm size in developing economies
874	0.999166272	2010 Figure 19 I Real prices of nickel, plywood and fish, Jan. 2000–July 09	2010 Figure 17 I Real prices of selected commodities, Jan. 2000-Aug. 09
875	0.999639144	2013 Table 1.3 I World exports of merchandise and commercial services, 2005–2012	2007 Table 2 I World exports of merchandise and commercial services, 2006
876	0.999677116	2010 Table 4	2012 Appendix Table 6 I Leading exporters and importers in world trade in commercial services (excluding intra-EU(27) trade), 2011
877	0.999677116	2010 Table 4 I World exports of commercial services by major category, 2009	2011 Table 6 I Leading exporters and importers of commercial services excluding intra-EU (27) trade, 2010

Notes: Gray words show the repeated occurrences of titles, and khaki means different content in a sentence pair

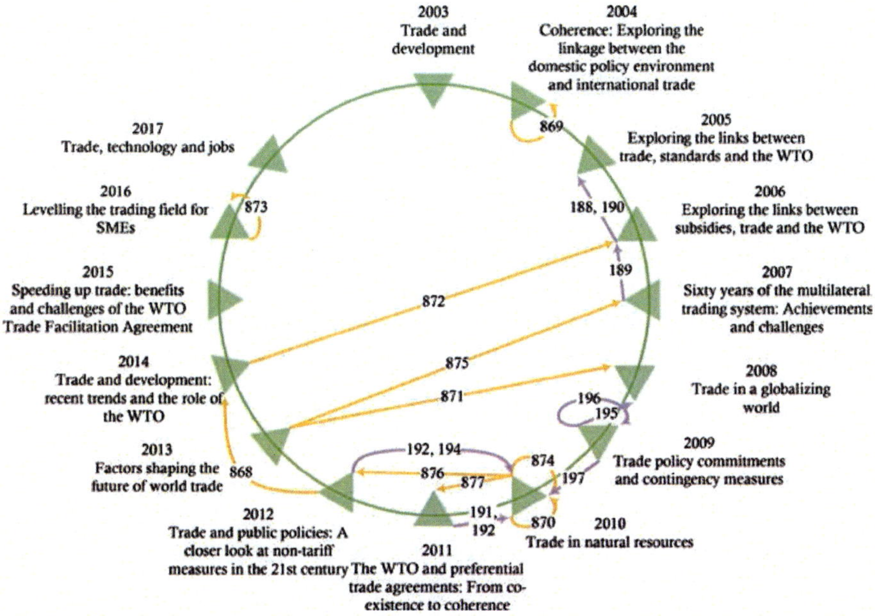

Figure 6.5 A correlation of different years' reports with Top/Bottom 10 similar sentence pairs that come from the titles of tables, charts, and boxes in the World Trade Report from 2003 to 2017. Note: This figure is in line with Table 6.1. (Source: Author drew the results produced by a learning model automatically)

distance is 0) excluded. It is easy to see these similar sentence pairs are all titles used in the Appendix, which indicates a correlation among reports of consecutive years. The bottom ten refers to the most dissimilar sentence pairs among a total of 877 pairs. One sentence pair may involve two different years, implying a correlation among reports of contents. For this reason, I drew figure 6.5, according to Table 6.1. Regarding content indications, the findings relating to the most dissimilar sentence pairs are more interesting than the most similar ones.

While being rough a little bit, it is enough to be an evident result for data products. The reason is that this product aims to purse the profits rather than self-/ or consumption. Whatever the results good or bad is not very vital, the key lies at the available analyzable data with both breadth and depth of what data technologies can reach. From a technical viewpoint, this product is more likely to receive on-demand outsourcing than the simple visualization one. Next, I can publish this on websites, such as GitHub, to await—wait for picking up a task if someone makes a request. Also, we can send to someone who needs. Sharing the original work may earn points from some open source communities, and it is possible to receive extra practice with a benefit for this product's subsequent development.

Tongue Diagnosis in the Future Scenarios

Before proceeding, we must gain a better understanding of the dataset. The value of the data is in two aspects. In one aspect, this continuously growing dataset relating to tongue diagnosis is extensible, comparable, and compatible. Data collection processes are standardized to acquire data by considering the requirements and expectations of not only patients but also various researchers. The dataset pursues at least three purposes. First, it covers almost all possible indicators of tongue diagnosis in Chinese and Western medicine and adds the content of face consultation additionally. Second, it aims to adopt the epidemiological method of investigation by (1) limiting the target population as an Asia population with a fixed age range, and (2) scheduling the collection time as the first day of hospitalization. Thirdly, the data can be easily linked to EMRs from different systems, such as CT (computerized tomography) scans or MRIs (magnetic resonance imaging) and clinical laboratory indicators, relying on more than 20 years of previous HIS (hospital information system) experience.

In the other aspect, the data is labeled by clinicians with rich clinical experience. A total of 16 physicians participated in manually labeling the data with the admission diagnosis. Each patient's diagnosis is determined through a panel of subject matter experts. The data will be updated if the patient is readmitted to the hospital. The dataset meets the requirements for use as a training set and is suitable for artificial intelligence and machine learning. Some preliminary results are able to correct false medical information or misleading claims concerning tongue and face consultation on the Internet and social media.

Developing a mHealth app is possible to consider by most people on reutilizing what learned from the dataset of tongue diagnosis. A scene from solely disease treatment to commercial healthcare is often natural, accessible, and acceptable. Integrated into an allergy-related mHealth app, for example, can help diagnose oral allergy syndrome. The tongue is affected by oral allergy syndrome; the others include the mouth, lips, and throat. Like allergic rhinitis, oral allergy syndrome is due to the body that cross-reacts to certain proteins: the former caused by pollens, while the latter is about foods. Proteins are found in all organic matter. The immune system identifies particular proteins to target bacteria, viruses, and other unwanted germs. However, it sometimes identifies everyday proteins, such as pollens, as harm full as well. For many, the key signs of oral allergy syndrome are swelling and itchiness of the lips, mouth, tongue, and throat immediately after eating certain fruits and vegetables, especially when raw. Humans likely know very little about bodies, especially a child who needs parental supervision. We often do not know what kind of food may cause allergies, how the allergic symptoms develop, and whether the case needs intensive care. An initial mild allergic reaction developed to be severe may only take a few hours. The severe swelling of the tongue or throat can cause suffocation. At this time, a mHealth is needed to get effective information in time.

I once designed such a mHealth app, of which the purpose is simple: to help decide whether to go to the emergency room now, according to existing reactions. Despite this, the app we called the Allergy Screener is based on continual analysis of vast amounts of available adverse sensitivity data, and requires heterogeneous or complex data integration as well as result multiple filtering. We at Allergy Screener want to make lives easier for people with adverse sensitivities, which is defined as allergies, hypersensitivities, and intolerances by helping them identify, document, and manage reactions and causative agents. Our app will provide information about users that self-identify quickly the signs and symptoms of a severe allergic reaction, nearby providers (e.g. hospitals, insurance companies) that have availability to see the patient, or wait times (i.e. skip to the prioritized queue track of a specific insurance company) at local emergency rooms to help a patient get back to a stable condition. This personalized treatment can exclude a potential fatal allergen, offer tips caution about foods or drugs, save patients' time in searching a good doctor, and reduce expenses if linked to their health insurance network. Therefore, the tongue diagnosis data can be a supplemental relating to other data. A visual description of my solution is shown in figure 6.6.

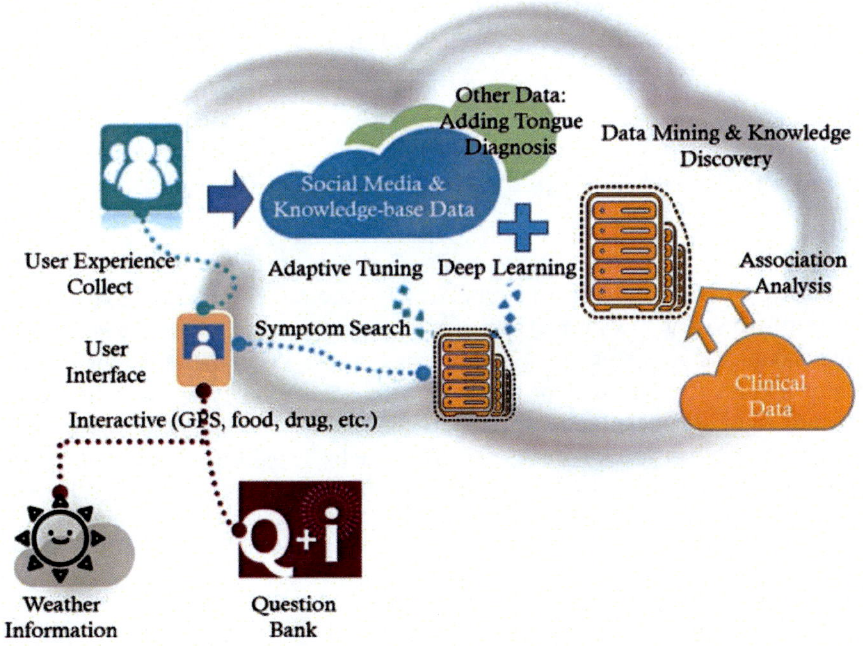

Figure 6.6 A visual description of the Allergy Screener with the data relating to tongue diagnosis. (Source: This is one of author's data products)

In sum, applying in mHealth apps is a pretty good future scenario. Regarding more user adoptions, however, my suggestion is to use tongue data across the field in the catering business. The reason is the plain truth—not everyone is allergic, but everyone has to eat. Eating with habits is well characterized by the proverb "you are what you eat." Chinese cuisine is an integral part of Chinese culture. There exists a series documentary *A Bite of China* showing this very well. British journalist Oliver Thring called this documentary as "the finest food TV ever." In one episode that "medicine and food spring from the same origin," advocates mindful eating: think before you bite. Bumps, patches, and spots in your mouth can be harmless. But sometimes, they can give clues to what's going on with the overall health. Infections, stress, medication issues, and even aging can make their marks on our tongue. Finding out what our tongue is telling us before eating, specifically from the health professionals, may be helpful to help.

Many food apps exist related to ingredient consumption for various purposes. Some pay attention to thousands of restaurant menus to offer useful information to both the supplier and consumer. Some recommend recipes that combine food materials consumers likely have on hand, with auxiliary sale products. Some provide QR code (quick response code) or bar code scanning tools to help consumers acquire nutrients and evaluate nutritional values. The others suggest where to buy cheaper food materials, how to burn calories after eating, to encourage better consumption. Then there are supporting services, such as food ordering service, food (cuisine) delivery service, and queue outside restaurant service, restaurant appraisal service, and so forth. It is possible to add tongue diagnosis information to the services mentioned above.

What's Different on Data Capital's Metamorphoses?

Thomas Piketty wrote one chapter in his *Capital in the Twenty-First Century* to dedicate the metamorphoses of capital (see Footnote 19). He considered various forms of wealth and examined their development over time, involving net foreign capital, other domestic capital, housing, and agricultural land. His examination specifically included the cases of Britain and France since the eighteenth century. In a more detailed paper also by him in 2014,[25] as shown in figure 6.7, Piketty added "net foreign debt" and additionally listed three countries. These countries are Germany, the US, and Canada, with differential timelines but the

[25] I redrew the two figures according to Piketty's paper instead of his book due to the lack of value of "net foreign debt" in his book. There exist other three countries in this paper; and they are Germany, the US, and Canada. I didn't use them because of inconsistent coordinate axes. ***See:*** Piketty T, Zucman G. Capital is Back: Wealth-Income Ratios in Rich Countries 1700–2010. The Quarterly Journal of Economics. 2014;129(3):1255-1310. Doi.org/10.1093/qje/qju018.

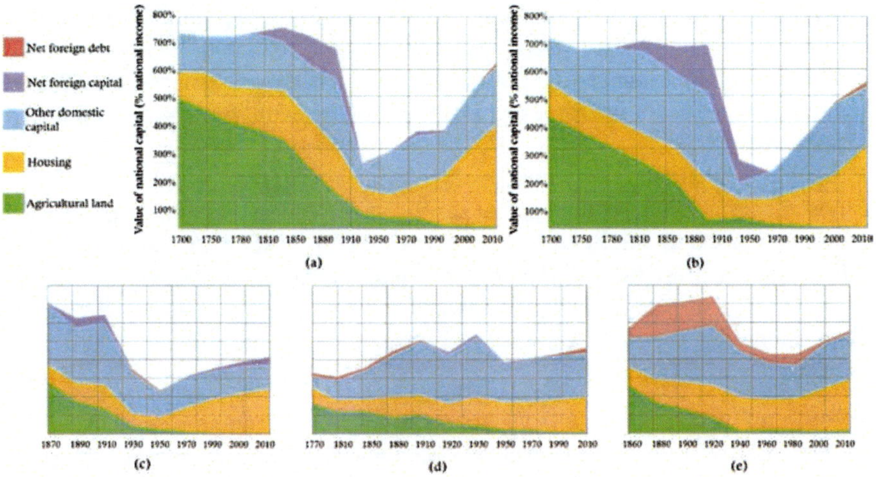

Figure 6.7 Capital in five countries from 1700 to 2010. (Data source: Author redrew them according to a Piketty T *et al.'s* paper (see Footnote 25) in 2014)

same ordinate as the capital/income ratio. The capital/income ratio followed quite similar trajectories in three European centuries, although the subfigure of Germany has no about the first 170 years' data. In all Britain, France, and Germany, basically, the total value of national capital fluctuated between 6 and 7 years of national income before 1910. But after World War I, the capital/income ratio sharply plummeted, and it continued to fall during the Depression and World War II. In contrast, the two American countries have some increases. According to Piketty, thus, an impressive "U-shaped curve" indicates (see Footnote 19)

> *the violent military, political, and economic conflicts that marked the twentieth century. Capital, private property, and the global distribution of wealth were key issues in these conflicts. The eighteenth and nineteenth centuries look tranquil by comparison.*

Note that national capital defined by Piketty is the sum of private capital and public capital, and his domestic capital can be broken down into three categories: farmland, housing, and other domestic capital. I've already summarized them in Chap. 2. So, Piketty considered the metamorphoses of capital combined with spatial scope, time, and capital classes. The foreign capital of both the UK and France, for example, is shaped by the turbulent history of these two leading colonial powers over the past three centuries. As shown in figure 6.7a, b (see Footnote 19),

> *the net assets these two counties owned in the rest of the world increased steadily during the eighteenth and nineteenth centuries and attained an extremely high level on the eve of World War I, before literally collapsing in the period 1914– 1945 and stabilizing at a relatively low level since then.*

Another example is the total value of farmland. In the beginning, farmland of all five countries represented 3–5 years of national income or nearly one-third to a half of total national capital. By now, farmland's worth is less than 10% of national income in all five for less than 2% of total wealth. But housing has risen largely after 1910.

Different data capital is. First, data capital as a large capital class covers almost all current intangible types of capital and even the tangibles' digital part, so it only can be divided and has not evolved. Next, data resources existing in cyberspace are presented as "data nature," without any fixed geographic boundaries. Then, data production tires down domain barriers to entering the traditional industries to make a converge between traditional industries and new forms of business in an instant. I explain the term "barriers to entry" as below.

Definition 6.4 *Barrier to Entry* Barriers to entry, an economic term, describe the existence of high startup costs or other obstacles that prevent new competitors from quickly entering an industry or area of business. Barriers to entry may be natural (high startup costs to drill a new oil well), created by governments (licensing fees or patents stand in the way), or by other companies (monopolists can buy or compete away startups).

Lastly, data capital depends on ecosystem productive forces rather than the scale of the data, and insufficient productivity can make data capital, an owner has likely on hand, disappear. Make no mistake: the data widely affecting the method is not negative in itself. We do not consider a dataset that is small enough to no carry out big data decision-making.

Ecosystem Productive Forces with Its Vanishing Act on Data Capital

We now know that data nature has been forming in cyberspace. This is an artificial ecosystem with a volatile and complex. Since there is no geographical shackles, data nature can offer greater "cross-border"[26] convenience for engaging in transactions and exchanges with very different cultures and societies, according to Max Weber's statement "capitalism rested on the elimination of differences."[27] In data nature, a very tiny part is evolving into data capital we can master. One in two key particularities of data is about value. Given two piles of

[26] The meaning of "border" not only is national or continental, but also refers to any restriction on various industrial sections or domains. In addition, there is no geographical distance in cyberspace.

[27] This viewpoint is referenced from the new biography of Max Weber wrote by Jürgen Kaube. ***See:*** Kaube J. Max Weber: ein Leben zwischen den Epochen. Berlin: Rowohlt Berline. January 17, 2014. 2014.

data, I'm guessing that comparing the amount of wealth held by each is NOT significant. If the data is not too small to carry out big data decision-making, the differential amount of wealth is not on the scale of data capital. The amount of wealth in data capital is subject to data technology. That is to say, the "wealth" magnitude of data capital is of how much value these data can create and create. This explains why many companies are still a single and non-profitable data business. In fact, some data organizations built before the notion of "big data" appeared in 2012, such as digital libraries. While holding a large amount of literature data, we cannot say these libraries have a lot of wealth due to the single type of data product. Instead, they spent a lot to collect, store, and manage the data.

In addition to determining the "wealth" magnitude of data capital, data technology can break the barriers formed by the existing division of labor in the current economic system. Then it helps develop the newly productive relationships in an innovative way, which is to replace the original chain or hierarchical structures by the networks; and a more adaptive network is possibly much closer to the "perfect competition." As described in the prior section about the typical processing mode of exchanging data through a service, primary data production has been exhibiting a characteristic of cross-industry (and across domains, realms, regions). This represents a gravitational effect—better service efforts mean more data; vice versa, more data with advanced technologies say much better service.

Some believe that the most significant achievement of twentieth-century economics was to discover the uncertainty. The uncertainty is the "human predicament" in our economic systems. American economist Frank H. Knight argued in his published Ph.D. dissertation *Risk, Uncertainty and Profit* that, the uncertainty differs with economic risk. Situations with risk were those where the outcomes were unknown but governed by probability distributions known at the outset, he carefully distinguished the two to explain that, whereas the "uncertain" ones were not only the outcomes, but even the probability models that governed them, were unknown. Knight argued that uncertainty gave rise to economic profits that perfect competition could not eliminate. But there are some benefits. Even though its probability likes human evolution, the uncertainty makes all prophecy to be unreliable, whatever apocalypse or fairy tale. Imagine that human evolution is possible to let flowing water in a swimming pool to shake together delicate watch parts for all eternity and not get that adjustment of the wheels and springs that makes to a watch.

There is no direct evidence so far that prediction is necessary for any uncertain situation. Perhaps we can cope with these precarious situations using gravity balance from classical physics. One may say that influence factors are electrically charged bodies, while links between them are springs. Springs pull similar factors together, and the charge does not let the bodies adjoin and push factors apart if there is no link between them. Originally, all such electrified bodies (factors) are randomly scattered in space. Springs are stretched, repulsion energy is high—the system is far from being at equilibrium. Then the factors start moving under the influence of the forces exerted, and in a while coming to a halt—forces of attraction now become equal to forces of repulsion, the system has reached its equilibrium. It is precisely that state that is we need when understanding economic systems and accurately interpreting the data economy.

The above interpretation shows data productivity should be ecosystem productive forces that can drive an "ecological balance"-like equilibrium in our real life. The ecosystem productive forces can also make data capital to be "disappeared." Assuming an extreme case of two parties, one has the data but no advanced data technology, we called the owner; the other can mine data in-deep but has no data, we called the producer. In the first stage, the producer wants the data that can move from the owner, but the owner is no willingness. The next is the owner cannot move the data due to the amount of data is enormous so that the producer no longer wants to have the data but to use the data in the owner site. In the last stage, the owner wants to share the data in any approach, but the data producer doesn't want these data because they found an alternative. Thus, data capital, the owner has on hand, devalued, and will disappear due to the lack of use.

Instantaneous Cross-Industry Effects

I'm going to explain the characteristic of cross-industry (and also across domains, realms, regions) represented by data production. The current data resource development, such as target marketing, online to offline (O2O), intelligent navigation, and ride-hailing, is in shallow applications. But they all show instantaneous cross-industry effects, and such results will become more and more visible soon.

Data productive powers can tire down the intrinsic economic division of production of the traditional "H" hierarchy, and this chain structure is the root reason for various barriers in our current economic system. The hierarchical structure of the division of production that we understand today was invented in the last century, and it is an incredible invention. It can direct and coordinate the actions of thousands of enterprises and entrepreneurs making and selling thousands of products or services across thousands of miles, and do so effectively, efficiently, and profitably, week after week after week. The industrial value chain "H" hierarchy partitions the production process within the industry into a series of stages, each further refining the specializing the product of the previous stage, beginning with raw materials, and ending with the completed product delivered to the customer. Still, it can only extend vertically or horizontally. Enterprises and entrepreneurs at the high-end of a value chain have the discourse of power and play a role of determinants on the pricing, the allocation of resources, and income distribution. Being at the top of the value chain means often a monopoly or at least an oligopoly. The further down the value chain, the more intense the competition could be, and the lowest end of the value chain should be very close to the "pure or perfect competition" studied by economists. Horizontal and vertical integration as the two ways businesses can expand depends on technological advancements and innovations, reflected in the deepening and refinement of the division of production. Data productivity may transform the traditional "H" hierarchy into a network-like structure. Because data technology tends to understand the complex world of interconnected entities we live in from comprehensive (i.e., different and complementary) viewpoints, with entities being modeled as nodes and their connections as edge, comprising large networks. These networked choices are a new and rich source mostly hidden with the value chain. Just as one principle of economics, "the more options that mean the lower cost of actions, the higher value could be created." Also, it is difficult to distinguish between the top and bottom of the value chain. The key factors that shape benefits is relevant on data technology.

In the future, one way to master data capital is to use more advanced data technologies. Data capital exists in data nature, and more and more data producers are willing to use data force given by data technology can help possess and protect data capital; thus, they could genuinely become a data sovereign.

Part III
Hopes and Challenges Brought from Data Capital

Abstract Data as capital can bring with it both immense hopes and equally tremendous challenges. That means not only industrial circles should be more open-minded to meet changing in the face of radical data-rich markets, but also the government and academia need to pursue different policies. Here, three selected issues are discussed, respectively; and they are the market, public choice, and economic indicator.

In Part II, I focused on analyzing the composition of data capital in order to understand its two components, as memories and time. Also, I examined two particularities of data itself and the productivity-based potential economic impacts. But I did not look directly at the hopes and challenges brought by data capital. In the next few chapters, I will show these them, and followed with potential responses.

Chapter 7
Radical Data-Rich Markets

The market is a mechanism of social innovation to help people divvy up scarce resources efficiently. Naturally, it equals purchasing power adding with purchasing desire. This is a simple statement but has an enormous impact. Markets have globally enabled us to feed, clothe, and help the needy program of about eight billion humans, and to improve significantly our life expectancy as well as life quality, making us superbly well aligned with human nature. That's why markets seem so natural to most of us and are so deeply ingrained in society's fabric. They are the building blocks of the economy.

Unlike the ideal market in economic analysis, however, each market can be sluggish; and a slow-growth market has always been a problem. And even some economists believe that it holds great peril for the world, especially for wealthy societies, which "threatens the legitimacy and stability of the values and institutions."[1] The mess they perceived, called the "crisis of the liberal order," has three components (see Footnote 1).

> *First, inequality within wealthy countries, and in particular the US and UK, has been growing dramatically with especially the share of national income going to workers falling. "Neo-liberals" like Ronald Reagan and Margaret Thatcher promised that in exchange for such inequality their "trickle down" economics [refers to the economic proposition that taxes on businesses and the wealthy in society should be reduced as a means to stimulate business investment in the short term and benefit society at large in the long term.] would stimulate growth and create jobs. Yet, second, at the same time economic and productivity growth rates have fallen dramatically and employment has declined. This combination of inequality and stagnation, which we call "stagnequality", has largely discredited existing economic policy in the eyes of much of the public. At the same time, democratic politics has struggled to respond to conflicts between minorities and majorities within countries, as well as those between international cooperation and national sovereignty, leading to increasingly polarized politics.*

[1] ***See:*** (1) Posner EA, Weyl EG. Radical Markets: Uprooting Capitalism and Democracy for a Just Society. Princeton, NY: Pricetion University Press (October 8, 2019). 2019; 383 pages; (2) Retrieved from http://radicalmarkets.com.

© Springer Nature Switzerland AG 2021

C. Tang, *Data Capital*, https://doi.org/10.1007/978-3-030-60192-8_7

Responding to both these crises of legitimacy, dangerous and incompetent populist leaders and policies have become prominent and threaten hard-won social progress.

The main reason is that traditional markets is basically fragmented due to the existence of various barriers to entry. For example, the automotive markets can be divided into auto parts or vehicle markets, new or used-car sales; luxury, mid-level, or superminis; and domestic or oversea markets. These divisions are mainly related to purchasing power and purchasing desire. This is why marketing research as it exists today. The marketing research is to understand the customer, so as to reasonably avoid other competitors to do market positioning concerning their products, and market positioning also means market segmentation.

Purchasing power, relating to the price, is the value of a currency expressed in terms of the number of goods and services that one unit of money can buy. Purchasing power is also related to inflation, which has the effect of a decline of a currency's purchasing power and an increase in prices. Many western economic theories we are familiar with are about this, such as quantity theory of money, purchasing power parity (PPP), interest rate parity model (IRP), exchange rate fluctuations, differences in productivity growth rate, and even the Keynesian IS-IM model (stands for investment-savings and liquidity preference-money supply).

Purchasing desire research is studied in both subjective and objective perspectives. Objectively, it refers to consumer demand. Demand and supply should go hand in hand. Supply is the amount of product available for consumers to purchase. Consumers want to pay the lowest price while the supplier attempts to maximize their profits. In the perfect scenario, the suppliers maximize profit while still satisfying the consumers. If the suppliers have too much supply, they won't enough demand to sell their products. If the suppliers run out of supply, they won't make enough profit to continue their business. An example of a poorly managed supply and demand is the iPhone 7 announcement. Apple stopped taking pre-sale orders for the new version of their high sales phone due to the inability to meet the demand. In response, the price of the company's stock decreased, and consumers were left unsatisfied. Not a happy ending for both Apple and the consumer. One well-known theory is William H. Hutt's consumer sovereignty, which reflected the rise of the figure of the consumer in the public space.

The subjective studies come from psychology which is impulsivity and compulsivity. Impulsive purchasing behaviors we often see the consumer desiring objects that they feel will please them. And compulsive consumers, on the other hand, desire the shopping experience itself, leaving the meaning of the purpose behind.

It's been a long while of the radical tradition in western economics with the countless theoretical attempts, led by thinkers like Adam Smith, Jeremy Bentham, John S. Mill, Henry George, Léon Walras, and Beatrice Webb. In terms of radical means, however, whether socialist regulation or liberal indulgence, it cannot exist independently from the other party. A recently published book (in 2019) called *Radical Markets* (see Footnote 1) gives a unanimously praised idea to ignore the traditional left-right debate, namely expanding the scope of markets to use "market power" (i.e., the power that comes from the market itself). The idea the authors claimed was adequately realized by actual practices of shallow use cases work with the data. Take the C2B (customer-to-business) model used in e-commerce as an example; C2B, in which consumers (individuals) create value and businesses consume that value, can be roughly achieved by two methods of shallow data application in the market. One is demand aggregation (e.g., group/collective buying), which provides "visibility" to the purchasing requests generally through the process of coordinating and consolidating individual requirements. The other is offer-based cooperation in social networks, such as reverse auction that is a type of auction in which sellers bid for the prices at which they are willing to sell their goods and services. Technically speaking, the two methods are on the need for eliciting decentralized information from individual privacy data, so require more active partition, analysis, and locking of the consumer.

Another example is targeted advertising. Targeted advertising many people understand is a form of online advertising that focuses on the specific traits, interests, and preferences of a consumer. Here's one of the most common methods:[2]

> *Let's say you go to Amazon to look for a new Paul McCartney CD. In order to keep track of your visit, Amazon creates a file called a cookie on your computer. Later, when you're reading an article on Slate, automated advertisements read this cookie and generate ads for items related to your Amazon visit, like other Paul McCartney albums or a Beatles T-shirt.*
>
> *The good thing about targeted ads is that you'll see ads for products you actually want. However, this system isn't perfect. Even if you only visit a page once, you may see ads for that particular product for quite some time. ... Aside from cookies, advertisers also learn about you in other ways, like checking your search engine history and finding your personal information on social media.*

But this understanding is only a half. The other half is to design auctions for advertising slots on webpages. As early as the end of 2008, when a winter storm loomed for the Internet sector after the financial crisis, many websites companies changed the strategy of selling their webpages' advertising slots. Their previous expectation is to let the viewer reach the advertisement as several industrial giants asked (e.g., from the automotive, real estate, and IT industries). The updated one is to obtain feedback directly or indirectly from the viewer as the new advertisers' (i.e., many SMEs) requirements, which is the purpose of targeted advertising. The new strategy can be realized technologically as real-time bidding (RTB). Simply, RTB is the automated means by which ad inventory is

[2] ***See:*** Retrieved from https://edu.gcfglobal.org/en/thenow/what-is-targeted-advertising/1.

bought and sold on a per-impression basis, via an instantaneous programmatic auction that matches the buyer and seller.

The idea from the book Radical Markets to unleash the power of markets is acceptable. Still, it might not be necessary to envision the radical markets as institutional arrangements using the auction because the rise of markets rich with data is originally radical. Any business practice of shallow use cases work with the data is worth more than the market value of 12 digits. For example,[3] location-targeted mobile advertising in 2018 spending in the US amounted to US$21.4 billion; and an industry forecast made by Statista is to surpass US$32 billion by the end of 2023. Smart advertising is another example. Smart advertising is rapidly gaining popularity and is emerging as one of the most preferred forms of advertising for enterprises across multiple industry verticals such as automotive, gaming, entertainment, consumer goods, and electronics. Based on the product, the market for smart advertising has been segmented into the interactive kiosks, digital posters, digital billboards, and others. Compared to other conventional forms of advertising, smart advertising enables marketers to effectively target audiences based on multiple attributes related to interests, age, and location, among others, thereby increasing user engagement. According to a study done by Grand View Research,[4] the global smart advertising market size was valued at US$655.1 million in 2018 and is expected to register a CAGR (compound annual growth rate) of 19.4% from 2019 to 2025. As Shoshana Zuboff wrote in her book *The Age of Surveillance Capitalism*,[5] market positioning on taking "the consumers' private information and data without the consumers' permission and … do not generally disclose to the user that this information will be sent to advertising companies" is too radical to be a little scary.

In this chapter, I'm going to connect the dots with potential responses. They are: the difficulties of traditional markets (with some barriers), the error of the financial market' pricing mechanism caused by its power structure, and the new questions faced by data-rich markets (e.g., data-intensive labor). I believe that a reboot of the market fueled by the data will lead to a fundamental reconfiguration of the global economy, which is supremely creative and wildly provocative, with new approaches to address "the crisis of the liberal order."

---

[3] ***See:*** Guttmann A. Location targeted mobile advertising spending in the United States from 2016 to 2023. Statista.com. November 8, 2019. Retrieved from https://www.statista.com/statistics/274837/local-and-national-mobile-us-ad-spending-since-2009.

[4] ***See:*** Smart Advertising Market Size, Share & Trends Analysis Report By Component, By Product (Interactive Kiosk, Digital Billboard, Digital Poster), By End Use, By Region, And Segment Forecasts, 2019 - 2025. Grand View Research. August 2019. Retrieved from https://www.grandviewresearch.com/industry-analysis/smart-advertising-market.

[5] ***See:*** Zuboff S. The Age of Surveillance Capitalism: The Fight for a Human Future at the New Frontier of Power. New York, NY: PublicAffairs; 1 edition (January 15, 2019). 2019; 704 pages.

A Possible Solution to Traditional Markets' Difficulties

Before proceeding, let's have a glance at a real case.

Case 7.1: Greenleaf Dairy Farmer Left with a Lagoon of Dumped Milk

GREENLEAF, Wis. (WFRV) It's enough to make any dairy farmer curdle, massive amounts of milk are being dumped because of the COVID pandemic. By: Rhonda Foxx, Posted: Apr 3, 2020/ 05:21 PM CDT.

Mark Mueller, owner of Mueller Dairy Farm in Greenleaf says that on April 1, a driver from the Dairy Farmers of America informed them they had to dump 56,000 pounds of milk because no one is buying it.

That massive amount of milk was dumped in pit on their property.

The milk harvested from the nearly 1000 cows at Mueller's Dairy Farm will rot in that pit, because the COVID pandemic is preventing dairy farmers, like him, from getting their milk to market.

Mueller says, "the closing of restaurants is a problem. Some feel that they don't have a place to haul our product."

This act was not to pour milk for the first time. The more famous one is the 1933 Wisconsin milk strike.[6] In the 1930s, Wisconsin was the largest producer of milk in the US, with more than 125,000 dairy farms. According to the 1930 decennial census,[7] 63% of all land in Wisconsin was farmland, and 71% of that land was used for dairy farming. The 1933 Wisconsin milk strike was a series of strikes (the three main strike periods occurred in 1933) conducted by a cooperative group of Wisconsin dairy farmers in an attempt to raise the price of milk paid to producers during the Great Depression. Newspapers reported that farmers lost US$10 million during the strikes. President Franklin D. Roosevelt helped lay results with the Agricultural Adjustment Act of 1933 but it was only a slight improvement.

Pouring milk in Case 7.1 occurred recently. Due to the coronavirus pandemic, stay-at-home orders and social distancing mandates have severely limited operations or closed restaurants, hotels and schools, leading to dairy farmers in Wisconsin, California, Florida, and other Gulf states, with no market—"Even

[6] ***See:*** The entry of "1933 Wisconsin milk strike" on Wikipedia. Retrieved from https://en.wikipedia.org/wiki/1933_Wisconsin_milk_strike.

[7] 1930 Decennial Census. According to 1929 Agricultural data tables, 11.052 billion gallons of milk were produced in the United States in 1929; Wisconsin was the top producer, producing around 10% of milk. ***See:*** Retrieved from https://www2.census.gov/prod2/decennial/documents/03337983v4ch10.pdf.

though there is no one to buy it, cows still need to be milked."[8] Other agricultural products also included. While millions of people are worried about having enough to eat and lines at food banks grow, farmers have been plowing under vegetable fields, dumping milk and smashing eggs. Some farmers acknowledge the efforts are "just a drop in the bucket"[9] of what they can't sell and are destroying instead.

How are economists able to explain such an issue? They would believe "the issue isn't necessarily the virus, instead it has to do with demand" (see Footnote 8). Economists use the term "demand" to refer to the amount of some good or service that consumers are willing and able to purchase at each price. Demand is based on purchasing power (needs) and purchasing desire (wants)—a consumer may be able to differentiate between a need and a want, but from an economist's perspective, they are the same thing. Demand is also based on the ability to pay. If a person cannot pay for it, this guy has no sufficient demand. What a buyer pays for a unit of the specific good or service is called "price." The total number of units purchased at that price is called the "quantity demanded." A rise in price of a good or service almost always decreases the quantity demanded of that good or service. Conversely, a fall in price will increase the quantity demanded. Economists call this inverse relationship between price and quantity demanded as "the law of demand." When the price of a gallon of gasoline goes up, for example,[10]

> *people look for ways to reduce their consumption by combining several errands, commuting by carpool or mass transit, or taking weekend or vacation trips closer to home.*

The quantity demanded at each price over some period can be in the form of graph, which is called a "demand curve." When economists talk about demand, thus, refers to the curve, and about quantity demanded is a specific point on the curve.

The "supply" economists mean is the amount of some good or service a producer is willing to supply at each price. In this way, price is what the producer receives for selling one unit of a good or service. The total number of units sold at that price is called the "quantity supplied." A rise in price almost always leads to an increase in the quantity supplied of that good or service, while a fall in price will decrease the quantity supplied. The positive relationship between

---

[8] ***See:*** Schneider L. Dairy farmers dumping milk amid COVID-19: Pandemic's impact on the dairy industry. abcNews.go.com. April 21, 2020. Retrieved from https://abcnews.go.com/US/dairy-farmers-dumping-milk-amid-covid-19-pandemics/story?id=70268302.

[9] ***See:*** Cokery M, Yaffe-Bellany D. 'We Had to Do Something': Trying to Prevent Massive Food Waste. NY Times.com. May 2, 2020. Retrieved from https://www.nytimes.com/2020/05/02/business/coronavirus-food-waste-destroyed.html.

[10] ***See:*** Retrieved from https://opentextbc.ca/principlesofeconomics/chapter/3-1-demand-supply-and-equilibrium-in-markets-for-goods-and-services.

price and quantity supplied, called the law of supply. When the price of gasoline rises, for example (see Footnote 10),

> *it encourages profit-seeking firms to take several actions: expand exploration for oil reserves; drill for more oil; invest in more pipelines and oil tankers to bring the oil to plants where it can be refined into gasoline; build new oil refineries; purchase additional pipelines and trucks to ship the gasoline to gas stations; and open more gas stations or keep existing gas stations open longer hours.*

Similarly, a supply curve is a graphic illustration of the relationship between price, shown on the vertical axis, and quantity, shown on the horizontal axis.

Due to the fact that the graphs for demand and supply curves both have the price (on the vertical axis) and quantity (on the horizontal axis), the two curves for a particular good or service can appear on the same graph. Together, demand and supply determine the price and the quantity that will be bought and sold in a market. When two curves on a diagram cross, this intersection usually means the "equilibrium." In other words, the point where the supply curve and the demand curve cross, designated by point, is the "equilibrium price." The equilibrium price is the only price where the plans of consumers and the plans of producers agree—namely, where the amount of the product consumers want to buy (quantity demanded) is equal to the amount producers want to sell (quantity supplied). This common quantity is called the "equilibrium quantity." At any other point, the quantity demanded does not equal the quantity supplied, so the market is not in equilibrium at that price. The word "equilibrium" means "balance." If a market is at its equilibrium price and quantity, then it has no reason to move away from that point. However, if a market is not at equilibrium, then economic pressures arise to move the market toward the equilibrium price and the equilibrium quantity. Imagine, for example (see Footnote 10),

> *that the price of a gallon of gasoline was above the equilibrium price – that is, instead of $1.40 per gallon, the price is $1.80 per gallon. ... At this higher price, the quantity demanded drops ... This decline in quantity reflects how consumers react to the higher price by finding ways to use less gasoline.*

In sum, economists study whether a market is equilibrium or not from a perspective of demand and supply. A market is connected by dynamic supply-demand activities to pursue added value for products and services. If the price is below the equilibrium level, then the quantity demanded will exceed the quantity supplied. Excess demand or a supply shortage will exist. If the price is above the equilibrium level, then the quantity supplied will exceed the quantity demanded. Excess supply or a surplus will exist. In either case, economic pressures will push the price toward the equilibrium level. Of course, such economic pressures typically contain the element of manual intervention (e.g., time and effort on the part of marketing). For example, many remedies can operate in Case 7.1 (see Footnote 9).

... the Department of Agriculture will begin spending $300 million a month to buy surplus vegetables, fruit, milk and meat from distributors and ship them to food banks.

The federal grants will also subsidize boxing up the purchases and transporting them to charitable groups.

Nationally, the Dairy Farmers of America, the largest dairy co-op in the United States, has diverted almost a quarter of a million gallons of milk to food banks.

... New York will give food banks $25 million to buy products made from excess milk on farms in the state; the state is working with manufacturers like Chobani, Hood and Cabot to turn the milk into cheese, yogurt and butter. Some of the state subsidy can also be used to buy apples, potatoes and other produce that farms have in storage.

A group of university students has started an online site, FarmLink, seeking to connect farmers with food banks. James Kanoff at Stanford and Aidan Reilly at Brown founded the group ... with donations from family and friends.

Manual intervention in the markets requires a supplier to try the best to understand various demands from the customer to avoid a supply surplus. The intervention includes analysis of product differentiation, target audience, and market feedback. Product differentiation is to understand the competitors and then to expect a completive position. The target audience analysis provides a comprehensive way to consider unknown customers' demand based on their interests, behaviors, and performances, to aggregate alternative groups to increase product or service sales. Market feedback generates and reports marketing results, based on the degree of product differentiation and target audience, and to adjudge the decision-making and implement a strategy. Apart from the above, manual intervention also involves business insights. Gaining business insights starts with market survey and leads to brand confirmations, advertising and promotions, and performance examinations, aiming to find business opportunities in a vast but fragmented market. A brand is not only a trademark; it concerns reputation, products, corporate culture, and the overall operations. The brand confirmation includes the analysis of brand distribution, brand competitive dimension, and brand sales forecasting typically, in order to obtain market attention and performance. Advertising and promotion has been always in vogue; it is used to emphasis the content or theme of specific products and services. Performance examination is somewhat similar to market feedback, but is the inspection carried out by an entrusted third-party for the purpose of accessing the key information on the markets, to identify returns on investment of manual intervention. At the same time, supply chain management (e.g., inventory, logistics) should be considered in manual intervention. Take Case 7.1 as an example; the shelf life of dairy storage is a problem despite having no upstream supply (if no considering cow feed).

Too much manual intervention is essentially the difficulties faced by traditional markets with the barriers. The heavy work caused by manual intervention consumes too many economic resources on the supply side, resulting in a considerable cost burden for the enterprises. Also, too much intervention may lead to a market failure that is the economic situation defined by an inefficient distribution of goods and services in the market; in market failure, the incentives for rational behavior cannot produce rational outcomes.

To exert their charms, markets require enough information; and the information depends on the easy flow of data, and the ability to discover the data for decisions. The capacity to make decisions, defined by Viktor Mayer-Schönberger in his book *Reinventing Capitalism in the Age of Big Data*, is how people transact on markets after having referred to multiple factors. These factors involve nearly all aspects that we should consider in the above manual intervention. But the capacity requires that everyone (both the supply and demand sides) has easy access to massive data about what's available, which could be only solved by data-rich markets. In the past, communicating such information in traditional markets were difficult and costly. We thus adopted a workaround to condense all of the information into a single metric: price. And we conveyed that information with the help of currency. We must admit it works, to a certain extent. But when all information is compressed into the price, details and nuance existed in other dimensions get lost, resulting in suboptimal transactions. We tolerated such inadequate solutions for millennia because no better alternative was available. I use a hypothetical example from *Reinventing Capitalism in the Age of Big Data* here.[11]

> *Suppose, you are looking for a new frying pan. An adaptive system, residing perhaps on your smartphone, accesses your past shopping data to gather that you bought a pan for induction cooktops last time, and also that you left a so-so review of it. Parsing the review, the system understands that the pan's coating really matters to you, and that you favor a ceramic one (it also notes your preferred material for the grip). Equipped with these preferences, it then looks at online markets for optimal matches, even factoring in the carbon footprint of the delivery (because it knows how worried you are about that). It negotiates automatically with sellers, and because you are ready to pay by direct transfer it is able to get a discount. With a single tap, your transaction is complete.*

But before explaining, I would like to change something since this example uses too much private data that belongs to a shallow data application. An instance with less user privacy is: first, the algorithm performs real-time analysis on relevant review data on the Internet and social media when searching the pan. For example, the comments on Facebook are generally related to the gift involving logistics, and Amazon's remarks are about product specification about user experience. Secondly, best matching the demand (e.g., the pan's coating and size; note that a container might not meet your imagination when just looking at the picture) in the market based on consumer habits from the location of your living area. Compared with our own search, the algorithm can refer to more variables and to evaluate more products and services. Neither does the

[11] ***See:*** Mayer-Schönberger V, Ramge T. Reinventing Capitalism in the Age of Big Data. Now York, NY: Basic Books (February 27, 2018). 2018; 288 pages.

algorithm tire easily (as we humans do when searching for something both offline and online), nor is it distracted in its decision advice by price, derailed by cognitive bias, or lured by clever marketing. While still using money as a store of value as well as seeing price as a part of the information, we are no longer focused on price broadens (that may help us get rid of vanity). Our perspective yields better matches, a more efficient transaction, and we believe less fragmented markets. For instance, considering healthcare, living area, and carbon emission when buying a frying pan may help combine e-commerce markets with energy markets, where natural gas consumption may be on the time of cooking, transportation and logistics markets, and health care markets together.

Therefore, solving the dilemma of conventional markets doesn't lie in achieving equilibrium between the demand and supply. Still, it is on how to the data with its driver. Data translates into "too much of an improvement in transactions and efficiency" (see Footnote 11), which can help reduce the frequency of market failure.

Threatened Power Structure Behind Financial Markets

Unlike price-based traditional markets, financial markets are based on money. According to Encyclopedia Britannica online,[12] money (e.g., metallic or paper money) is a commodity accepted by general consent as an anonymous medium of economic exchange in which prices and values are expressed. The money market, as one vital component of financial markets, is "a set of institutions, conventions, and practices, the aim of which is to facilitate the lending and borrowing of money on a short-term basis" (see Footnote 12). The definition of money for money market purposes is not confined to banknotes but includes a range of assets that can be turned into cash at short notice, such as short-term government securities, bills of exchange, and bankers' acceptances. Note that the money market is different from the capital market; the latter is concerned with medium- and long-term credit. Commercial banks are at the center of most money markets, as both suppliers and users of funds, and in many markets a few large commercial banks serve also as middlemen. Central banks' work is to control the quantity of money in an economy, in which one of the basic methods is the reserve requirement. Well-developed money markets exist in only a few high-income countries, such as the US, British, French, and so on. A process of continuous change during the twentieth century has produced a genuinely national money market in the US, which carries out the largest volume of

---

[12] The topics of money and money market at Encyclopedia Britannica online. ***See:*** (1) Retrieved from https://www.britannica.com/topic/money; (2) Retrieved from https://www.britannica.com/topic/money-market.

transactions in the world. In Great Britain, the money market consists of a number of linked markets, all of them concentrated in London. The 12 specialist banks, known as discount houses, have the longest history as money market institutions. The French money market is fairly well established, but its size is restricted by one fact, which is, by France's regulations that nonfinancial private sector is excluded from the market.

Another critical concept combined with money is credit. According to Joseph Schumpeter, the first known advocate of money credit was Plato. But the earliest modern thinker to formulate a credit theory of money was recognized as Henry D. Macleod. Macleod's work was expanded on by Alfred M. Innes. In Innes' 1914 article *The Credit Theory of Money*, he argued:[13]

> *The Credit Theory is this: that a sale and purchase is the exchange of a commodity for credit. From this main theory springs the sub-theory that the value of credit or money does not depend on the value of any metal or metals, but on the right which the creditor acquires to "payment," that is to say, to satisfaction for the credit, and on the obligation of the debtor to "pay" his debt and conversely on the right of the debtor to release himself from his debt by the tender of an equivalent debt owed by the creditor, and the obligation of the creditor to accept this tender in satisfaction of his credit.*

Since this round of globalization (i.e., the collapse of the Bretton Woods system in 1971), financial capital rapidly replaced industrial capital. The scale of global liquidity finance sharply exceeded trading in the late 1970s, from hundreds of billions US dollars to tens of trillions of US dollars. In the 1980s, the radical economics represented by Thatcher and Reagan saw free market values once again that became the progressive ideal with striving for while government intervention was marked as a backward concept to be shunned. In essence, this is a victory for financial markets. Countries around the world competed to set up stock markets, bond markets, forex markets, and derivatives markets. These over 50 countries include developing countries having socialist systems, such as China, Vietnam, and Russia, low-income countries in Latin American and Africa, such as El Salvador, Honduras, and Malawi, and some countries in the Middle East or the Caribbean are no exception. And even Iceland, with a population of only 300,000, established stock exchange in 1985. Besides, investing in stocks of emerging economics from developed countries has grown fast, from nearly zero before 1980 to hundreds of billions of US$ just through a decade (in the 1990s); and the current investment scale has exceeded US$ 1 trillion.

The contemporary global economy deeply influenced by the global financial market can be roughly divided into five levels.[14]

---

[13] There are two references. ***See:*** (1) Innes AM. The credit theory of money. Banking Law Journal 1941. Retrieved from http://wfhummel.net/innes.html; (2) The entry of "Credit theory of money" on Wikipedia. Retrieved from https://en.wikipedia.org/wiki/Credit_theory_of_money.

[14] ***See:*** Xiang, S. The New Capitalism: Rising, Crisis and Future of the Global Financial Capitalism. Beijing: China Citic Press; 1 edition, December 1, 2014, 616 pages.

- The first level is markets and the price system of the global final goods and services.
- The second level is markets and the price system of the global strategic resources (e.g., oil, natural gas, iron ore, etc.) and other factors of production (science technology and human values included).
- The third level is markets and the price system of the global corporate equity (e.g., listed companies' stocks, private equity funds, venture capital, mergers and acquisitions)
- The fourth level is markets and the price system of the global bonds (e.g., issued by corporations as well as by municipalities, states, and sovereign governments to finance projects and operations)
- The fifth level is markets and the price system of varieties of money and credit systems.

> *Among the above five levels, ... The fifth level is the most decisive at the top of the global pricing system pyramid. ...*
>
> *The global monetary transmission mechanism is: (1) the price of the reserve varieties of money determines or affects credit systems (i.e., interest rate and treasury yield), (2) all interest rates and treasury yields determines or affects the price of the global corporate equity, (3) all interest rates and treasury combined with the price of the global corporate equity determine or affect the price of the global strategic resources, (4) all interest rates and treasury combined with the price of the global strategic resources determine or affect varieties of foreign exchange rates the global inflation rate, namely the price of global final goods and services.*

However, this monetary transmission mechanism is not as fair as it looks. Thomas L. Friedman commented in his known book *The World Is Flat: A Brief History of the Twenty-First Century* in 1996:

> *There are two superpowers in the world today in my opinion. There's the United States and there's Moody's Bond Rating Service. The United States can destroy you by dropping bombs, and Moody's can destroy you by downgrading your bonds. And believe me, it's not clear sometimes who's more powerful.*

Friedman believed that in today's world, holding the pricing power of financial markets means to manage global liquidity, and even to control the rise and fall of a specific sovereign government. Then, who holds the reins of the worldwide bond market and pricing the capital markets? The answer is the reserve US dollar issued by the US Federal Reserve System and the Big Three credit rating agencies that are Standard & Poor's (S&P), Moody's, and Fitch Group. The pricing of the reserve US dollar determines or affects other currencies. The US Treasury remained at the top of credit ratings; its credit rating is the reference standard for rating other bond markets around the world. S&P and Moody's are based in the US, while Fitch is dual-headquartered in New York City and London, and is controlled by Hearst. As of 2013 they hold a collective global market share of "roughly 95%,"[15] in which Moody's and S&P having approximately 40% for each. Without a good credit rating, no government, enterprise, or institution allows to enter the international bond markets for financing; namely, the global credit rating is a pass for any bond issuer to access any financial market.

Once one of Big Three announces the downgrade of a country, the domestic financial market of this country may fall into turmoil. And this pattern has been seen many times before. To cite just one more case, on April 27, 2010, S&P Global Ratings Services lowered its long- and short-term credit ratings on the Hellenic Republic (Greece) to non-investment grade BB+ from BBB+, a "junk" rating; and on the same day, Poor's downgraded Portuguese debt to A− from A+, a downgrade of two levels with a negative outlook. Both the Greek and Portuguese stock markets fell on April 27, 2010, with the ratio contracting 6% and 5%, respectively. Other stock markets in Europe, such as Germany, France, and the UK, also generally declined about 2–3%. By the next day (April 28, 2010), the total market value of the global stock market had evaporated by more than US$ 1 trillion.

Someone believes that the Big Three that holds the pricing power of financial markets is not just bystanders. For example, together with investment banks, the three credit rating agencies were key enablers—"essential cogs in the wheel of financial destruction"[16]—of the 2007 US subprime mortgage crisis. This crisis could not have happened without the rating agencies. The subprime mortgage crisis eventually evolved into a more severe financial crisis than the Great Depression worldwide. But I think that such an economic disaster falls victim to the same flawed decision-making caused by our long-term reliance on money (and price). This is "a toxic combination of opaque information and deeply flawed human decision-making" (see Footnote 11), described by Mayer-Schönberger; and a more detailed explanation is below (see Footnote 11).

> *Money eased the exchange and evaluation of market information for many centuries, by collapsing a great deal of it into price. But in large part, those coveted greenbacks paper over the fundamental challenge of taking highly condensed information and translating it into transaction decisions. Money-based markets are fraught with inefficiencies, and these are felt in how well or how badly the market fulfills its promise of coordinating human activities to everyone's best interest.*

---

[15] ***See:*** (1) Alessi C, Wolverson R. The Credit Rating Controversy. Archive.org. February 11, 2013. Retrieved from https://web.archive.org/web/20130727223220/http://www.cfr.org/united-states/credit-rating-controversy/p22328; (2) The entry of "Big Three (credit rating agencies)" on Wikipedia. Retrieved from https://en.wikipedia.org/wiki/Big_Three_(credit_rating_agencies).

[16] The ten-member commission appointed by the United States government with the goal of investigating the causes of the financial crisis of 2007–2010. ***See:*** Conclusions Of The Financial Crisis Inquiry Commission. Retrieved from http://fcic-static.law.stanford.edu/cdn_media/fcic-reports/fcic_final_report_conclusions.pdf.

Someone may be curious: how do data-rich markets threaten financial markets. I would say, in addition to recent innovations based on massive data and novel data technology, financial capital has been declining itself. The financial capital decline is not a simple replacement by other types of capital, which is similar to industrial capital replaced by financial capital. Instead, the formerly single source of financial capital will be diluted by the data from other sources (e.g., another sector, field, domain, etc.); and all these data will be merged together to form a new financial data capital. Attendant, the data-rich market will gradually disintegrate the current power structure behind financial markets.

Financial Capital Decline

In the money-based markets, a critical reason for the popularity of financial capital is a fungible factor of production. When necessary, it is easy to be changed for a much-needed resource, and vice versa. That means enabling efficient resource use in the value dimension—the freedom of choice as well as relative power. The rapid growth of financial capital is another driver, and one way is from an outside investment. The investment involves early-stage venture capital, midterm mergers and acquisitions, and corporate equity transactions in late stage. An outside investment, besides the value dimension as the cash inflow itself, has the information dimension such as reputation. When a prestigious venture capital firm invests in a start-up, it's akin to conferring a peerage by British Queen. The recipient gets immediate name recognition and obtains additional market value as a result. Another example is in the stock market. I once mentioned (see Chap. 2) that news (or other signals; e.g., gossip) and financial data are twins. Therefore, existing financial capital has two dimensions: value and information.

As markets embrace diverse information streams, the two dimensions of financial capital are no longer necessarily intertwined. Somewhat, they will more frequently be separated. The information dimension can be characterized by multiple attributes and measurement to show its diversity via data mining, instead of a single signal, such as corporate reputation and timely news or gossip. With the richness on the information dimension, resulting in the demise of price and money as the market's preeminent conveyor of information, the value dimension of financial capital also can be replaced by the data (i.e., data as currency).

When people discuss currencies, they tend to think of paper notes—US dollars or Euros. Printed money, however, is only one kind of currency. Throughout history, currencies have appeared in many forms, from the storied stone wheels of the Yap islanders to cowries. These mollusk shells became a popular means of exchange in China more than three millennia ago. Currencies have evolved over time from stones and seashells to the sophisticated forms of legal tender that enable today's global financial transactions, such as bitcoin, holy dust used in the online game World of Warcraft. Currency is how we create and exchange economic value across geography and through time. It is anything that can serve as a medium of exchange. Data has each of these essential characteristics, as cellist Zoë Keating noted in suggesting that instead of sending her royalties, streaming music services should provide her with data about her audiences.[17]

Some Innovative Practices in Financial Markets

Apart from financial capital decline itself, data-rich markets will cause an extreme shock to the financial system, which is related to numerous disruptive data innovations. I'm going to list several existing innovative practices in financial markets, involving settlement approaches, loans, and investment services.

While an HD (high definition) movie takes several minutes to download in a 4G (i.e., the fourth generation mobile communication) network and will use a handful of seconds using 5G, banks' payment solution such as settlement of cross-border currency transactions still takes days. That lag is evident wherever looking, from institutional finance to retail.[18]

> *T+2, or "trade date plus two days," is still the standard timeframe for most stock trades, meaning that settlement occurs two full business days after the order has been executed -- the rule of thumb for even the most reputable brokerages. Meanwhile, expectations for cross-border payments must be similarly tempered, with the best-known international money transfer services requiring 3 - 5 business days to coordinate a cash wire across countries.*

Banks are under pressure to meet their clients' heightened needs, but the clients are in search of alternatives. Some fintech startups focus not on lower fee but innovative additional services. These startups are both targeting niche markets and undertaking a highly customized service bundle of payment with accounts that comes from banks (see Footnote 11).

> *... Conconut [in the United Kingdom targeting freelancers] offers customers the ability, whenever receiving or making a payment, to swiftly (re)calculate taxes and to put money aside to pay them. Holvi's [in Finland, targeting small businesses] services include free integrated invoicing and bookkeeping. ... [Other] mobile-payment startup such as Stripe and Square, are having the banks for lunch.*

---

[17] ***See:*** Harris D. Data Isn't Just the New Oil, It's the New Money. Ask Zoë Keating. Gigacom, November 20, 2012. Retrieved from http://gigaom.com/2012/11/20/data-isnt-just-the-new-oil-its-the-new-money-ask-zoe-keating.

[18] ***See:*** Xu H. It's About Time: Disruptive Settlement Technologies Are Coming to Banks. Finextra.com. June 11, 2020. Retrieved from https://www.finextra.com/blogposting/18871/its-about-time-disruptive-settlement-technologies-are-coming-to-banks.

Besides, the race is on to reinvent settlement based on the next-gen technology in the form of distributed ledger technology (DLT) (see Footnote 18). Note that deeply decentralizing technologies (e.g., blockchain) for transferring and holding value prevent the need for the centralized service banks are offering. I hope the solution can associate with "data as currency" in the next step.

> *Entrants into settlement on the DLT side include the high-profile Ripple, which is looking to solve the sector's problems with the RippleNet global payments network. The emerging DLT competition can also be seen in players like Baton Systems and Copper, the latter of which recently announced its ClearLoop system for instant settlement of crypto trades off-exchange. Roxe is a new entrant that goes beyond DLT and fully leverages blockchain technology to create a unified clearing and settlement network that connects banks and the global financial system at-large. Roxe will purportedly enable instantaneous, and more reliable, global asset transfers for banks.*

A loan with no credit check (i.e., Peer To Peer Lending, P2P) is an innovation, which is called "Internet banking" in China. The business attracted investment exceeding US$19 billion globally in 2015 alone. Loaning has progressed from a matter of personal trust. Deciding on whether someone gets a mortgage or not by the bank is based on a single statistic: an individual's score. Reducing personal trust to a number may help banks choose their borrowers easily. But as we know, the reality is more sobering. Credit scores are slowly on past credit transactions without much data. If missing a few data points, errors in that data get amplified, then a credit score may be overstated or understated. New loan providers spurned up the scores by adding more data.

> *SoFi, a fintech start-up, which originally focused on student loans, factors many data points into its prediction of creditworthiness, allowing it to offer low interest rates to individuals with limited credit history; Kabbage offers a similar service to small businesses.*
>
> *... Another new entrant, Upstart, is a fintech that uses educational data in addition to traditional credit scores to assess credit risk and capture the risk end of the credit market. And fitches such as Avant and ZestFinance (founded by former Google CIO Douglas Merrill) combat payday loan-sharking. ... In 2016, Chinese Internet giant Baidu invested in ZestFinance, ...*

The digital investment advisor might not be a human being and is essentially an algorithm. The algorithm can be run on our smart terminals. In this way, privacy is guaranteed by creating a private environment. That is, we can authorize the algorithm to access our behavior data relating to very personal investments if necessary. Then, based on rich data, the algorithm can help detect any proper investment opportunities without the need to share it with anyone else. Also, the algorithm can be designed to fit any requirement bundling with existing, high-quality data products. It may pick up the most optimal solution from various providers or combine them. For current data technology, it is no difficult to realize the digital investment advisor I envision. Also, I believe such an algorithm is possible to seize some market share from the Big Three.

Data Labor, Data Partnership, and Digital Errand Runner

The previous content in this chapter, so far, can be summarized as follows. Data-rich markets are inherently radical. Because It can provide more information based on analyzable data, compared with price-based traditional markets as "a paucity of information (i.e., nearly no data)" and money-based financial markets as "a single information (i.e., two dimensions)." The more information here means more sources, more characters, and more matching degrees between the demander and supplier. The word "data-rich" in essence means big data. I once mentioned before; big data does not only refer to the data, but includes technology and application. Of course, the information within the markets has no value without human society. In this section, I'm going to talk about the "people"—no matter who they are: the real data labor and data partnership, and the virtual one as a digital errand-runner.

Data Labor

"Data, everywhere," a common saying is that we live now in a data-driven age. There are many buzz words to describe this new reality: data science, big data, data mining, machine and deep learning, even artificial intelligence—I don't like the last one; it is too capricious to gain new connotation every decade. The aim is to inspire young and new minds to consider data science as a necessary intellectual pursuit for an educated citizen in the digital age, even if one does not necessarily aspire to become a data scientist. Chief economist at Google Hal R. Varian argued that "data are omnipresent these days and that what is scarce are the talent … needed to make sense of these data." It is an interesting time to be a data scientist. After all, *Harvard Business Review* announced that this is "the sexiest job of the twenty-first century."[19] Inspiring young minds to consider data science as an intellectual pursuit for an educated citizen in the digital age is important, even if one does not necessarily aspire to become a data scientist.

Now the first question is how to define a data scientist. A popular description goes something like this: "A data scientist knows more computing than a typical statistician and more statistics than a typical computer scientist, but can communicate better than either of them."[20] There is an article on TheBalanceCareers.com, a known career website, which provides more specific job skills for a data

---

[19] ***See:*** Davenport T, Patil D. Data scientist: Sexiest job of the 21st century. Harvard Business Review. October 2012. Retrieved from https://hbr.org/2012/10/data-scientist-the-sexiest-job-of-the-21st-century.

[20] ***See:*** Lazar N. Data, Data, Everywhere … Harvard Data Science Review. May 27, 2020. DOI: 10.1162/99608f92.a6e7a24e.

Figure 7.1 Important job skills for data scientists. (Data source: Author redrew this figure according to the content of Footnote 21)

scientist consisting of six categories and 90 items.[21] I drew a radar chart figure 7.1 to carry out it simplified. It does not as people hope, a single, simple solution could enable everyone to be a data scientist. On the official website of the Initiative for Analytics and Data Science Standards (IADSS), a recently created body formed to try to create a broad set of standards for data science qualifications, the current situation is described as:[22]

> *Almost every company in the industry has a unique way of defining roles and assigning titles in data analytics related positions which have resulted in a chaotic market that is confusing to employers, academic and training institutions, and candidates; with a large number of unqualified candidates calling themselves "data scientist," "data architect," "data engineer" or "analytics professional."*

With all of the different types of data and data-related issues, it is no surprise that "data scientists are in high professional demand" (see Footnote 20). As shown in figure 7.1, four of the six categories are related to a technical concept.

[21] ***See:*** Doyle A. Important Job Skills for Data Scientists. TheBalanceCareers.com. June 3, 2019. Retrieved from https://www.thebalancecareers.com/list-of-data-scientist-skills-2062381.

[22] ***See:*** (1) Fayyad U, Hamutchu H. Analytics and data science standardization and assessment framework. Harvard Data Science Review. May 1, 2020. DOI: 10.1162/99608f92.1a99e67a; (2) Retrieved from https://www.iadss.org.

But the job requires a person, who is open-minded, easy to communicate enough to analyze, interpret, and present data in meaningful, understandable ways across domains—a wide span involves science, business, education, government agencies, and industries of all variety.

Of course, data labor is not limited to data scientists. Just as Victor R. Fuchs, a founder and leader of the field of health economics, named the physician as "the captain of the team" in his book *Who Shall Live? Health, Economics And Social Choice*, data scientists are the captain of the data economy. The assumption is that the data revolution will unfold much as the Industrial Revolution did: in the new economy with highly automation, initially displacing many people but eventually leading to a variety of new jobs. So a positive mentality is that people can rejoin the workforce if they acquire new skills. But new skills did not represent new job opportunities. The ever-changing in the skills means that what may be much in demand in 3 years may no longer be so sought—after in another 3 years. I think keeping up with times is not equivalent to learning greedily to "bite off more than one can chew."

Data Partnership

I envisage that data partners or data partnership will take a significant proportion in the future data-rich market, far more extensive than the number of data labor. This phenomenon, at least for a bit, makes the social hierarchy tends to be flat—people can work together "more equally." This may be not in line with conventional convictions about the labor market, let's research on it.

When money and price prevailed, it may have been appropriate to understand human labor in terms of wages and financial benefits. Labor, as the ability of a "social person," provides identity, community, and a sense of belonging for many people. Based on this logic, finding suitable work to keep it will continue to be a central role for society, and even only for a family. So the labor market emerged to make human labor as commodities. A critical problem in the labor market is why some inherent differences exist; for example, some laborers earn more money than others and their worth or position are higher. Some conclusions are:[23]

> *(a) that the labor market tends to be fragmented along occupational lines, (b) that the processes governing wage attainment vary from one occupation to another and (c) that occupational differences in these processes can be predicted from the explained in terms of the forces which lead to occupational segmentation of the labor market.*

---

[23] ***See:*** Stolzenberg RM. Occupations, Labor Markets and the Process of Wage Attainment. American Sociological Review. 1975;50(5): 645-665.

The occupational segmentation of the labor market is reflected within a corporate, which is the hierarchical structure of management—allowing a better one to participate in decision-making.

"The best Amabots" is a story given by Mayer-Schönberger in his book *Reinventing Capitalism in the Age of Big Data*, to fully embody "companies and control;" namely, the "hierarchally organized, command-and-control structure" unique to the firm. The robot, here, refers to human labor finely managed like a machine.

> *In 2011, a former Amazon engineer named Steve Yegge garnered international attention when he shared his thoughts on his ex-boss [Jeff Bezos] in rant on [a social media] ... "Bezos is super smart; don't get me wrong," Yegge noted. "He just makes ordinary control freaks look like stoned hippies." ... [This caused resonance among other Amazon's employees.] ... "If you are a good Amazonian, you become an Amabot, "one employee told ... The best Amabots claim to put in one-hundred-hour workweeks in order to pore over the data and answer any and all questions posed to them about how the information should influence decisions. Others grow frustrated, burn out, and move on – or, if they are among the bottom 10 percent of the company's performers, get a warning or get fired. The metrics allow Bezos to control staffing up and down the hierarchy without observing performance firsthand. ... As the Economist noted, "... Far from being an outlier, it would seem that Amazon is the embodiment of a new trend" – what the magazine branded "digital Taylorism," after the scientific management principles of Frederick W. Taylor. It seemed that new technologies were ushering in a supercharged version of command-and-control, fueled by data about employees, processes, products, services, and customers.*

The Taylorism championed by Frederick W. Taylor around the 1890s was often resented by employees, who felt that quantifying every aspect of human labor turned workers into mere cogs in the industrialists' machines. But it was sometimes effective—at least in the era of data scarcity, the information age included. Centralizing information flows and decision-making as a tool of comprehensive control is the key concept of the firm over the past few centuries. Jeff Bezos obviously accepted the practice: embracing the centralized structures, rules, and behaviors of the firm, to manage the vast majority of his business empire. In the post-information age, as more and more detailed information floods to the center, the firm must ensure them to be translated into good decision. Because it has always been a deeply manual task, the executive finds ways to spread the task across more individuals so that an increasing number of people are involved in it. This operation is so-called "delegation of decision-making," and delegation of decision-making is a delicate balancing act: ensuring that the information flows to the power center for execution, and the advantages of centralized information flows are not lost. So, newly empowered decision makers must be trained to make choices that align with the central leadership. If they don't the "noise" they create in the firm causes inefficiencies. Some management strategies can be used to avoid such "noise," including the establishment of standard operating procedure (SOP) and other rule books that guide decision-making in organizations. Workflow-based OA (office automation) systems can also help. By and large, the firm has become a highly effective way for humans to coordinate efficiently.

Data-richness may fundamentally disrupt this best practice. Market information is more up and flows faster via cutting-edge data-focused automatic tools. When the running speed of information in a market exceeds in one company, inefficiencies would cause the company to lose its competitive advantages in the market. To tip the competitive balance between the mechanism of markets and firms, as markets evolve, the traditional firms should handle the speed of internal information flow, such as easing the crushing quantity of decision and upgrading executive decision-making. This will inevitably lead the firms to choose, whether to focus on cost or organizational structure. The former is a bet on the "past," willing to disrupt its workers (i.e., replacing human labor with machines as digital errand runners; I will be back here in the next section), rather than its structure. The later seems like a bet on the "future:" willing to sacrifice historic hierarchies to keep pace with a reinvigorated market. Regardless of the choice, the number of employees will both decrease. The former is due to layoffs (replaced by the machines that are better than humans at obeying executive orders), and the later results in organization split—Internally, the firm splits into multiple small teams that face the markets directly. There exist two types of teams with or without the boss. Mayer-Schönberger introduced one type without a boss, called "squads", when explaining a Swedish startup Spotify that operates music-streaming service (see Footnote 11).

> *"Squads" are small teams responsible entirely for certain aspects of a product (such as the search function or the user interface) or certain business activities (such as sales in a given market). A squad don't have a boss, only a so-called product owner, whose task is to ensure that all the members of the team have everything they need to do a good job. The product owner also keeps an eye on the self-defined goals and deadlines of the team, but unlike traditional team leaders, this one has no enforcement power. Neither does the squad's "agile coach," whose task is to foster and facilitate collaboration within the team. [Both product owner and agile coach are moderators.]*
>
> *... don't ask your manager, because you don't really have one. Experiment with data to create evidence. Then share it with your teammates, the teams working on connected issues, and anyone in the company you consider knowledgeable on a given matter. Get feedback, then make the decision yourself (or as a squad) and implement it as swiftly as possible. If there are problems, it is up to you to fix them.*
>
> *To avoid silos and foster collaboration across squads, specialists with connected expertise meet in "chapters." Squads working on similar projects form "tribes," which should have no more than 150 members. Above the tribe level, there are "guilds." The main goal of this structure is to facilitate the flow of information and knowledge throughout the company. To ensure coherence of business strategy and decision-making, ... [It's] fluid organizational chart denotes the ranks of "system owner" and "chief architect." But in practice even those two top management positions do not have the power to give orders. ... they operate [also] like moderators. They use the soft power of persuasion and the hard facts of data-driven evidence if they want to see their opinions turned into action.*

Another type has a boss who is responsible for group members' salary treatment, similar to the "principal investigator" for the independent grant project in science. If having collaboration across groups, full time, or onsite work is no necessary. There exists an interest swap between the bosses. But mostly, it depends on the individual's willingness. In general, labor and benefits are subject to negotiation; no fixed work time required.

Both two types of team rely on a company brand to undertake and complete products and services. These teams can freely use data capital owned by the company, and products or services carried out by these teams will be shared within the company. In essence, the company has evolved into a market with relatively small competitions. Therefore, the phenomena of hiring "shared employees" or the internal talent marketplace I mentioned above (see Chap. 3) conform to this trend. Besides, crowdsourcing (e.g., Amazon Mechanical Turk, MTurk) is an alternative for these teams to achieve their goals. Taking on an employee as a partner is then achieved. In the past, the bundle of work benefits was not flexible, and the monetary dimension prevailed. An increasing number of high-tech firms has been focusing on the nonmonetary elements of employment when competing for talent now. We will appreciate the rich diversity of preferences for what a person wants from work soon—and embedding this appreciation, the employer may allow us to leave salary and money behind as the main determinants of work. The elements of a job's benefits bundle with career planning, such as meaning, experience, and identity, thus, will become a crucial feature of an organization's overall human resource strategy.

Note that finding a data partner aims no longer to speed the internal information flow and is an optimal matching issue based on data technology complementation and data resource sharing, and thus to form a strategic defense. It is a component for a virtual fortification built by a future data sovereign or the cluster of data sovereigns.

Digital Errand Runner

One option of "betting on the past" to meet the challenges from data-rich markets, I mentioned in the prior section, is also acceptable. That means choosing layoffs replacing human labor with machines, instead of disrupting the hierarchical structure of management used skillfully. Take Fukoku Mutual Life Insurance, a Japanese giant headquartered in Tokyo, as an example; Fukoku decided to use IBM's Waston to process insurance claims at the end of 2016, and this decision made around one-third of staff in claims-assessment department redundant. The machines are the "digital errand runner" I coined here. The term "errand runner" is a buzz word used on the Internet and often shows in the films and TV shows played by the secondary or tertiary leading actors. An errand runner refers to the person to be used as the tools—always on call to hard work but not treated equally in emotion. In reality, as people get older to lost capable of managing daily chores, such as grocery shopping, picking up prescription medicine, and more. Senior citizens usually hire errand runners to do the tasks they no longer can or want to do. The purpose I call these algorithms or data models as "digital errand runners" is to distinguish them from human labor and real robots.

The digital errand runner has a wide range of coverage, such as IBM's Waston system, Tesla's Autopilo (the semiautonomous driving system Elon Musk boasted on Twitter in later 2016), the digital investment advisor I envisioned above, and other automatic office clerks. I'm going to compare two tax digital errand-runners, see Case 7.2;[24] and they are tax e-filling bots vs. chatbots, which may help navigate and comply with complex tax regulations.

Case 7.2: Two Commonly Used Tax Digital Errand-Runners

Case IV.2 Two commonly used tax digital errand-runners

- Tax e-filling bots

Tax e-filling bots can help SMEs file taxes accurately with less manual effort through the following capabilities.
Extraction of financial data from primary sources
Tax compliance is dependent on information from primary data sources such as invoices, balance sheets, profit and loss accounts sheets, etc. To fill in tax forms, we have to separate out expenses and revenues, which involves a lot of manual data sorting. Tax bots extract primary data sources from PDFs and spreadsheets into our tax management software and sort the information accordingly.
Automatically calculate tax adjustments
Calculating the adjusted gross income (AGI) is a key component of knowing your actual tax liability (net tax). Tax e-filling bots automatically calculate AGI by subtracting deductions and expenses from the total tax as it transfers this data to the tax form from primary sources such as invoices. This helps to eliminate the need to manually key in these fields to calculate tax adjustments, saving time, and reducing the risk of error.
API integration with tax management tools
Through APIs, tax e-filling bots can integrate with other accounting or tax-specific tools to capture primary details related to taxes. For example, if our accounting tool contains information about your total earnings in terms of revenue and profit, all this information can be coded into the tax bot to fill in your tax forms.

- Tax chatbots

Chatbots for tax management serve as knowledge portals that enable SMEs to ask tax-related questions.
NLP(natural language processing)-based character user interface data processing
The primary focus of any chatbot is to answer customer questions whether for sales, marketing, taxes, or another topic. Just like Siri or Google Now, a tax chatbot converses with a user through written text in the question box or through audio and provides an answer in a similar format.
Tax knowledge management system
To answer user queries related to taxes, chatbots need to be integrated with a knowledge management system. Once a user asks a question, the chatbot matches the text string with the data in the knowledge management system.
Self-service tax kiosks
Since chatbots automate the process of answering tax-related queries, they often don't need any human intervention to complete basic tasks such as scheduling tax filing reminders. As virtual assistants, tax chatbots serve as self-service kiosks that can help users resolve their tax queries without any human assistance.

[24] *See:* Singh A. Tax bots vs. Chatbots: Which tax automation technologies will help your small business file taxes? GetApp.Com. January 31, 2019. Retrieved from https://lab.getapp.com/tax-automation-technologies.

Numerous deeply manual, occupational tasks that are previously the most tedious, time-consuming, and non-artistry can be replaced by the digital errand runner. These occupations include, but not limited to, government officials, office clerks, bank or insurance counter operators, sales, lawyers, investigators, and other intermediaries. This is a pretty brighter future, but what is the result eventually?

> *... machines, and their seemingly infinite ability to accomplish complex tasks at low or practically no cost, will recycle the resources we have forever, essentially taking us into a true utopia. In it, humans, freed from daily chores, will enjoy life and have means to live it to the fullest.*

Proponents of the view use the term "fully automated luxury communism" to describe the idea, is that "we all can work less and still enjoy whatever we want." But Mayer-Schönberger opposed, and his thoughts on this are (see Footnote 11):

> *Advocates of abundance have built that vision on a fundamental fallacy. They focus on physical resources and misunderstand that the market is not just a way to allocate scarce physical goods, but a way for human to coordinate with each other effectively and efficiently. As humans coordinate, we face the scarcity of time. Without perfect immortality, we need markets to coordinate with each other in the limited time we have.*
>
> *Rather than hoping for a timeless world that does not exist, we need to embrace the reality we live in. We do not dispute the existence of severe economic woes throughout the world. We, too, see them as acute challenges. But the market isn't the root cause of the problem. Instead, we view it as our best hope for overcoming them. ...*
>
> *Our future as human is not about centralizing power, nor is it about consumption and opulence; rather, it is about interacting with each other so that we can spend the very resource that we cannot replenish – our time – most meaningfully. But a comprehensive and sustainable freedom ... rests on the market as a decentralized social mechanism coordinated by humans.*

But from my perspective, why it is not a "communism" if having plenty of matter and spiritual freedom.

Driving Health Data Profit Return: A Brainstorm

The healthcare market, especially its digital part, always attracts new market entrants. Data innovation is the name of the game here, with the primary goal of streamlining physicians' work, improving patient outcomes, reducing human error, and extracting insights that can be monetized to support opportunity areas including population health. These diversified data innovations are the necessary elements for the re-creation of the healthcare market. I will bridge two of my mHealth products to report the fact that, data and techniques, when combined, makes us formidable buyers and sellers. Not because we will win every negotiation but because we'll think right and act efficiently, relentlessly optimizing outcomes based on our preferences. They will not only benefit participants but may significantly improve the healthcare market as a whole and make the market the most efficient place for human health and well-being.

Introduction to a Bone Health-Related App

Osteoporosis is "a pediatric disease with geriatric consequences;"[25] that is, in need of an early prevention campaign. If we think of bones as a bank account, where people make "deposits" and "withdrawals" of bone tissue, then childhood is a critical time as much more bone is deposited than withdrawn during this span of life as this is when the skeleton grows in both size and density. This means that the more children can put away in their "bone bank" account when they're young, the longer it should last when they get older. Thus, the bone mass attained in childhood and adolescence is "an important determinant of lifelong skeletal health." I'm going to introduce the mHealth app of promoting bone health in children as follows.

1. **One-Line Pitch (120 characters)**
   Our exergame improves child bone health via customized exercise routines by combining traditional Chinese medicine and big data analytics.
2. **Idea Description (500 characters)**
   Performing group-based physical activity consistently is an effective therapy for osteoporosis and a reliable way to increase happiness. Gamified digital health technologies are to augment fitness with fun by increasing engagement and motivation. We envision a gaming app that translates the Wu Xing Bone-Strengthening Exercises into an accessible form of exercise to increase enjoyment during workout sessions in children through engagement.
3. **Your progress to date (e.g., the idea formed, proof-of-concept completed, preclinical studies lined up, entered the clinical research and soon)**
   We chose to use the Wu Xing Bone-Strengthening Exercises, a complete workout program derived from Chinese Kung-Fu, as the source of our exercises. It has shown promise in our prior research on rehabilitation in osteoporosis awarded the third Place of 2017 Shanghai Science and Technology Awards for Popular Science, China. We currently have a functional prototype that is being piloted in China. Children and their parental guardians interact with our app using separate but integrated user interfaces. The children's interface has four functions (i.e., normal mode, enhanced mode, My

[25] ***See:*** Hightower L. Osteoporosis: Pediatric disease with geriatric consequences. Orthopaedic Nursing. 2000;19(5): 59-62.

Growth Manual, and the game hub) while the parent app interface has three (i.e., Child App Settings, Child's Growth Manual, and Social Hub). Both interfaces are fully functional. We are continuing our community outreach to the Chinese market and have early adopters in 18 local communities in Shanghai.

4. **The market (Who are your customers and competitors? What's the market size? How do you plan to enter the market?)**
   Our exercises specifically targeted children, age 7–15. There are no other apps that focus on the needs of an individual child's bone health. During our pilot study in 18 local Shanghai communities, we received valuable feedback from our users in the 7–15 aged demographic and their parents regarding the effectiveness of our app and who would benefit from it. The pilot study has indicated that our customized exercise routines have comparative advantages compared to standard broadcast gymnastics prevalent in primary and secondary schools and demonstrates that our approach is reasonable.
   Chinese elements are frequently added in today's highly globalized world to increase promotional appeal in audiences interested in Chinese culture. In the field of exergaming (games related to physical motion and fitness), an estimated global market of US$6.4 billion (in 2017), we believe that an app that incorporates these Chinese elements would succeed. We hope to derive growth by systematically increasing user adoption, by adding value to our stakeholders, and by expanding our current efforts in China and also extend to the US. Thus, during our first phase of US expansion, we plan to target Chinese immigrants, Chinese living abroad, and Chinese Americans. A significant group of Chinese people living abroad is teenaged students; part of our target age demographic. In fact, according to the latest data from China's Ministry of Education, 523,700 Chinese students went abroad in 2015, a 13.9% increase over the year before. As of 2014, almost 30% of all Chinese students abroad enrolled at the secondary school level. Interestingly, these students often live with grandparents overseas. Our app offers group-based physical activities that are beneficial for all ages (including the elderly) and a remote fitness monitoring component for parents in the homeland.
   We hope to utilize the specific market distribution of the local New England region. We will work to collaborate with Partners Healthcare to combine our app with clinical data from Partners. This market distribution will allow us to participate in various cross-promotional activities, such as giving talks, handing out the app as a gift to patients, and local demonstrations of the Wu Xing Bone-Strengthening exercises.
5. **Intellectual Property status (Have you filed or intend to file for a patent? Who owns the IP, your or your institution? Do you have the permission to use it?)**
   Our prior research on exercises is for relieving older adults osteoporosis summarized in the book, Wu Xing Bone-Strengthening Exercises, published in 2014 and has a print distribution of more than 50,000 copies.

A Brief of the Platform for Data Profit Sharing

Data transaction has a history for far too long. The former Reuters correspondent and the current residence at Harvard Adam Tanner discussed "how the hidden trade in our sensitive medical records became a multibillion-dollar business, but has done little to improve our healthcare outcomes,"[26] in his newly published book (in 2017). He said that (see Footnote 26),

> *Hidden to consumers, patient medical data has become a multibillion-dollar worldwide trade industry between our health-care providers, drug companies, and a complex web of middlemen. This great medical-data bazaar sells copies of the prescription you recently filled, your hospital records, insurance claims, blood-test results, and more, stripped of your name but possibly with identifiers such as year of birth, gender, and doctor. As computing grows ever more sophisticated, patient dossiers become increasingly vulnerable to reidentification and the possibility of being targeted by identity thieves or hackers.*
>
> *Paradoxically, comprehensive electronic files for patient treatment – the reason medical data exists in the first place – remain an elusive goal. Even today, patients or their doctors rarely have easy access to comprehensive records that could improve care. In the evolution of medical data, the instinct for profit has outstripped patient needs.*

The practice on how we can best balance data profits between advance medicine and improve lives while preserving the rights and interests of every patient is critical. I'll brief the platform for medical data profit sharing as follows.

1. **What's the problem you are looking to solve? Describe the market need and venture opportunity. Character limit is 800.**
   The starting question we want to solve is to help healthcare researchers to get their desired data easily. While existing vast amounts of data that are open access allowed them to apply, the data required by medical researchers that often targets specific individuals is too purposeful to be collected. For example, pharmaceutical researchers need genetic data for a particular disease. Paying the fee to the data owner is a possible solution. However, on the one hand, the increasing gap between healthcare costs and outcomes is one of the most critical issues; on the other hand, individuals are not willing to sell their data for little money. Also, people cannot get any profit via data ownership because some bad data brokers are making money with their personal data spread on the Internet (796 of 800 characters).
2. **Describe your solution to the problem you have specified: What will you make or do? Why will it be effective? Character limit is 800.**
   We aim to create a machine intelligence (MI) based global leading health data platform to share data, profit, and economy, to help medical research. So

[26] ***See:*** Tanner A. Our Bodies, Our Data: How Companies Make Billions Selling Our Medical Records. Boston, MA: Beacon Press; 1 edition (January 10, 2017). 2017; 248 pages.

far, we have two core data techniques: (1) data self-governing openness that has been applying for a patent and was invited to give a poster at the first Harvard Data Science Initiative Conference in November 2018; (2) visual data mining process that is our previous research work. We will optimize the platform using the blockchain technique to connect (possibly) heterogeneous existing systems. Also, we propose an advanced concept related to this venture that is "no raw data in the transaction." That is to say, the individuals or institutions can enjoy risk-free profits only depend on data ownership without responsibility (782 of 800 characters).

3. **Who is your target customer/user for this solution? Character limit is 600.**
   Our target potential customers can be divided into two classes of buyers and sellers. The purpose of the buyers is for medical research, involving medical schools (with their departments), hospitals, insurance companies, medical suppliers, and drug companies. Most of the sellers are individuals who are willing to authorize us to help them (1) find their data through search engine we offered, and (2) obtain their own EMR/EHR data using our services via hospitals' open APIs. Also, we will provide free data tools to hospitals/the others to mine their data for research or administrative purposes (600 of 600 characters).
4. **Who are your existing competitors, and who might become a competitor? What makes your venture a better option? Character limit is 600.**
   We did marketing research in 2017 to list several existing competitors, and they are: Covance that took 46–50% market share with 10.2 billion US$, iQVIA that took 36–40% market share with 8 billion US$, and the others (e.g., Hu-umanity, Optum, IMS, and Symphony) that made the rest market share around US$1–2 billion. However, the data they provided is also without a target. So, they need to collect and store data in advance accounted for a large part of the cost. We will collect data according to customer demand to reduce data inventory, and offer more affordable and precision (588 of 600 characters).
5. **How do you plan to go to market? How are you thinking about piloting your venture? Acquiring initial customers? Partnering? Achieving financial sustainability? Character limit is 600.**
   We step-by-step segment our business goal by multiple minimum viable products (MVPs) to pilot our venture. First, we designed the platform interface that adapts multi-user interfaces to meet customer needs including data sellers and buyers. Secondly, we defined all business modules. Thirdly, we launched a sample webpage of data ownership claim, which helps the individuals or institution to find and claim/deny their data on the Internet. Also, we are achieving an enhanced function of data ownership service, which is privacy service to hide negative links corresponding to customer request (594 of 600 characters).

Integrating the Two in a Cross-Border Market

It is possible to separate the above two data products facing different market segments in the healthcare market. But this is not what I want to discuss. I want to discuss an integration with you to see how we can move this forward.

The mHealth app for promoting bone health in children is a business between the post-information and data industry aiming to develop users and then to collect their data. It targets a group of population in childhood and adolescence clearly, age 7–15 years old, living in the US accompanied by one or more intergenerational guardians. Note that most parents of them live in China. The data the app needs is the content relating to the growth manual for target audience, which records bone or other information. The other information includes, but not limited to, workout habits (sedentary or not), gaming preferences (the proportion of time that spent on games, workout, and the others), family medical history, and social network constructed by family members and friends. The biggest disadvantage lies at app stickiness to retain customers; namely, make sure: the app is actually "fun" to use over time, or the customer continues to use it on a regular basis. The biggest advantage aims to a cross-border market.

The data profit-sharing platform is without a specific target audience. It aims to solve two pain points. One is how to make individuals (and other data owners) benefit from data products used their data partially. The other is how to let research institutions obtain their desired data easily. Even the world's best health systems are fragmented. While hard efforts were given by the academia uninterruptedly, various kinds of tangible and intangible barriers make it difficult for researchers to get free access to the data they need and want. These efforts are such as publishing numerous articles about data sharing (my paper included) and constructing international communities in favor of data interoperability (e.g., the Observational Health Data Sciences and Informatics, OHDSI). The core problem is nothing more than two: data and profit. On the one hand, no stakeholder in healthcare is willing to share the data to be free. Medical-related data acquisition is laborious and time consuming, and storage costs are expensive; meanwhile, freely sharing data means giving up an income—after all, it is a multibillion-dollar trading market share. On the other hand, the data itself is a problem. Many medical studies blindly pursue data targeting specific populations that are too purposeful to be collected. The relevant researchers need some recommendations from data scientists to use alternative data sources. For instance, the data on any patient-centered social network (e.g., PatientsLikeMe) can be used to replace them. Another fortunate thing is that APIs are starting to enter the EHR market because of the US government's Health IT Certification

Program authorized under three federal laws. They are[27] the Health Information Technology for Economic and Clinical Health (HITECH) Act, the Medicare Access and CHIP Reauthorization Act of 2015 (MACRA), and the Twenty-first Century Cures Act. So, The pieces are in place for a truly disruptive shift in "patient-led data sharing"[28]—to make patients access and use their own clinical data to improve their health. These are also the foundation on which this data profit-platform can be built. According to the report published by Allied Market Research,[29] the global Electronic Health Records (EHR) Industry generated $24.90 billion in 2017, and is expected to generate $33.29 billion by 2023, registering a CAGR of 5.0% from 2017 to 2023.

By linking the two products together in the same market, it not only is through the entire data industry chain (i.e., from data collection to data production, then to data profit sharing) but also breakthroughs geographical restrictions. Next, I'm going to analyze the two products related issues on data capital rights, income and output, and wealth distribution, in the cross-border market. To simplify, I use product B refers to the app of promoting bone health in Children and product P to the data profit-sharing platform in the following paragraphs.

First, both two products are for dual users. Product B is for kids and their parents—reporting data collected from the kids to their parents regularly. It may involve raw data transmission. Product P is for individuals (or other data owners) and medical research institutions—selling the data entrusted from the individuals to institutions after having processed as a data product (instead of raw data); such a data transaction may respond to a specific institution's requirement. Defining data sovereign is obviously useful because it helps to distinguish the users and data producers clearly. The data collected through product B is possible to be attacked by the governments (e.g., censorship) due to cross-border. And some claims as an attack may be received from the individuals; for example, a father complains his son spending money in the game without his permission, and requires compensation. Since the data collected through product P is based on personal entrustment, such attacks are less likely to occur. For instance, a man designated us to search relevant data about him on the Internet, or to obtain his medical records from multiple hospitals via APIs, respectively.

Second, technological depths in the two products are different. Product B belongs to a shallow data application involving more private data processing.

---

[27] ***See:*** Thornewill J, Esterhay B, Schneider EC, *et al.* Making Health Data Useful to Patients Through Open APIs. CommonWealthFund.Org. December 7, 2016. Retrieved from https://www.commonwealthfund.org/blog/2016/making-health-data-useful-patients-through-open-apis.

[28] ***See:*** Gordon W, Chopra A, Landman A. Patient-led data sharing – a new paradigm for electronic health data. NEJM Catalyst, 2018;4(6).

[29] ***See:*** Allied Market Research. Electronic Health Records (EHR) Market to Reach $33.3 bn, by 2023. April 13, 2020. Retrieved from https://www.globenewswire.com/news-release/2020/04/13/2015136/0/en/Electronic-Health-Records-EHR-Market-to-Reach-33-3-Bn-by-2023.html.

While having the analytic content based on personal data, such as suggesting the time, length, and sequence of workout via some recommendation algorithms, it is generally without advanced data techniques. Besides, data marketing is more required, including creating community engagement strategies, carrying out app promotion, and advertising. Product P is relatively cutting-edge, a self-governance model included (in which profit sharing from the data transfer can be realized by the data owners). The others involved the next generation search engine (that can return the search result based on data mining, instead of the list of original links) and automatic data deletion after the expiry date.

Third, the two products are not based on the same principle of sharing data profits. Product B requires more time and effort the user spent in exchange for the benefits (e.g., token). For example, a kid may receive two tokens (but another kid only one) because his growth manual is more integrity than another, or he spent more time on the exercise or a game. Product P tends to be simple and easy and has no user activity requirements. That will be awesome if logging in your account regularly to discover thousands of dollars. These operations are automatically implemented according to personal entrustment and algorithms. Of course, you can choose to exchange data for data. After all, money and price have no longer prevailed in the future data-rich markets.

Someone may wonder if we should only keep product P. While product P is more advanced, the phase of initial user accumulation may be much longer due to its too blood new idea. This phase will be much shorter if integrating with product B that features a high degree of medical focus and a great user base. Also, product B has a cross-border market. Cooperation between the two product facilities forms a sovereign data alliance, which is helpful for product B to improve defense level and increase user stickiness; in turn, product B provides a real application scenario for product P, making medical research more effective.

Chapter 8
Public Choice on Data Capital

Radical data-rich markets may cause "the Great Adjustment"[1] of the global economy. Then, should the governments intervene in this adjustment? How should governments do an intervention? When and Which means the governments intervene in the marketplace? What about the short- and long-term effects of government intervention in markets? All of them are determined by rational decision making on the public choice given by the government according to a country's economic foundation itself.

Take, for example, public choice in upfront infrastructure investment for data capital is different between developed countries and emerging economies, particularly between the US or UK and China on physical infrastructure. The data economy needs different sorts of physical and intangible infrastructure. For instance, the 5G infrastructure will require new transmission infrastructure, including thousands of cell towers and tens of thousands of antenna, deployed on utility poles and other urban infrastructure, meanwhile needs software architecture (e.g., network function virtualization). Just like explained by Jonathan Haskel and Stian Westlake in their book *Capitalism without Capital*,[2]

> *Such "infra" means "below," infrastructure consists of those structures that underpin the way society works. One way of thinking about infrastructure is the set of durable assets that are complementary to the capital employed by firms and households. ... Some infrastructure may be publicly owned (for example, a public highway); some may be owned by firms (for example, a private airport or telephone cable) – but whoever owns it, infrastructure has a public character in that it enables economic activity on the part of a large number of other people or firms.*

[1] ***See:*** Mayer-schönberger V. Delete: The Virtue of Forgetting in the Digital Age. Princeton University Press; Revised edition (July 5, 2011). 2011; 253 pages.

[2] I reference directly here from the original sentence published in Jonathan Haskei's and Stian Westiake's book of *Capitalism without Capital: The Rise of the Intangible*. ***See:*** Haskel J, Westlake S. Capitalism without Capital: The Rise of the Intangible. New Jersey, NY: Princeton University Press; Reprint edition (October 16, 2018), 2018: 296 pages.

© Springer Nature Switzerland AG 2021

C. Tang, *Data Capital*, https://doi.org/10.1007/978-3-030-60192-8_8

The fact is that the great adjustment caused by data as capital will affect nearly all human activities. These changes can be both gradual and dramatic, and are inseparable from social security. It seems to come naturally for politicians and governments to react to public choice. Whatever the emergency and thoughtful response on the public policy, it requires some institutional design principles. I explored the five institutional design principles that may dictate the shape of how State or Local governments and related organizations to cope with "the Great Adjustment." These principles are self-evident, including well-defined guidelines, collective-choice arrangements, monitoring, graduated sanctions and credible rewards, conflict-resolution mechanisms, minimal recognition of rights. In this chapter, I'll design three possible scenarios (i.e., political leadership, alternative currencies, and cyber encounters) for the governments during the great adjustment. When explaining these scenarios, I will mention the five principles.

Rational Choice of Political Leadership

Public choice[3] an economic tool often deals with the traditional problem of political science, concerning the study of political behavior—the self-interested agents (voters, politicians, bureaucrats) and their interactions. In other words, public choice is also close to decision making, relating to some approaches and the explanation of how the procedure results in outcomes that conflict with the preferences of the general public. So, political leadership is included. But there exists no chapter on leadership mentioned in Dennis Mueller's encyclopedic survey of this field (i.e., public choice; or more generally rational choice in politics). Because he believes that, the leadership issue is partially related to "political entrepreneurship;" and its another part is about "dictatorship," largely following Ronald Wintrobe's opinion.[4]

In this section, I'm going to explore whether various strands of rational choice theory can justify and explain the leadership, by looking at the different types of infrastructure that help for data capital. The leadership is manifested in how to solve problems, first is at physical infrastructure and cluster effects; and then at the softest of soft infrastructure: norms and trust. Two of five principles as "well-defined guideline" and "collective-choice arrangements" are mentioned here.

[3] ***See:*** (1) Brennan G, Brooks M. Rational choice approaches to leadership. In The Oxford Handbook of Political Leadership (pp. 161-75). Oxford: Oxford University Press. 2014; (2) The entry of "Public choice" on Wikipedia. Retrieved from https://en.wikipedia.org/wiki/Public_choice.

[4] ***See:*** Wintrobe R. (2000). The political economy of dictatorship. Cambridge, England: Cambridge University Press; 1 edition (September 25, 2000). 2000, 404 pages.

Regarding the meaning here, I give an example from mine. My mother had been complaining about the slow Internet speed used in the home for a prior period, which prevented her from watching HDTV series smoothly. But because of the price, she did not decide to change the data package. I helped her decide to apply for broadband acceleration. The installation needs to wait for a while, during which she still kept complaining. I began to test the current network speed and found the rate only to achieve one-third of the quoted one through my wireless adapter. Then, I called a statement claim. A staff from telecom consumer service suggested I reset the network speed from the router they provided instead of my wireless adapter. I did so to find the cause: it was due to the wireless adapter I installed myself. Thus, I changed the wireless adapter and canceled the new data package application. I only spent about 5% of the expenditure (i.e., on applying and using a new data package) to solve the problem my mother always complained about. As a result, my "leadership" increased at home. The leadership may lie in deciding both whether infrastructure is needed and how we reasonably invest under our choices. The justification of leadership also involves two things, one is on delegation (that is, why my mother wanted me as a representative to deal with the problem instead of doing it directly) and the other is about hierarchy (that is, why representative institutions like "home" are organized with "leader" at the top). In what follows, I will investigate what the public choice has to say in all aspects.

Physical Infrastructure and Cluster Effects

Over the past decades, there has been a veritable flood of interest in industrial localization, not least in the work of Alfred Marshall, Kenneth Arrow, Paul Romer, and Edward Glaeser. Industrial localization, of course, is nothing new. One way of doing this is to build or nurture dynamic local clusters like Silicon Valley. But clusters are absolute mint for policymakers and pundits because it brings the right places where innovative businesses and people are likely to come together and share ideas, especially in high-innovation sectors. For example (see Footnote 2),

> *Witness the dozens of Silicon-soundalike names that have been coined around the world in homage to northern California's tech cluster – from Silicon Roundabout in London and Silicon Wadi in Israel, to any number of more aspirational variants elsewhere.*

Cluster policy is appealing in part driven by politicians' self-interests—they would earn a reputation as a do-well leader if their country is at the forefront of a technological revolution. This is an impressive leadership. Often, the light-touch cluster policy can also be politically convenient (see Footnote 2).

> *Governments that for ideological reasons are anxious about seeming too interventionist can point out that they are only helping out clusters that are already there, not trying to create clusters from scratch; governments unable or unwilling to spend much money will find that when it comes to cluster policy, moral suasion and cost-effective networking events go a long way. ... The flipside of cheap, light-touch cluster policies is that it is difficult to show whether they have been effective. There are relatively rigorous ways of evaluating the economic effect of grants, tax breaks, or infrastructure investments. But when it comes to ongoing, subtle policies meant to accelerate clusters that are already forming, it is hard to test whether the policy itself has change anything. (For some policymakers, that a policy is difficult to evaluate may – as they say in the world's most famous tech cluster – be a feature, not a bug.)*

Similarly, the rational choice theory (RCT) regards the appropriate framework for the "leadership" as involving some discretion on the part of political agents. The cheap cluster policy does not require many fiscal resources. Decisions about how to use a government's fiscal resources are subject to disputes which require the collective-choice processes; this will ensure that the stakeholder has a say in how the support is to be used and managed. Fewer decisions on using fiscal resources mean less collective-choice processes, resulting in the rise of such discretion that is conducive to these relatively weak cluster policies.

Of course, the relatively strong cluster policy with efficiency metrics is different. This creates something of a dilemma for leadership. Physical infrastructure investment may matter in an age where clusters are becoming more critical. It is the same reason for SoftBank Group Corp that once poured more than US$13.5 billion into WeWork, aiming to invest in housing and working spaces with an artistic or creative leaning. When more stakeholders are empowered to participate in making and modifying the rules that govern resources, the discretion of leadership will decline. If such discretion is not weakened accordingly, meaning that the agents' interests do not coincide with their duty, the administration may evolve into the benevolent despot and even a dictatorship.

Fortunately, the industry formed by "data as capital" is biased a cyber-agglomeration, and might not or less require specialized concentrations in particular geographical locations. So the relatively cheap, light-touch cluster policies are applicable.

Good public policy on physical infrastructure should be creating the conditions for data and information to flow and process as well as for knowledge to spread and mingle. While a prediction that the data economy tends to a cyber-agglomeration, the markets being "our best hope for overcoming the future challenges" happen more communicative coordination in specific places where people congregate, especially in cities. This leads to proper urban planning and land-use policies to be necessary. There exist a vast literature on how we constitute good policy for cities. So, I only have two pieces of advice about this. One involves a mixture of judicious planning and organic growth; and the other is to allow diversification. Specifically, it features on humanizing benign planning as (*a*) cluster cohesion that is obvious to support competition and cooperation among people and enterprises; (*b*) a driving force combined with data capital and technology, making arrangements with science and research to participate in data innovation; (*c*) an accommodating workplace and housing, providing enough infrastructure for people get places to meet; and (*d*) a very diversified ecological community, tolerating messy and diverse areas.

Soft Infrastructure: Trust and Norms

Policymakers have long known that sufficient soft infrastructure can help decision-making, then to encourage investment. Conversely, the tyranny of small decisions[5] that is a situation in which many decisions, individually small and insignificant in size and time perspective, cumulatively result in a more substantial and significant lousy outcome. Among these both formal and informal institutional infrastructures, the principle as the softest of soft infrastructure needs "well-defined guidelines." These guidelines dictate how a government's strategic resources will be used and explain why the guidelines are essential.

Compared with the "expressive" leadership required by policy innovations in the previous part, the soft infrastructure goes further to ask for "political entrepreneurship." Political entrepreneurship tends to interpret leadership in a somewhat favorable light (see Footnote 3).

> *The notion of political entrepreneurship entered the public choice literature at the hands of Richard Wagner (1966) in a review of Mancur Olson's (1965) The Logic of Collective Action. Olson's 'logic' emphasizes the role that apparently 'incidental' private interest must play in any provision of public goods: public goods for a group are more likely to be provided if they come with incidental 'selective' private benefits available for contributors. Wagner conceives of the political entrepreneur as the broker—the one who conceives and delivers the peculiar package of selective private (and general public) benefits. Wagner emphasizes (in the Olsonian spirit) that such entrepreneurs are more likely to be forthcoming if they themselves will receive 'selective benefits'—either electoral advantage or rents from office.*

Apart from the guidelines, the key to soft infrastructure is trust and norms. Trust is related to the provision of trusted information, correctly and effectively. Since the data economy can assist increase profits of the others, smoothly flowing the trusted information given by the governments encourages the kinds of interactions that tire down barriers to entering the traditional sectors, to create synergies of market performance, then to accumulate newly data capital. Of course, these governmental functions that were previously difficult to achieve in

---

[5] ***See:*** The entry of "Tyranny of small decisions" on Wikipedia. Retrieved from https://en.wikipedia.org/wiki/Tyranny_of_small_decisions.

the money-based markets can be helped by digital errand runners in data-rich markets. For example, the updated industry directories and maps may make new and emerging sectors readable, both to outsiders and to firms within the industry.

Regarding norms, making rules to encourage people and businesses to participate in "data as capital" should consider three aspects, and they lie at legislation, market, and business practices. First, one option for a government to rely on data as capital is to tighten and extend the relevant laws. It could provide assurance for data producers while strengthening data protection and privacy regulations. For instance, designing a tolerance of noncompete clauses to separate the right to use and own data resources (e.g., the progressive data-sharing mandate[6]—the mechanism is to protect against excessive concentration of adaptive systems through the sharing of different data subsets, to enable all competitors using the same data input to build the systems), so that government administration could regulate the negative lists to reduce the costs of data resource development. Another example is to encourage new forms of debt finance or tax reform via legislation, making it easier for companies to borrow data products as a charge or lien and a tax credit for financing.

The second is to promote establishing data product exchanges in the market. A significant undertaking of the kind of transactions is to reform the existing approaches for valuing a data product; that is, to use public decision-making powers held by the government that coordinate all relevant stakeholders, including rights-holders, data product providers, and collection agencies. One goal is to ensure that individuals or other data owners can participate in profit distribution.

Lastly, since data capital is emerging, it relies on prior business practices (e.g., norms, rules, and standards) about how to bring them together. Some norms are technical, like interoperability protocols that allow different pieces of software to interact. Some rules are regulatory, like rules about what websites can enable the crawler to gather the data and what they cannot do with it, or the regulations that govern relationships between enterprises. For example, Google's crawler is shielded by Facebook. Another example is that YouTube and other video sharing platforms instructed how to "Fair Use" copyrighted video content. Some standards are professional norms, like intellectual property backed by patent tools. Most business practices are underpinned by social consensus about things work; meanwhile, establishing and enforcing these practices is time-consuming and costly: they require appropriately funded government administration and regulators to regulate reasonably (far more than the will of the lobbyists from interest groups) by expending political capital. Therefore, the data economy needs these carefully thought-out business practices, based on informed and somewhat stable social consensus.

---

[6] ***See:*** Mayer-Schönberger V, Ramge T. Reinventing Capitalism in the Age of Big Data. New York, NY: Basic Books (February 27, 2018). 2018; 288 pages.

The Scenarios for Alternative Currencies

When speaking about currencies, there exist numerous terms. To simplify, I won't clarify them in this book. I only partition them into two categories. One is sovereign paper-based currencies (i.e., coins and banknotes) described with the word "currency." The other includes the rest with the term "alternative currencies," and the future sovereign digital currencies included. Besides, I believe innovation is also a new currency because it offers the approaches to cover more ground and create distinct value which was supposed to be paid as the cost.

Public choice theory to explain the government in various alternative currency scenarios may involve "special interests" and "rent seeking." These are related to one of five principles as "graduated sanctions and credible rewards."[7]

> *Sanctions discourage people from breaking the rules. An effective application of sanctions ensures that no participants can routinely or flagrantly break the rules without suffering consequences. Conversely, rewards can be arranged for those who contribute to maintaining and enforcing rules for using resources.*
>
> *Without a system of sanctions and rewards, it's difficult to encourage behaviors that produce sustainable outcomes. Sanctions can include an array of incentives designed to influence behavior, the most severe including things like being fired or receiving a downgrade from a rating agency. But sanctions will work best in a series of graduated steps that start with a light touch and progress in intensity as a behavior increasingly deviates from the established and agreed rules.*
>
> *Governments have traditionally relied on heavy-handed methods of enforcing compliance with policies (e.g., fines and citations), but researchers have found that more subtle approaches can be more effective. For example, benchmarking information or other commonly accepted best practices can be used to demonstrate that financially sustainable decisions are widely practiced by other governments.*

Due to public decision-making powers held by the government, rent seeking behavior may occur around the authorities. Rent seeking is the activity of individuals or interest groups to influence public decision-making or government officials in order to seek their own economic profits. Although rent-seeking's initial motivation stems from the economic profits of non-government officials, the production of behavior is closely tied to the monopoly power from government administration; and its ultimate beneficiaries are often the government or government officials. So, rent seeking is a vital factor that leads to inefficient government regulatory mechanisms.

The "rent," or "economic rent" does not refer specifically to payment on a lease but rather to Adam Smith's division of incomes into profit, wage, and rent. The economic rent represents the balance to the actual income of an element

---

[7] ***See:*** Kavanagh S. Institutional design principles for sustainable communities. GFOA.Org (Government Finance Officers Association of the United States and Canada). August 2, 2017. Retrieved from https://www.gfoa.org/sites/default/files/GFR081728.pdf.

exceeding its opportunity cost. Under perfect competition, the income of any element must equal its opportunity cost.

When having developed a new product successfully, the enterprise can enjoy excess profits that are higher than the others. This behavior is called "rent creating." Rent creation is the driving force behind social development. Only when the profits exist, then entrepreneurs will innovate and improve productivity. Rent creation, therefore, is linked to innovation. When having seen the new product is profitable, other enterprises may flood into the market, thereby lowering product prices and dissipating excess profits gradually. This behavior is called "profit seeking." Profit seeking belongs to a regular market competition characterized by the pursuit of new added social economic profits, resulting in social welfare increase. But if pursuing vested socio-economic benefits, the behavior belongs to rent seeking. The direct infringement of property such as theft and robbery is the most primitive rent seeking behavior. In modern society, a common and advanced rent seeking is to maintain or redistribute vested benefits by the administrative means.

"Specific interests" are on the economic rent (see Footnote 3).

> *While a good government tends to be a pure public good for the mass of voters, there may be many advocacy groups that have strong incentives for lobbying the government to implement specific policies that would benefit them, potentially at the expense of the general public. For example, lobbying by the sugar manufacturers might result in an inefficient subsidy for the production of sugar, either directly or by protectionist measures.*

Of course, government agents may have specific interests and will seek the privileges provided by themselves in order to partake in the resulting monopoly rent. And alternative currencies is one of their specific interests. It now seems for the government to appear all three types of behavior (i.e., rent creating, profit seeking, rent seeking) in various scenarios of alternative currencies.

Major Counties' Stances on Facebook Libra

In 1976, Austrian-born British economist Friedrich A. Hayek published a pamphlet *The Denationalization of Money* to envision a system of private monies. Because he worried that political constraints in developed countries might prevent central banks from tackling the high inflation, so he argued that money-issuing should be opened to market forces.[8] And now Hayek's proposal was proved to be workable, after decades of being considered as a curiosity.

[8] ***See:*** Hayek F. The denationalization of money: An analysis of the theory and practice of concurrent currencies. In S Kresge (ed.) The Collected Works of F A Hayek, Good Money, Part 2. Chicage, IL: The University of Chicago Press, 1999.

Thanks to fascinating advances in cryptography within computer science field, it is straightforward to create privately issued money as a cryptocurrency. Private money is different from government-issued money (i.e., sovereign currencies) because it has no fiscal backing. Today, anyone with Internet access can use a bewildering array of cryptocurrencies as a means of exchange. While still representing a trivial fraction of all payments in the world economy, the shares may perhaps increase exponentially over the next few years. In particular, cryptocurrencies can become widespread in emerging economies with dysfunctional government monies.

The topic of currency competition among privately-issued fiat currencies sparked controversy in economics in the last 2 years. This topic raises positive questions.[9]

> *Will a system of private money deliver price stability? Will one currency drive all others from the market, or will several of these currencies coexist along the equilibrium path? Do private monies require commodity backing? Will the market provide the socially optimum amount of money? Can a unit of account be separated from a medium of exchange? ... Should governments prevent the circulation of private monies? Should the private monies be taxed? ...What is the best strategy to jump-start the circulation of a currency?*

Still, the controversy was quickly overwhelmed (in 2019) by another significant one relating to Facebook's Libra—digital forms of money issued by tech giants. So, the questions are changed to: what are the effects of cryptocurrencies on government monetary policy? How is monetary policy changed by the presence of alternative means of exchange? Can cryptocurrencies and a government-issued money compete? Should we revisit the role of governments as issuers of money? This controversy is no more than just from the academia, the industrial circle (e.g., the currency circle), as well as the world's central banks and nearly all countries' policymakers, included.

Before proceeding, I'm going to brief an introduction to Libra[10] I interested in first, which is more likely to improve data productive forces greatly.

> *Libra is a stablecoin-like digital asset and a blockchain-based financial infrastructure project that is led by Facebook. As per the promotional material, it largely targets the unbanked population (which accounts for around 1.7 billion adults across the globe, according to the World Bank) and will initially focus on cross-border remittance. .... Several media reports suggesting that Facebook ... will facilitate payments across its platforms – WhatsApp, Facebook Messenger, and Instagram – which have ostensibly amassed a combined 2.7 billion monthly users.*

---

[9] ***See:*** Fernández-Villaverde J. On the economics of currency competition. Voxeu.Org. August 3, 2017. Retrieved from https://voxeu.org/article/competition-between-government-money-and-cryptocurrencies.

[10] ***See:*** O'Neal, S. Facebook Libra Regulatory Overview: Major Countries' Stances on Crypto. CoinTelegrap.Com. July 16, 2019. Retrieved from https://cointelegraph.com/news/libra-vs-us-congress-all-there-is-to-know-ahead-of-hearings.

The new cryptocurrency will be governed by the "Libra Association [named Calibra]," a not-for-profit, Switzerland-based consortium, which includes Mastercard, PayPal, Visa, eBay, Coinbase, Andreessen Horowitz, Lyft, Farfetch and Uber among a total of 28 founding members, the majority of which are US companies. ... The Libra Association is itself controlled by the Libra Association Council. The council's initial members are the founding members, each of whom runs a validator node on the network and was required to make a minimum investment of $10 million to secure the position. Each $10 million investment grants an entity one vote on the council, per Facebook's rules, although no council member can acquire more than 1% of total votes (evidently, to prevent a monopoly).

According to its official white paper,[11] Libra uses a Byzantine Fault Tolerant (BFT) consensus approach, the open-source Blockchain technique, to help people and businesses transfer via the "unhosted wallet" mode—as convenient as sending a text message, for everyday use.

The purpose of Libra is for use. Stablecoins are cryptocurrencies designed to minimize the volatility of the price of the stablecoin, relative to some assets; for Libra, the assets refer to Facebook's and its founding members' billion-dollar reserve assets. The Libra network planned to bind a "basket" of major sovereign currencies (at least four as USD, EUR, GBP, and JPY). Libra uses BFT that does not support the purpose of mining coins. In other words, there is no blockchain innovation technically. BFT in the several criteria of comparison as "performance" is excellent, while for the BFT algorithms are with mathematical theorems that prove their correctness. Although it does not go decentralization, BFT represents multi-centralization. BFT supports to the multiple "validator node" alliance system, to a certain extent, and has the function of countering regulation from the bottom up. The Libra network unites the power from an entire industrial chain of human activities to have many scenes. Base on solving the pain point of cross-border payments (of which the average fee is around 7% accounting for tens of billions of US dollar each year), Libra aims to support about one billion adults who did not have bank accounts but use mobile phones to make payment via a wallet mode. This will be very popular in countries with hyperinflation (i.e., currency instability), such as Venezuela and Zimbabwe. These countries typically have no advanced payment instruments and a prohibition on the circulation of US dollars domestically.

[11] The first Libra Association white paper was published in June 2019. This link is the new version of Libra Association white paper, published in April 2020. The most significant departure that Libra 2.0 takes from the first version of the project is the movement away from a 'single token' model. Originally, the Libra network planned to operate with a single coin, the Libra Token, that would be tied to a 'basket' of sovereign currencies such as USD, EUR, GBP, JPY. Now, however, Libra laid out plans to launch a series of currencies. These include a number of stablecoins, each of them tied on a one-to-one basis to different kinds of fiat currencies; in other words, a USD Libra token, a EUR Libra token, a GBP Libra token, and a JPY Libra token. ***See:*** Libra Assoication White Paper. April 2020. Retrieved from https://libra.org/en-US/white-paper.

Since most of the founding members of the Libra association are US companies, Libra may represent a new type of globalism. It can, to some extent, help strengthen the US dollar because the current globalization partially limits the USD's physical attributes. So, the US' concerns only lie at compliance issues. In Europe, due to the existence of Euro (which is a super-sovereign currency having a similar to Libra—a sovereign country issues neither of them), it is somewhat no problem on European circulation. Next, let's take a glance at the major counties' stances on Libra.

1. The US: Negative, wary
   The US, as the most potent jurisdiction, when it comes to controlling Facebook, has expressed a lot of concern regarding Libra because of Facebook's previous data scandal.

   The congressional group wrote an open letter to Facebook founder and CEO Mark Zuckerberg, asking him to share details about the then-secretive project. In a series of questions, the Senate highlighted certain aspects of consumer protection, asking Zuckerberg how the company plans to protect users' financial information. The committee also asked the social media mogul whether Facebook shares or sells consumer data with unaffiliated third parties, something the company has been accused of before. Moreover, on June 18 [2019], the same day the Libra white paper was released, Rep. Maxine Waters, chairwoman of the House of Representatives' Financial Services Committee, called for Facebook to halt the development on its cryptocurrency. Specifically, Waters argued: "Given the company's troubled past, I am requesting that Facebook agree to a moratorium on any movement forward on developing a cryptocurrency until Congress and regulators have the opportunity to examine these issues and take action."
2. The European Union: Neutral
   The European Union government (not separate members) seems to treat Libra as a sign of inevitable evolution. In turn, spurred Brussels to get ready in advance.

   on July 7 [2019], European Central Bank (ECB) executive board member Benoît Coeuré said that financial regulators must act fast to prepare for Facebook's Libra. ... He concluded: "We [the financial regulators] have to move more quickly than we've been able to do up until now." According to Coeuré, the development of digital assets has exposed gaps in current financial regulations and underlines banks' slow rate of adoption of new technologies: "All these projects are a rather useful wake-up call for regulators and public authorities, as they encourage us to raise a number of questions and might make us improve the way we do things." Furthermore, on June 26 [2019], Reuters reported that EU banks could have an instantaneous payment system in place by 2020.
3. The UK: Open-minded
   The governor of the Bank of England, Mark Carney, has been keeping an open mind in regard to Libra. Specifically, Carney said Libra could have genuine use cases if it can conform to regulatory demands, saying, "Anything that works in this world will become instantly systemic and will have to be subject to the highest standards of regulations." Carney further suggested that Facebook's cryptocurrency can indeed solve some financial problems: "It's way too expensive to do domestic payments. It's way too slow, and that hurts consumers and businesses. It stifles innovation, and it's far too expensive to

send money cross-border, and there are huge financial inclusion issues related to that and costs related to that. So, while we are trying to address all these issues, we have to absolutely acknowledge the problem that they're trying to solve."
In addition, Christopher Woolard, the executive director of strategy and competition at the Financial Conduct Authority, Britain's financial watchdog, argued that ... the forthcoming cryptocurrency's "size and scale will pose questions for society and government more generally about what is acceptable and desirable in this space." Christopher Woolard said, adding, "Historically, this may have been a sector that has lived by the mantra of 'move fast and break things, but the issues raised here require deep thought and detail."

4. Russia: Neutral
   Russia has made it clear that it will not develop a separate regulation for Libra.
   The statement was made by Deputy Finance Minister Alexei Moiseev, who explained there will be uniform rules for all circulating digital currencies: "No one is going to introduce a ban. A large number of businesses ask when it will finally be possible to legally conduct an ICO [initial coin offering] transparently. This will definitely be regulated, permitted, and that is about it."
5. India: Hostile
   India, as the largest market for Facebook, has presented itself as one of the most hard-to-reach territories for Libra. In fact, Facebook has already announced it won't be offering its Libra product in India, at least in the near future.
   "There are no plans to offer Libra in India," a Facebook spokesman reportedly told the publication. "As you may know, there are local restrictions within India that made a launch of Libra not possible at this time." The company's representative was most likely referring to the cryptocurrency ban imposed by the Reserve Bank of India (RBI) in 2018. Previously, local publication The Economic Times reported that Libra may not be available in India due to the current ban of blockchain-based currency transactions.
6. China: Cautious
   Since introducing the blanket ban on cryptocurrencies in 2017, China has been favoring the "blockchain before Bitcoin" policy. China has recognized Libra as a potential threat to traditional forms of money. But China might use the concept behind Libra for its own monetary system.
   A former governor of the People's Bank of China, Xiaochuan Zhou, has already stated that Beijing may delegate the issuance of digital currency to commercial entities.
   Nevertheless, the former head of the PBoC has also stated that "China should take precautions" against the trend represented by Facebook's Libra. Indeed, according to media reports, China's central bank has started developing its own digital currency in response to Libra. Wang Xin, director of the PBoC research bureau, ... said that the bank decided to create its own digital currency specifically because of the unclear role of the U.S. dollar once Libra is issued: "If the digital currency is closely associated with the U.S. dollar, it could create a scenario under which sovereign currencies would coexist with U.S. dollar-centric digital currencies. But there would be in essence one boss, that is the U.S. dollar and the United States. If so, it would bring a series of economic, financial, and even international political consequences."
7. Japan: Cautious
   Japanese regulators seem to waste no time in regard to Facebook's forthcoming crypto project. According to Cointelegraph Japan, local authorities

have already started investigating the impact of Libra ahead of the aforementioned meeting of the G-7 finance ministers in France.

While it is currently unclear how Japanese regulators will treat Libra as a result of their discussion, the central bank has already expressed its concerns regarding the tech giant's project. Thus, a Bank of Japan official has reportedly stated that Libra will pose a risk to financial systems by "piggybacking for free on a financial system that takes heavy costs." The central bank representative added, "It will move money into an absolutely virtual world, so it is completely different than other forms of digital payment."

8. South Korea: Cautious

   South Korea's financial regulator, the Financial Services Commission (FSC), published a report titled "Understanding Libra and Related Trends" ... reportedly stresses that little information regarding the project has been revealed [in its official white paper], and there is a possibility that personal data might be leaked. Meanwhile, Libra has a higher chance of being successfully commercialized compared to other cryptocurrencies due to Facebook's unprecedented large audience, the FSC adds. Moreover, the watchdog argues that "Libra may turn into a money laundering solution," if the project is not regulated.

9. Singapore: Neutral

   Singapore is one of few countries that have ostensibly held discussions with Facebook about its upcoming Libra.

   According to Ravi Menon, the managing director of the Monetary Authority of Singapore, Libra wields a number of potential benefits, such as making payments cheaper or supporting the unbanked. However, Menon said the regulator needs to understand exactly how the tech giant's system will work. Additionally, Singapore would be seeking further assurances on security and privacy issues before making a regulatory decision, he added.

10. Switzerland: Relaxed

    According to Reuters, Switzerland's financial regulator, the Swiss Financial Market Supervisory Authority, has also been in touch with Facebook regarding Libra. Thomas Moser, alternate member of the governing board at the Swiss National Bank, reportedly said the tech giant has indicated it is "willing to play according to the rules." He concluded, "I think it's an interesting development and I'm pretty relaxed about it."

    ... Switzerland's position regarding the cryptocurrency should be crucial for the project, given that the Libra Association is based in Zug.

Regardless of the views from the governments, in fact, Libra is doing "rent creating" by numerous industrial applications via commercial innovation.

Options of Sovereign Digital Currencies

I mentioned in Chap. 3: the head of the IMF (International Monetary Fund), economist Kristalina Georgieva, listed sovereign digital currencies as the organization's top priorities in 2020.[12] An article Gopinath penned at the *Financial*

[12] ***See:*** Daily Hodl Staff. Head of IMF Highlights Digital Currencies As Top Priority in 2020 for Its 189 Member Countries. TheDailyHodel.com. January 16, 2020. Retrieved from https://dailyhodl.com/2020/01/16/head-of-imf-highlights-digital-currencies-as-top-priority-in-2020-for-its-189-member-countries.

Times emphasizes central bank digital currencies (and cryptocurrencies issued by tech giants) could help improve the "slow, costly and burdensome"[13] status quo of cross-border payments. Also, a recent research estimates:[14]

> *Cash may not fully disappear everywhere in the next twenty years, but cash will sooner or later be gone, or be discontinued by decree when maintaining the infrastructure of coins and banknotes for ever fewer and smaller transactions would make losses.*

Apart from Georgieva's pleadings, Coeuré from ECB Executive Board treated the 2019 Libra project as the finial "wake-up call" for sovereign digital currencies. An expected comprehensive changeover from traditional cash to safe and secure base money will be similar to the paper money reform of the nineteenth century, when national legal-tender notes replaced private or regional banknotes.

There are currently seven consensus options for introducing sovereign digital currencies (see Footnote 14).

> *Direct access to an individual central-bank account, custodial access to central-bank account balances, digital currency as mobile money, central-bank issued cryptocurrency, synthetic digital currency as a stablecoin based on central-bank money, indirect digital currency as e-money issued 1:1 against central-bank money, and indirect digital currency as 100%-reserve bankmoney.*
>
> *These categories result from (1) whether it is about an account-based solution or mobile central-bank money or cryptocurrency, (2) whether the access to digital currency is direct or indirect, and (3) if indirect, whether the respective PSP [payment service provider] or bank acts as a custodian of customers' digital currency or as an intermediary that changes paid-in cash and reserves 1:1 into their e-money, promising to re-convert the money on demand.*

While being feasible, the above solutions all belong to a simple changeover from traditional cash to its digital form rather than a reflection of advanced digital currencies. First, these solutions still allocate medium businesses to commercial banks as a traditional mode of money market. For most of us, there might not be an immediate difference in our daily life: we still pay with money. But banking is in dire straits. Banks have been already suffering from the increased competition from traditional credit and saving operations. Some examples are (see Footnote 6):

> *Banks ... shrunk ... margins in the United States from almost 5 percent (in 1994) to around 3 percent (in 2016). Researchers at the German central bank predict that the interest spread staying low, only one in five banks in Germany will earn a decent return on cost of capital in the coming years. ... In 2015 alone, one in ten private banks*

[13] ***See:*** Gopinath K. Digital currencies will not displace the dominant dollar. Financial Times. January 6, 2020. Retrieved from https://www.ft.com/content/e5dd66b8-2ca0-11ea-84be-a548267b914b.

[14] ***See:*** Huber J. The Age of Digital Money. InternationalMoneyReform.Org. January 31, 2020. Retrieved from https://internationalmoneyreform.org/news/2020/01/theory-of-dominant-money-part-ii-the-rise-of-sovereign-digital-currency-cbdc.

in Switzerland vanished, including centuries-old brands such as Bank La Roche (founded in 1787). The sector's crisis is affecting well-known names, too: Germany's Commerzbank will cut one in five positions by 2020, and Italy's large UniCredit bank will close 26 percent of its branches.

But, "as if they aren't bad enough," as Mayer-Schönberger commented on this, another "element looms on the horizon. It may topple even some of the supertankers in banking, and it has to do with the role of money."

With the rise of data-rich markets, our view of the economy may evolve. Rather than equating markets with price and money, and the economy with finance, the market will be understood by rich data flows. For example, the revised Payment Service Directive as an EU Directive, administered by the European Commission, to mandate that any bank must, if a customer requests it, provide competitors as well as third parties with the customer's digital data that banks hold. The directive aims to transfer more dimensions of information instead of a credit score and the balance when bank customers are switching banks. If having access to rich banking data, the new bank can continue to assist in decision-making following consumers' preferences (like or more than the prior bank). It is the example of a disentangling of traditional banking and value generation based on sufficient and comprehensive information. The advantage of Libra is not an alternative payment choice but lies in many application scenes created by data-rich markets. Designing the newly sovereign digital currencies should consider rich information through a new way to convey it. I hope the world's central bank can utilize more dimensions of information when revising the above solutions for issuing sovereign digital currencies; otherwise, such a profit-seeking behavior is not significant.

Government Roles in Data as Innovative Currency

Governments are one of the largest providers of generating data—and one of the few that deliver data to the public free of charge. Governments already regulate how organizations may use personal data and myriad other issues related to data. For example, a United States social security number as the data in question is a similarly complex and sensitive data element. This number is originated by the Social Security Administration, a government entity and data provider, who generates the number upon birth or naturalization of a person. The person is also the data owner of the social security number, as this personally identifiable data is bestowed upon them. The person must utilize the social security number when accepting employment, appointing the employing entity as a second-level data owner over that data within their systems. Local, State, and Federal regulations govern how this personal information must be protected by the organization, and may require data localization to prevent the data from being stored or transferred internationally.

Governments are also one of the most diverse customer segments as data consumers in the data economy. In the past, governments spent a great deal of time, and efforts, and other fiscal resources to product data in better serving its citizen-customers, from demographics to statistics on people's behavior, from traffic patterns to web search trends. With accurate data available in the commercial and other markets, more and more governments tend to explore the benefits of the purchase of data and data products. For instance,[15] one could combine census data with data on consumer shopping trends to analyze public health issues, such as nutrition and obesity, and then connect the results to other health and social issues, such as localized infant mortality rates or high school dropout rates. The trick will be to strip out personally identifiable information (PII) in order to protect privacy.

When governments switch between the two toles, the practices, such as more open and transparent access to the data governments provided as well as more data products governments bought, directly challenge the clientelist and rent-seeking practices built around governments' decision-making. This is because the practices are based on data as innovative currency. As a new form of money, data offers the promise of new wealth for the private and public sectors alike, and more opportunities to government rent-seeking.

The US government's work on open data access is the thing people take delight in talking. Open access as it originated from academia was an attempt to break the monopoly of scholarly publishing. In November 2004, the US Congress passed a bill for the National Institute of Health (NIH) to force each sponsored researchers revealing the finial peer-reviewed manuscript freely available to the public. American Cures Act further made open access compulsory to agencies that perform biomedical research. Open access, then, targets at public administrative data or governmental affairs data. In January 2009, on his first day in office, President Barack Obama signed the Memorandum on Transparency and Open Government. The memorandum ordered federal agencies to provide their mountains of data to the public through open application programming interfaces, paving the way for dynamic enterprises that organize public data for social good. Enterprising citizens can build real-world solutions out of data. Data from sources as disparate as crime records, reports of power outages, and personal accounts of corruption tell a story to those who can translate it (see Footnote 15). We did observe a radical trend towards networked behavior such as crowdsourcing and co-creation, driven by the emergence of the open-source software community, the general use of social networks, and increased availability of open government data. One app, for example, designed by Silicon Valley-based Palantir, matches patients to clinical trials. Another,

[15] ***See:*** Eggers WD, Hamill R, Ali A. Data as the new currency. Deloitt Insights. July 24, 2013. Retrieved from https://www2.deloitte.com/us/en/insights/deloitte-review/issue-13/data-as-the-new-currency.html.

from the University of Rochester, overlays disease incidence data from the Centers for Disease Control, plus related tweets, on a map to track the spread of illness. A similar solution traces the path of a recent salmonella outbreak. Maya Designs has used the US Department of Agriculture Food Environment Atlas to highlight sources of cheap vegetables in America's "food deserts," areas lacking supermarkets or large grocery stores. One Code for America fellow in Boston noticed that homeowners shoveled snow from their sidewalks but left fire hydrants buried. This led to Boston's "Adopt-a-Fire Hydrant" app, which allows citizens to commit to clearing snow from a fire hydrant, to keep it clear for fire department access. Because Code for America's programs is open source, other cities have adopted the app; Honolulu uses a updated version of "Adopt-a-Fire-Hydrant" to keep citizens checking batteries on its tsunami warning system, Seattle to get them clearing storm drains, and Chicago to organize volunteer snow shoveling. In San Francisco, a website called Crimespotting is an interactive map for visualizing criminal records with the real-time updates, which situates crime scenes to alert tenants or people searching for apartments and also those who go home late at night.

In 2009, the UK government made its first public sector data assets available. At first, the initiative gave citizens and the media the chance to uncover poor performance and behavior. Despite the market is relatively immature, a research based on statistics for more than 37,000 data sets from three of the largest official open data portals in the UK revealed that (see Footnote 15),

> *the quantity and quality of open data in the UK had reached the critical mass necessary to trigger a step-change in attitudes. Businesses in all industries can now find relevant open data and use it to improve their products and services. Thus, new business models are beginning to emerge: Suppliers, aggregators, developers, enrichers, and enablers. And new businesses, like Placr, ELGIN, Locatable, and Mastodon C, are delivering new products and services predicated on the insight they deliver from open data.*

Besides encouraging commercial and social innovation, governments might also use benefit from the data, in particular, to replace some of their traditional data gathering activities. For instance (see Footnote 15),

> *the US government spent $13 billion on the 2010 US Census.21 That included salaries paid to more than 565,000 temporary workers who conducted in-person interviews with millions of households. But data aggregators such as Acxiom, ChoicePoint, and Rapleaf already offer much of the information the Census Bureau needs, including demographic, lifestyle, financial, and other personal data on individual households, not to mention lists that match individuals to address, age, gender, household makeup, country of origin, and race.*
>
> *To be sure, enlisting a private aggregator to help with a census count could draw challenges focused on everything from privacy and legal constraints to issues of transparency, data accuracy, and reliability (not to mention some public uneasiness with the data aggregation industry itself). Nevertheless, as these databases become larger and more reliable (Acxiom's database includes 126 million households and 190 million individuals in the United States alone 23), it is worth exploring whether governments*

could conduct a more efficient and less costly census by tapping into third-party data. If nothing else, such data might be used to prefill census forms that citizens would edit as necessary.

... Much of the information that agencies need already resides on the servers of other government agencies. Better sharing among agencies could reduce costs by eliminating redundant data collection. A ... study by the London-based think tank Policy Exchange concluded that the United Kingdom could save more than $56 billion a year by making better use of personal data that citizens had already volunteered to various government agencies. By promoting better inter-agency information sharing, the UK could eliminate the national census, according to the study, creating approximately $800 in savings per citizen per year and troubling the population with far fewer data requests.

According to the statistics,[16] the number of initiatives relating to open government data has grown from 2 to over 300 in the period 2009–2013, and membership in the Open Government Partnership has gone from 8 to 59 countries in 2 years. Over 280 government data catalogs have been published and over a million datasets have been released by governments around the world, spawning new businesses and social projects. For instance, the European Commission launched an Open Data Strategy for Europe in December 2011, including directions regarding charging rules, licensing, search on data portals, and interoperability. It is expected to deliver around 53 billion EUR boost to the EU's economy each year. In May 2013, the Obama administration made open and machine readable the new default for government information, which is expected to benefit to a wide range of domains, including health, energy, education, public safety, finance, and global development.

While making significant progress in open data access, it should notice that for now, governments have focused mostly on its scale and network effects without any feedback effect on utilizing the data. Both governments and users didn't know what a particular advantage of datasets may quickly reveal, whether two or more datasets are complementary, and how to compare the value among multiple datasets. Services based on feedback can "buy" more innovation at diminishing cost as the user base grows. The input may include: how many algorithms or data models did you run the dataset? How much or how little were your needs met when using this dataset (and what if yes or no)? Would you go back to change another dataset so that there was more valuable for you (and why)? Such feedback we may refer to the articles of data descriptors published in emerging data journals like Data in Brief. We also can use literature mining as an automatic method to develop the question bank of the feedback. All comments as feedback may help governments to evaluate opened data, and will be useful for future users. Some firms that analyze these comments may thus spawn new businesses. We need to consider the principles of "graduated sanctions and credible rewards" when designing feedback.

[16]***See:*** Jetzek T, Avital M, Bjorn-Andersen N. Data-driven innovation through open government data. Journal of theoretical and applied electronic commerce research. 2014;9(2), 100-120.

Cyber Encounters: More Role of Government

In the prior scenario, I introduced crucial tow roles of government play in the data economy as a data provider and data consumer. Governments also can do a data producer and regulator. So, in this section, I'm going to talk about a data sovereign when meeting a government as the latter two roles. Appling public choice on the scenario of cyber encounters needs to consider "political stance" and "bureaucracy." The founder of public choice theory James M. Buchanan suggested that public choice to be interpreted as "politics without romance" (see Footnote 3), an important truth that politicians are "no less selfish than the rest of us."[17]

A bureaucrat is a member of a bureaucracy. Bureaucrats are the executor of government public policy, who are usually appointed by politicians elected by the voter. Therefore, bureaucrats are directly responsible to politicians rather than the voter. Politicians and bureaucrats implement division of labor based on functional specialization following the principle of separation of powers between decision-making and executive. This makes bureaucrats relatively independent in implementing public policies. While individual voters cannot influence a specific bureaucratic behavior directly, they may do this by lobbying politicians to enable bureaucrats to make arrangements on public policies that benefit some individuals and interest groups. Compared with the agent of private-sector organizations, a bureaucrat is subject to much less institutional constraints because the bureaucracy features giant monopoly and difficult supervision without property rights constraint and cost analysis. It may result in procrastination and inefficiency of work styles, to highlight inside lag, and even to show net welfare loss caused by the excessive provision of public goods. Of course, the application of a rational actor also can model to the bureaucrat. Bureaucrats seek to maximize their personal interests during the tenure, and one vital way to achieve the goal is to maximize public budgets. Under the pressure of interest groups, therefore, every governmental department is possible to emphasize and even exaggerate the importance of themselves to achieve the department budget maximization. For example, the defense ministry can exaggerate the potential threat to increase the defense budget. Another example is the environmental protection agency. It may emphasize people's concerns about the environment to require additional governance funds.

Here may involve two of five principles as "monitoring" and "conflict-resolution mechanisms."

For clarity, I'm going to give a practical example first.

---

[17] ***See:*** Roberts R. Michael Munger on EconTalk's 500th Episode. EconTalk.Org (Podcast). Library of Economics and Liberty. November 23, 2015. Retrieved from https://www.econtalk.org/michael-munger-on-econtalks-500th-episode.

Effective monitoring discourages participants from breaking the rules. For example, experiments have shown that participants are more apt to cheat on a test if the lights are dimmed, or if the subjects are simply wearing dark glasses while taking the test. Conversely, they cheat less when there is a cartoon picture of an eye nearby. The point is that people are more prone to dishonest and self-interested behavior when they believe they cannot be seen and their reputations are not on the line.

Practicing constructive conflict-resolution methods means that conflict becomes a way to solve problems, not exacerbate them. The best way to avoid destructive conflict is to prevent it from happening in the first place. Research shows that if participants in a decision-making process believe that the process was fair, they are more likely to support the decision, even if the decision is not in their own interest. But disagreements will sometimes run deeper and be so intractable that conflict is unavoidable. In these cases, having structured mechanisms to help parties resolve disagreements constructively can help the organization move forward.

For clarity, I'm going to give a practical example first.

Food is always a pursuit. The global online food delivery services market grows fast, of which the most significant segment is Restaurant-to-Consumer Delivery, with a market volume of tens of billions US dollars. Delivery orders owned by the costumer are often placed through a mobile app (e.g., Uber Eats, Ele.me). These data are of use to these app providers for contact tracking the valuable insights restaurants need, such as improving their bottom line of services, expanding the customer base, and even optimizing their menus. Also, the data can be used to shorten the distance between farm and table, particularly in helping dairy farmers in the COVID-19 era. Cookie-based related to food (and ingredient) consumption for the costumer belong to a shallow data application; and people-based products for restaurant franchise optimization use more advanced data techniques. These app providers are not only to collect transaction data, but provide information via data products for decision making to all stakeholders, and themselves included. As a data producer and data sovereign, app providers have the responsibility to protect the consumer data they controlled; at least, the data includes similarly complex and sensitive data elements such as the mobile number location. Local, State, and Federal governments as regulators govern how this personal information must be protected by the data sovereign, and may require data localization to prevent the data from being stored or transferred internationally. Governments has a legitimate reason to keep this data private and to not allow it to be part of the global data economy.

So far, governments at all levels only have one role as a regulator. Suppose that the local (e.g., provincial) government purchased some restaurant-related data without individual privacy from those app providers. Based on combining some government data, the provincial government highlights the featured restaurants in a live map, such as the ranking information of employment and taxes. The provincial government, thus, can be a data producer. Note that not all data producers can be identified as a data sovereigns because there exists a prerequisite that is the ability to possess data. If satisfied with this, the provincial government also can be a data sovereign like the app providers. Our issues, thus, are (1) how the government treats the other data sovereigns with the assumption that the government is a data sovereign, and (2) why the government needs to coordinate data sovereign-involved disputes in data-rich markets.

Government as a Data Sovereign

A typical case of government as a data sovereign is Datapalooza. Someone may question it is just a conference. Actually not. The Health Datapalooza hosted in Washington, DC, was born out of the US Federal government's commitment to open data (in June 2010), aiming to "catalyze the development of an ecosystem—an ecosystem that leverages data to improve health" (see Footnote 15). Todd Park said when propelling the organization (see Footnote 15). Park was the chief technology officer (CTO) for the US Department of Health and Human Services (HHS) at that time, and then served as US CTO appointed by the Obama administration since 2012.

> *Each program, if successful, promises to save or improve lives. Health care data could add billions to the nation's economy, says Park, and he wants to attract more innovators to use it. As founder of successful health care management start-ups, he knows an opportunity when he sees one.*

Having released troves of data, the HHS uses Datapalooza to debut some of the best healthcare-related web and smartphone apps driven by open government data. Datapalooza continues to enjoy the greater and deeper engagement from the HHS' leaders and agencies than any meeting of its type. It also gave rise to the Health Data Consortium, a public-private partnership that now manages the Health Datapalooza and works to promote the availability and innovative use of open, machine-readable health data. And it later became a model for the White House—to showcase how freely available government data might fuel innovation and jobs in other fields, including energy, public safety, and education. In 2016, Datapalooza became a Patients Included accredited event. That means that Health Datapalooza is committed to incorporating the experiences of patients as experts while ensuring they are neither excluded nor exploited. This point is unique because some norms often block the connections between health professors and patients in non-medical scenes. Take social media as an example; physicians and patients have two social networks, respectively, with little or no intersection.

Datapalooza, as a data sovereign via the efforts of a decade, is with hundreds of mHealth data applications, enabling thousands of individuals to self-manage their diseases (e.g., asthma), and health systems to provide real-time, data-informed clinical support for their patients. When facing other data sovereigns (e.g., healthcare providers), Datapalooza offers a collaborative approach to help work with the state and federal governments together, itself included. And the relationship is evolving to eliminate outside threats or scattered regulations and then to make it easier to merge conventionally fragmented healthcare markets. This kind of market behavior is organized by a dialogue with various level governments, covering more than a thousand agency leads, academics, innovators, advocates, and others. Datapalooza also helps to check the difficulty of competition between new entrants and traditional healthcare providers. It, to a certain extent, reduced the lobbying of interest groups and improved the efficiency of the bureaucracy, and further prospers the market radically. So, Datapalooza is a great model.

The current deficit financing and high debt are a common problem faced by governments worldwide. This is a stubborn illness brought by financial capital, and both the cause and results of the global debt and economic crisis. The non-money-based wealth, as well as data capital, created by the government as a data sovereign, may facilitate reducing fiscal deficits globally.

Government Being Supervisory Role

The case of government as a data sovereign may have adverse effects. Especially in emerging economies, the government's role in this process is like both an athlete and a referee, possibly resulting in markets to lose the right features. Perhaps the most important of the markets' principle is the diffused nature of decision-making. If this principle is violated, a market cannot coordinate human activity efficiently. In any case, the primary function of governments lies in the supervision under proper conflict-resolution mechanisms, such as arbitration of disputes and review of lousy market behavior.

An archive is an accumulation of historical records via physical facilities in which they are located. Professional archivists generally understand archives as "the secretions of an organism,"[18] which are distinguished from documents written or created to communicate a particular message to posterity. Archives contain primary source documents of regular legal, commercial, administrative, or social activities.

The Internet also has its archives. In 1996, Brewster Kahle founded the "Internet Archive" which has begun to archive and preserve the World Wide Web in large quantities. The Archive is a nonprofit operating in the US, and has an annual budget of US$10 million derived from various partnerships, government grants, periodic funding campaigns, donations, and the Kahle/Austin Foundation. The Archive now expands its collections beyond the Web archive,

[18] *See:* Galbraith VH. Studies in the Public Records. In Series: University of Oxford. Ford Lectures. London: T. Nelson (1948). 1948: 163 pages.

starting with the Prelinger Archives (that is a collection of films relating to US cultural history). According to its website:[19]

> *Most societies place importance on preserving artifacts of their culture and heritage. Without such artifacts, civilization has no memory and no mechanism to learn from its successes and failures. Our culture now produces more and more artifacts in digital form. The Archive's mission is to help preserve those artifacts and create an Internet library for researchers, historians, and scholars.*

The collections of digitized materials held by the Internet Archive include "over 20 million books and texts, 3 million movies and videos, 400,000 software programs, 7 million audio files" (see Footnote 19), and "450 billion web pages (downloads of URL [as Uniform Resource Locator] linked pages and metadata)."[20] In addition to its archiving function, the Archive is an activist organization, advocating a free and open Internet.

A recent specific historical website captures the Internet Archive to store .GOV—Government on the Internet (see Footnote 20).

> *For nearly 20 years, governments have used the web to share information and communicate with citizens and the world. .GOV is an archive of nearly two decades of content from .gov domains (the US federal, state, local) organized into a database format that is nine times larger than the entire print content of the Library of Congress (90 terabytes, [9×2*^{40}*]).*
>
> *The Internet Archive also curates sub-collection: .GOV. .GOV contains approximately 1.1 billion page captures of URLs with a .gov suffix (from 1996 through Sept 30, 2013). At the federal level, this includes the official websites of elected officials, departments, agencies, consulates, embassies, USAID [the US Agency for International Development], missions, and much more.*
>
> *.GOV offers four types of data from each webpage capture: the link data (the page URL and every other URL/hyperlink found on the page); the parsed text of the page; the full content of the page (the text including HTML [Hyper Text Markup Language] markup language; images; video files etc.); and the CDX index file that is used to access the page via the Wayback Machine.*

Apart from archiving, the Internet Archive also does data cleaning and subsequent data production. Some existing applications involve government attention to the financial crisis, terrorism, and climate change.

The Internet Archive as a data sovereign recently received an attack from the writers' union and digital publishing.[21] The US National Writers Union (NWU) presented a public informational webinar on "What is the Internet Archive

---

[19] ***See:*** (1) Internet Archive. Retrieved from https://archive.org; (2) The entry of "Internet Archive" on Wikipedia. Retrieved from https://en.wikipedia.org/wiki/Internet_Archive.

[20] ***See:*** Gade EK, Wilkerson J, Washington A. The .GOV Internet Archive: A Big Data Resource for Political Science. March 16, 2017. ThePoliticalMethodologist.Com. Retrieved from https://thepoliticalmethodologist.com/2017/03/16/the-gov-internet-archive-a-big-data-resource-for-political-science.

[21] ***See:*** Hasbrouck E. What is the Internet Archive doing with our books? Nwu.Org. April 16, 2020. Retrieved from https://nwu.org/what-is-the-internet-archive-doing-with-our-books.

doing with our books?" on April 27 and May 5, 2020, to speak out against the Internet Archive's ongoing and expanding book scanning and e-book bootlegging practices.

> *The Internet Archive's book scanning and e-book bootlegging have never been limited to the practices its supporters have described as Controlled Digital Lending, or those it now describes as a National Emergency Library. The Internet Archive's actual uses and re-distribution of unauthorized copies of images (and audio generated from them) of pages scanned from books are, and have long been, more extensive, less controlled and more damaging to authors' incomes.*

The NWU listed five distribution channels to indicate how are authors (and publishers) harmed by this. These are (1) downloads via OpenLibrary.org of e-books assembled from page images, (2) audiobooks generated from images of scanned pages, (3) viewing of page images on OpenLibrary.org, (4) Viewing of page images on Archive.org, and (5) APIs for automated downloads of page images.

The answer from Kahle appears to support the defenders of the Internet Archive is because "many of these books are no longer available for sale in the original book form." It may be mostly true; there's no effect on the sales of the original books if they are no longer available for purchase in that format. And of course, people aren't likely to buy a copy of a book if they want just an excerpt, photo, or illustration.

> *The real lesson though, is this: If these works are being distributed and read primarily as individual pages rather than as entire books or e-books, and most of them aren't available in the original book formats, the only reason to talk about "book sales" is to set up a straw man that can easily be knocked down – and to distract attention from the real issues of clickstream diversion and undermining of revenues from digital rights.*

It is unknown whether any government will intervene or how they resolve this dispute. But I recommend that such government intervention is necessary. In essence, this is a unavoidable collision between government authority and data sovereignty. The collision is challenging in not only the organizational capacity, the staffing, and the required investigative and enforcement authority, but also the appropriate expertise on data science. As Mayer-Schönberger likes to point out, data-rich markets are vulnerable to a dangerous concentration of power. Government intervention may partially eliminate the excessive concentration of power from data sovereigns. Take the dispute between the Internet Archive and the writers' union as an example; five distribution channels proposed by the NWU almost involve data usage. "On the Web, clicks are money" (see Footnote 21). Pageviews can be monetized in many ways; for example, generating revenues from Web content through advertising and fees for downloads, subscriptions, or licensing. The fact that free to the readership does not conflict with creating income for the author and publisher. If government agencies want to be responsible for active intervention, an efficiency-enhancing regulatory framework, and more advanced data technologies with supporting policies, are required. Whether an existing agency is asked to expand its jurisdiction or a new one will be created depends on the national context. This recommendation does not call for a bureaucratic expansion lightly.

Chapter 9
Rethinking Economic Indicators with Data-Informed Insights

There exists a consensus about the 2007/2008 global financial crisis that shook not only the global economy but the foundation of economics itself. In 2011, a group of Manchester economics students as rethinkers, who were dissatisfied with the inertia of the economics discipline after the crisis, founded the Post-Crash Economics Society.[1]

> *[They] went to study economics hoping to understand why the financial crash happened; why was unemployment so high; why were public services being cut? The reality was a narrow, dispassionate exercise in some fairly basic calculus, coupled with a requirement to perform multiple-choice exams twice a year. Theory was not applied to the real-world, students worked with hypothetical apples and oranges rather than real case studies, past economic events were ignored and, perhaps most importantly, the theory that had led to the surrounding economic chaos was being imparted as if it was gospel, with no opportunity to evaluate its validity.*

And now the rethinks have developed to a global body of students, academics, and professionals.

Each economic crisis can stimulate people to reexamine our view of the world and methodology as well, especially economists. But in fact, economics was once in success. The success stems from a restriction of its objective: it endeavors to explain the regularities in the behavior of objects. But when its aim has been expended on attempting to explain society, economics are a bit powerless. This is in line with the great success of physics; physics does not endeavor to interpret nature. As a result, some economics argued that,[2]

> *the quantity theory of money is not used to understand inflation or deflation globally, and even for an individual country only. Purchasing power parity does not figure to*

---

[1] ***See:*** (1) Rethinking Economics. May 5, 2014. Retrieved from http://www.rethinkeconomics.org; (2) The entry of "Rethinking Economics" on Wikipedia. Retrieved from https://en.wikipedia.org/wiki/Rethinking_Economics.

[2] ***See:*** Xiang, S. The New Capitalism: Rising, Crisis and Future of the Global Financial Capitalism. Beijing: China Citic Press; 1 edition, December 1, 2014, 616 pages.

© Springer Nature Switzerland AG 2021

C. Tang, *Data Capital*, https://doi.org/10.1007/978-3-030-60192-8_9

predict the impact of exchange-rate fluctuations on a country's economy. Interest rate parity cannot govern the relationship between interest rates and currency exchange rates in various countries. Fluctuations in exchange rates and differences in productivity growth cannot explain the changes in global trade and the balance of payments. The Keynesian IS-IM (investment-savings and liquidity preference-money supply) model cannot guide to national macroeconomic policies. Risk diversification cannot show the direction of international capital flow. Marginal productivity theory cannot suggest wealth distribution and income inequality within worldwide countries.

While existing flaws, these Western theories we are familiar with are useful in interpreting some rules under certain assumptions. After all, the paradigm of a scientific assertion is "If so … then so."[3] We often forget the conditional nature of scientific laws is that those are stated in a shorthand manner, with the former part of "If so …" abbreviated or dropped. Because of this, the driving force behind the 2000 dot-com bubble is the global financial capital represented by Wall Street, such as private equity firms, venture capital, and mutual funds. They pursued the maximization of shareholder value; that was, to find one startup with the Tobin's Q ration less than 1 (i.e., Q Ratio equals the market value of a company divided by its assets' replacement cost). And, these funds conducted value revaluation to increase tech stock price and market value, and to create wealth myths. The cost of pursuing value with such an extreme simplified standard is huge:[4]

On January 14, 2000, the Dow Jones Industrial Average peaked at 11,722.98, a level it would not return to for more than six years. The tech-heavy Nasdaq peaked on March 10, 2000, at 5,048.62, a level it would not reach again until March 2015. From that March 2000 peak, all the way down to the trough it reached on October 9, 2002 (the bear market bottom would be 1,114.11), the Nasdaq would lose nearly 80 percent of its value.

At the very beginning of explaining economic regularities, simplification and modeling work smoothly. Simplification means reducing the amount of computation. Experience shows that the amount of computation involved increases at least as fast as the square level as the size of the problem increases.[5] For example, epidemiological methods often draw upon statistical mechanics, creating simplicity by dealing with people in the structureless mass, as interchangeable units, and taking averages. And the law of medium numbers can be applied, of which the importance is in the scope of its application instead of its power of prediction. Diseases have medium numbers of patients; clinical notes have

[3] ***See:*** Menninger K. Menninger Clinic monograph series. Theory of psychoanalytic technique. New York: Basic Books. 1958. https://doi.org/10.1037/10843-000.

[4] ***See:*** McCullough B. An eye-opening look at the dot-com bubble of 2000 – and how it shapes our lives today. Ted.Com. Retrieved from https://ideas.ted.com/an-eye-opening-look-at-the-dot-com-bubble-of-2000-and-how-it-shapes-our-lives-today.

[5] ***See:*** Weinberg G. An Introduction to General Systems Thinking. Wiley Series on Systems Engineering and Analysis (Book 4). New York: Wiley; 1 edition (January 1975). 1975: 279 pages.

medium numbers of vocabulary words. Translated into our daily experience, the law of medium numbers can be Murphy's Law:

Anything that can happen, will happen.

So, "the fruits of science are simple fruits, or more precisely, fruits of simplification" (see Footnote 5).

Modeling is a starting point of the process by putting "objects of interest" under "well-defined conditions" to delimit its application and to magnify its power of prediction. By finding general laws, people attempt to think in medium number systems to help recall and initial understanding. But faced with a complex system, some turn to the living system as a model. Knowledge about living systems was applied by analogy to other systems, to gain some points of leverage on the complexity. For instance, Thomas Hobbes viewed the state as a "body" of a giant, with the various "organs" representing the government agencies. Sometimes we can gain from modeling on a poorly known system. This is like Hobbes' use of the metaphor "body," but his knowledge of physiology that may be incomplete. "Every model is ultimately the expression of one thing we think we hope to understand in terms of another thing we think we do understand" (see Footnote 5).

If simplification and modeling are not done well, it may cause a false model or a model involving distortions. But some idealized models such as Galilean idealizations are ones that involve deliberate distortions: economists assume that agents are omniscient; biologists study isolated populations; physicists build models consisting of point masses moving on frictionless planes; and so forth. Neoclassical economics, especially Keynesian economics, for example, has not beautified the market. That means they view the market as an ugly woman, who needs cosmetic surgery by the government.

Simplification and modeling are obviously the methods to apply to insufficient data. A research article on *Science* indicates that,[6] the big data era began in 2002 because more digital data was stored than analog data from the year, despite analog data showed at the beginning of keeping records by human tying knots.

We estimated the world's technological capacity to store, communicate, and compute information, tracking 60 analog and digital technologies during the period from 1986 to 2007. In 2007, humankind was able to store 2.9×10^{20} optimally compressed bytes, communicate almost 2×10^{21} bytes, and carry out 6.4×10^{18} instructions per second on general-purpose computers. General-purpose computing capacity grew at an annual rate of 58%. The world's capacity for bidirectional telecommunication grew at 28% per year, closely followed by the increase in globally stored information (23%). Humankind's capacity for unidirectional information diffusion through broadcasting channels has experienced comparatively modest annual growth (6%).

[6] ***See:*** Hilbert M, López P. The World's Technological Capacity to Store, Communicate, and Compute Information. Science. 2011;332(6025): 60-65. DOI: 10.1126/science.1200970.

Telecommunication has been dominated by digital technologies since 1990 (99.9% in digital format in 2007), and the majority of our technological memory has been in digital format since the early 2000s (94% digital in 2007).

More data means less inference; further, the data can widely affect methods' use. In view of this pressure, it reminds me of a general system law—the *Used Car Law* (see Footnote 5). Consider the problem facing a car owner. His operating expenses as the cost of regulation consist mainly of gas and oil. If keeping a careful record of these expenses, the car owner may notice that the gas mileage is declining. It eventually will be profitable to have the engine turned due to the resulting savings on gas. He may also begin to find oil consumption, at which point it may pay him to install piston rings, and even buy a new engine. The reason for taking these drastic steps that are adaptations is economically justified. By adapting, the cost of regulating is reduced. But there may come a time where nothing short of buying a new car will suffice to keep down operating costs in the face of the ravages of time. So, the law states that: (1) a system that is doing a good job of regulation need not adapt. (2) A system may adapt in order to simplify its job of regulating.

Examples of the Used Car Law are ubiquitous. The straight-A student might not consider the path he will choose for his life. Only when he finds that dissatisfaction with his profession that is pressing his toward changes. We might, therefore, wish to rephrase the used car law to put specific emphasis on the pressure from big data: (1′) the economics no putting excessive stress from big data, does not need not be changed. (2′) The economics may be changed to reduce the stress from big data.

In other words, why do we continue expending mammoth efforts to repair old viewpoints for the world? Sometime we need to trade in the old. In this chapter, after glancing at a brief history of economic indicators, let's consider how to do replacement.

A Brief History of Economic Indicators

We live in a world defined by economic indicators as to the signals. The leading indicators brought the modern concept "economy." We treat our economic statistics as absolute markers of our success or failure. Most of them didn't exist before the 1950s. The content they measured is what their designers wanted when being invented. But the world is no longer what it was the same. Some media criticized that,[7] for example, the US labor department's economic numbers was at risk of giving misleading signals during the COVID-19 pandemic.

[7] ***See:*** Irwin N. The Economic Data Is About to Get Weird. NY Times. April 18, 2020. Retrieved from https://www.nytimes.com/2020/04/18/upshot/economic-data-distorted-coronavirus.html.

Another example is from some experts' arguments; they believed that,[8] much of the friction between China and the United States in recent years revolved around trade statistics and GDP (i.e., gross domestic product) figures. Nevertheless, leading indicators among them have profoundly affected numerous economic theories and models. Examining the evolution of these indicators is vital for understanding economic methods, especially for people (my included) without a deep understanding of economic theories.

A few leading indicators appeared in the 1920s. With the rage for hard data and statistics and these indicators in hand, people went about inventing more. The indicators invented before 1945 aiming to deal with the twin crises of depression and war, and the time after was marked by how to measure affluence. Various indicators made the global economy that more like a three-dimensional chessboard of countries, companies, and individuals, and then moved towards their capable of contributing to the private sector, productive sectors, scientific field, etc.

To know economic indicators, we need to go back even further in history to track the first attempts that understood the world in numbers. The main tone of this section I referred to Zachary Karabell's *The Leading Indicators: A Short History of the Numbers That Rule Our World.* Since Karabell is a far better writer than I, the interested reader is strongly advised to read his book—even if that means skipping this section.

Mathematics and the Census

While population estimates remain an essential function of a modern census, the census of thousands of years ago mostly involved land use because the land is the vital economic resource in the ancient. Figure 9.1 shows a redrawn version of the results map of China's earliest census for the land in the Xia Dynasty of Yu the Great.[9] China's earliest preserved census for the population was held in ad 2 during the Western Han Dynasty, and is still considered by scholars to be quite accurate. The population was registered as having 57,671,400 individuals in 12,366,470 households.[10]

The known *Domesday Book* finished in 1086 is also resulted from a land census. Its name derives from the Day of Judgement (see Footnote 8).

---

[8] ***See:*** Karabell Z. The Leading Indicators: A Short History of the Numbers That Rule Our World. New York: Simon & Schuster; Reprint edition (December 30, 2014). 2014, 304 pages.

[9] ***See:*** Shui Y. Expert opinions: Ancient China's census is first for the land, and then for the population. September 23, 2015. In Chinese. Retrieved from http://www.crnews.net/40/4343_20150923113150.html.

[10] ***See:*** Bowman JS (ed). Columbia Chronologies of Asian: History and Culture. Columbia University Press. 2000.

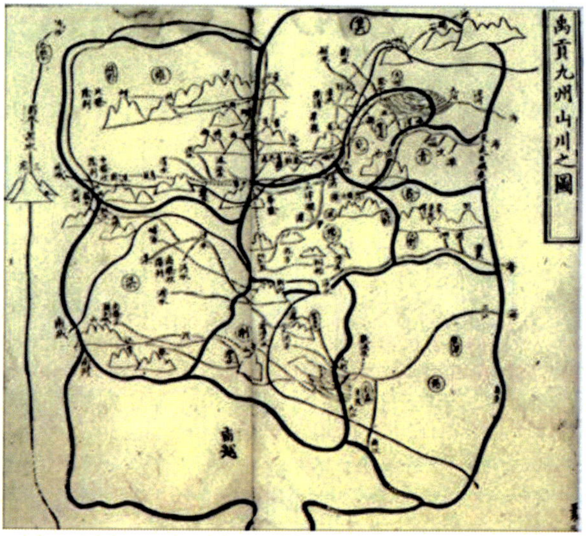

Figure 9.1 "Yu-Gong" (i.e., 《禹贡》) is the results map of an earlier census for landscapes in ancient China. Notes this version was redrawn in the Song Dynasty. (Sources: this picture was found on the Internet, see the reference (see Footnote 9))

> *"For as the sentence of that strict and terrible last account cannot be evaded by any subterfuge," explained a later courier, "so when this book is appealed to on those matters which it contains, its sentence cannot be quashed or set aside with impunity."*

In this book, material goods were the order of priority, and people, secondary. And the biggest flaws it suffered is that the Church was in various parts of the country omitted altogether. Nonetheless, the book was useful to allow King William the then conqueror understand the imperative of counting what he had (see Footnote 8):

> *how many weapons, how much grain, and how much his subjects could be taxed, what could be levied, and what armies could be raised.*

The main difference in censuses between millennium ago and today is that then, the flaws lay with insufficient models and an incomplete understanding of the distinction between counting and adding versus creating statistics. The changes need an opportunity that is to wait mathematics branched off from philosophy. When Isaac Newton and Gottfried Leibniz were exploring calculus, some seventeenth-century people began to think about the nature of probability and the necessity of understanding it to measure accurately a material world that was ever in flux, with wars and famines. They included Blaise Pascal and Pierre de Fermat. Pascal and Fermat investigated in their letters about the nature of gambling and almost all probability games of rolling the dice. With their shared love of the dice, their musings ranged widely, from how to determine the population of a dense city to how to account or adjust for inevitable errors in some tabulations. These letters, along with other nearly 80 mathematicians', became a part of Marin Mersenne's legacy when he died in 1648. French theologian

Mersenne was an unknown math enthusiast, and in 1635 he set up the informal, private Académie Parisienne. Académie Parisienne was a precursor to the French Academy of Sciences, where many of the leading mathematicians and natural philosophers of France shared their research. One hundred years later, Pierre-Simone Laplace entered the French Academy of Sciences as well. Laplace then penned his *A Philosophical Essay on Probabilities*, which contains some content from these letters (see Footnote 8).

> *In a concise essay of a few hundred pages, Laplace laid out the basic principles of why probability matters to everything from the selection of juries to accessing how long people live.*

Take "average value" as an example; Laplace showed how complicated seemingly simple figures can be (see Footnote 8).

> *For mortality figures, it is not enough just to take birth registers and death records, and then add them up and divide. Why? Because that will vastly overstate the chances of death at an early age. The logic is clear but easy to overlook. Far more infants and toddlers in those years failed to make it to adulthood. Their mortality rate was significantly higher, and so including them in an average significantly lowered average life expectancy. But once a child made it past those first dangerous years, life spans were considerably longer.*

In addition to the average value, Laplace illuminated another challenge of compiling accurate statistics. He suggested using a large sample to ensure assessing the full range of probabilities and vagaries of life and death. If only having a small sample from one area, the sample may be contaminated due to distorted results caused by regional factors such as disease, plague, or drought. Mathematics intervened in censuses resulting in not only fewer biases but a subject change—from material goods to people. But such a change lasted about seven centuries. A signal is the US Constitution embedded at least one statistic: proportional representation. That is on the idea that "each citizen should have his voice heard in national deliberations via an elected representative" (see Footnote 8). As a result, a clause was inserted into the Constitution for a regular census for the population to be taken every 10 years, in order to determine how many voters there were in each state.

> *... in Article 1, Section 2: ... Representatives and direct Taxes shall be apportioned among the several States which may be included within this Union, according to their respective Number, which shall be determined by adding to the whole number of free Persons, including those bound to Service for a Term of Years, and excluding Indians not taxed, three fifths of all other Persons. The actual Enumeration shall be made with three Years after the first Meeting of the Congress of the United States, and within every subsequent Terms of ten Years, in such Manner as they shall by Law direct.*

The "All other Persons" in this clause was to count slaves as "three-fifths" of a person.

Of all official statistics, the census remains the most time consuming and laborious to assemble. These initial census endeavors relied entirely on the door-to-door questions that individuals dispatched across the country to ask. The US first census began on the first Monday of August 1790. The federal government employed 650 marshals who fanned out across the 3 colonies. The operations of gathering and assembling data cost US$45,000, lasting 18 months. It evolved over the nineteenth century into far more than just a count of people. Those asked and obtained content are (see Footnote 8):[11]

> *for example, the number of people employed by the railroads, the number of shoe factories in Massachusetts, the average size of a household in Michigan, the education and literary level in Kansas City, and the number of free blacks in Baton Rouge in 1850.*

All the data generated from the US census was the foundation for the indicators that became ubiquitous in the twentieth century.

After the nineteenth century, people began to do more applied work based on the analog census data, and then augmented by sampling methods, by different questions, and by mathematics. Americans turned to mechanized agriculture, for example, and that sparked the need to figure out how much, and what, the industry was producing, on markets, grain exchange, prices, above all, yields. The first rudimentary census of agriculture was taken in 1840, followed by the Homestead Act in 1862. That was also the year of the creation of the Department of Agriculture. The new department went much further when including information about crops. The first commissioners traveled to Europe regularly to meet counterparts in London, Paris, Berlin, and Vienna, and to share methods and approaches. This was the transnational efforts at establishing common standards of statistical measurement. Nineteenth-century agricultural statistics were the precursor of contemporary economic indicators. The department only depended on hundreds of agents in the middle of the nineteenth century that increased to several thousand by the end of the century. That meant to let many people have a job.

The period was the history of mathematical intervention, from basic addition and subtraction, to mean calculation, then to probability statistics. But people at that time only relied on a large sample generated from the door-to-door questions, without any sampling methods. Over the centuries, the census turned its subject shifted from material wealth to demographics, added more content that was the foundation for contemporary indicators, and formed international cooperation between North American and Europe. Also, more censuses led more people find their job.

[11]*See:* US Departmentof Commerce, Bureau of the Cnesus. 200 Years of US Census Taking: Population and Hosing Questions, 1790-1990. Washington, D.C., 1989. Retrieved from https://www.census.gov/history/pdf/200years.pdf.

Vibrant Statistics Based Leading Indictors

Compared with the ambitions of the people who worked hard for the census assiduously, statistics at the end of nineteenth century were marked dubious quality and consistently inaccurate (see Footnote 8).

> *Almost from the moment official statistics began, the goal of drawing a perfect map of the world through numbers fell far short of the actual numbers assembled.*

Just like Jacob Dodge's understanding as the long-serving statistician of US Agriculture Department, "pure mathematics and immaculate judgment combined cannot cure the inaccuracy of erroneous original data." The statistical experts knew and know the flaws. As a result, some statisticians began working to let the numbers moved out of the obscure bureaus that assembled them and into the political and social limelight, to meet public demand of simplicity. When considering domestic inflation and future economic growth, for example, politicians weren't interested in sampling errors or multiyear revisions. They wanted a number to indicate "the truth" or at least a proxy for it.

But it is difficult to find a number. When the Great Depression began, no one truly knew how bad things were because most of the data were unavailable. There existed an official joke about employment and unemployment. Then-President Warren Harding convened a meeting to ask unemployment after a severe recession in 1920–1921. Too many divergent opinions about the number resulted in the attendees put it a voter because "the low end was three and a half million; the high end was five million."[12] A vote on a fact or on something that should be a fact. Yet the reality is that the answer is not a simple case of counting who has a job and who doesn't. Even though it is simplified to the task as simple as it seems, the questions include: what does it mean to be employed? Full-time work, part-time, temporary, seasonal? How about an actor between jobs or a ranch hand in fallow months? Suppose to ask a classical economist about unemployment, and the answer might be surprised. Until the nineteenth century, the concept of unemployment was alien. In any society, there are always jobs that need doing by someone at some price. Theoretically, there is no unemployment; there is only an individual's choice to work or not to work. That began to change in Europe with stream power that facilitated the growth of more massive factories and railroads that made possible the mass transportation of finished goods. And then jobs and wages became more central features of society. Although the government had long been keep track of trade and agriculture, almost every country lacked critical employment information.

With the passion of measuring newly America, the late nineteenth century was rife with the creation of countless associations, from academic guilds such

---

[12]***See:*** Card, D. Origins of the unemployment rate: the lasting legacy of measurement without theory. American Economic Review. 2011;101(3), 552-57.

as the American Statistical Association (1839), the American Economic Association (1885), and the American Political Science Association (1903), to professional interest groups such as the National Association of Manufacturers (1885). The tools provided by those associations and other groups had a large impact on defining the contours of the modern economy. Massachusetts was the first state to create a labor department focused on collecting assembling information about employment and working conditions in 1869. A dozen other states followed suit over the next decades. Their impetuses are similar. That is to collect information that would support their contention, such as poor conditions, insufficient wages, and nonexistent safety, and indifferent companies. The federal government was then spurred. President Chester A. Arthur signed into law the creation of the Federal Bureau of Labor, along with its statistics office, in 1884. By the time Ethelbert Stewart arrived, the Bureau of Labor Statistics (BLS) had been in existence for nearly 30 years on accumulating various work information. Still, in terms of measuring employment, it could just as well not have existed (see Footnote 8).

> *Stewart had little truck with homilies and ignorance, and saw in statistics as a way to force society to deal with pressing issues of justice, fairness, and decency. ... As an advocate for a minimum wage law long before that was fashionable, Stewart saw the issue as one of simple social utility: unless people had sufficient means to meet their needs, their lives would be diminished, as would the strength of the country.*

The voices from the others who advocated for better statistics on unemployment, similar to Stewart, had been largely ignored or taken less than seriously outside of the American Statistical Association and the burgeoning file of economics. In the ensuing Great Depression—a profound and essentially permanent rupture with agricultural America, an employment crisis along with a severe contraction of price and production, broke out suddenly. Later estimates indicate that more than 20% of the workforce had lost their job, and with farm foreclosures adding substantially to the national misery. As a result, Herbert Hoover was trounced by Franklin D. Roosevelt and failed to win the re-elected presidency after 1932. With the support of Roosevelt, Frances Perkins served the labor secretary as a rare woman in high policies. Perkins appointed a respected statistician named Isador Lubin to replace the aged and retiring Stewart at the BLS. Lubin oversaw the birth of what we now know as an indicator of the unemployment rate. Unemployment meant first determining the size of that workforce at any given time, next defining the difference between being part of the labor force and outside of it. Starting in the 1930s, the BLS defined unemployment and the definition has been evolving continuously since then. The BLS drew on two primary sources. One is a survey of businesses and their payrolls. The other is intermittent Census Bureau surveys of households. The most significant change was the widespread introduction of sampling techniques. That is to say, they relied entirely on sampling, or counting only a fraction of whatever is being measured. Employment numbers today, for instance (see Footnote 8), are based on samples and surveys of six-four thousand businesses and four hundred thousand households, which is less than 0.5% of the population. The trick in sampling is obviously to make sure that the "sample" set is representative of the whole set (see Footnote 8).

A sample of a household with six people, including three wage earners, would show a much higher income than a household of six people with one wage earner.

Sampling error is a major problem, remaining one of the weak links in the ongoing attempt to define the world in numbers. So, the following much work is to refine simple sets and to assess whether they are sufficiently representative to draw conclusions.

In this historical period, statistics represent the fusion of mathematics with data collection and analysis. Most advocates and changemakers were statisticians, thinkers, and policymakers, looking for answers to their questions. Early in the twentieth century, the field of statistics was still forming but is vibrant to develop and refine today's leading indicators. With statistics to be a distinct discipline, combing sophisticated mathematics, probability theory, and sampling methods, as well as accounting, statisticians tend to focus on how to improve measurement, estimation, sampling, and revisions, while leaving the policy questions to others. So when statisticians changed their minds, history opens the next phase.

The Concept of "Economy" Under Physics Envy

In the early decades of the twentieth century, statisticians' attempts to measure what had been unmeasurable happened to intersect with the desire of governments and politicians to quantify the effect of their policies. The first pass is at what became the gross domestic product (GDP). GDP as a new invention of the 1930s, much like the unemployment rate, was created to meet specific needs at a particular time—coping with the Great Depression and World War II. Of course, the figure was largely wonky and academic at the beginning. Later, GDP, gross national product, and the national accounts on which they are based on evolved because of the dual demands of war and Depression. In the 1950s, to meet the demands of the United Nations that new countries measure their economics, GDP became the litmus of economies everywhere. By the end of the twentieth century, GDP growth could determine the fate of nations and the arc of elections, along with its cousin, the unemployment rate, and inflection. In the United States, most of the credit on GDP goes to Simon Kuznets. During the year of spending in the new National Bureau of Economic Research, Kuznets worked at the rigorous collection, formulation, and analysis of statistics. Mentored by one of the masters of his profession Joseph Schumpeter, in less than 5 years, Kuznets' work was catapulted out of obscurity. He established the first US national accounts data and found the elusive formula for growth—a "bell curve" called the Kuznets curve.

National accounts, GDP, and GNP are not perfect. At the outset, they were limited in what they measured. To assess prosperity of a country, they were designed to adequately understand every aspect of life. It was not only domestic work and hobbies omitted, but the way that output was gauged as a net positive regardless of the nature of that output. For example (see Footnote 8),

> *... residents of the southern United States buy lots of air conditioners, which they do, to offset the crushing heat of summer, that will show up as a positive for GDP (assuming those air conditioners are made in the United States, which would have been the case until late in the twentieth century and is no longer). So too will the money that people spend on electricity bills. Vermont, with less arduous heat, likely sees fewer purchases, and thereby would show a lower GDP than Alabama, at least insofar as air conditioners are concerned. But such numbers say nothing about the relative prosperity of the two states or about the overall quality of life in eight places.*

Kuznets and his colleagues understood utterly about what was included and what was not, which would have a significant impact on conclusions drawn. But when becoming touchstones of public policy and media references, the necessary vigilance to simultaneously understand both the definition and limitations of these numbers was lost.

The indicators that should be discarded in the year after the war and Depression but came to rule our world, is a story tied to a man who saw the numbers' power of capturing the imagination of millions. This man is Henry Luce. The magazines, such as *Time*, *Fortune*, and *Life*, were able to retool under Luce to become new organs of American economic beliefs in the twentieth century, and then to push the rage for data and statistics in the US New Deal era. Marketing was crucial; without this, the numbers might have remained useful but obscure. The indicators invented after 1945 was marked by how to measure affluence. Not just Americans but also people throughout the world participated in define indicators. These indicators can be used to measure perhaps content to keep people alive or satisfied, such as military, food, and security. The proliferation of indicators after the war was also driven in part by industry. Private indicators were developed by professional associations (e.g., US Institute for Supply Management), academic institutions, and nonprofit organizations (e.g., Conference Board). Meanwhile, various government departments mentioned above (or not) continued their analytical work. So, the indicators evolved in scope, scale, and complexity. The final element was the vast proliferation of information collected, collated, and analyzed by newly created global institutions (e.g., World Bank) that further developed a hunger verging on a fetish for statistics. These forces combined to create the modern concept "economy." One tried to look up the "economy" on Google's Ngram Viewer site that displays a graph showing how phrases occurred in a corpus (e.g., of books). Until the middle of the 1930s, the term economy appeared hardly but is with a soared use after that. In short, economic indicators invented the "economy."

The economy as a concept is a close system with physical boundaries that define a nation-state. Evolving as it did when it did, the economy is inseparable from the nation-state. That may have made sense before World War II, but now it seems to affect the development of human society seriously. The invention of the "economy" with numbers to measure it was a vast improvement when lacking a set of theories to explain an unruly-rule material world. But it now is causing harm. To control economic activity genuinely inside geographic borders, the government's efforts for all possible means include as follows (see Footnote 8).

> *Currency rates were zealously guarded by governments, and gold acted as a common denominator to value those currencies against one another. Tariffs and taxes, even if they were relaxed in the name of free trade, were barriers that states erected to defend themselves or keep out competition. Central banks were (and in most respects still are) national entities established by national governments to manage national economic affairs and the printing of national money.*

As we saw, even today, in a world that is unarguably global and more like a three-dimensional chessboard of countries, companies, and individuals.

The leading indicators revealed so far very little about the economic lives of people. But people are at least an important variable. A non-economist George Katona developed a micro indicator: consumer sentiment, based on his interests on human psychology with his fascination with markets. He aimed to measure something much more emotional, personal, and harder to capture with numbers: what people felt about their economic lives and what they believed the future had in store. He argued in his book *War Without Inflation* that (see Footnote 8),[13]

> *Inflation was as much a product of consumer expectations as it was of the typical factors that economists emphasized, such as money supply. In order to prevent consumers from assuming that inflation, the government had to create "a gestalt," which would temper those fears and instead create expectations of price stability.*

For most economists, such a hybrid approach of focusing on both emotions and behavior was alien and unwelcome, because the "rational actor" principle has assumed that individuals utilize all of the information at their disposal to maximize their self-interest. Compared with the classical economics of Adam Smith and David Ricardo, Katona saw individual decisions as flowing from a myriad of causes, not only the desire to maximize gain but also the interplay of complex and powerful emotions ranging from fear, to greed, to hope. Selling the kind of micro indicators proved to be increasingly lucrative as the twentieth century progressed. Katona's efforts, along with other substantial statistical endeavors, branded Michigan University as a global leader in the field. Katona

[13] ***See:*** Katona G. Psychology and Consumer Economics. Joural of Consumer Research. 1974;1:1-8.

remained Michigan after having ended his government stint until he died in 1981. There is an entire substory about the beginnings of commercialized social science research that created income streams for higher education. One reason for getting profits is, the release of these numbers can move markets, so hedge funds and traders are willing to pay money to have the first read.

The government is in favor of sentiment metrics. The growing sentiment metrics filled a glaring hole (i.e., a human element) in numbers, such as trade, agriculture output, unemployment, and GDP. The sentiment surveys speak to the broader questions of happiness and satisfaction and whether the state and society are structured to maximize those, which helps understand, for example, unemployment. The surveys indicate people's feelings, and that, in turn, raises the policy question of "whether the system in place societally is succeeding in providing the most potential to the greatest numbers" (see Footnote 8). Another reason comes from fears of inflation that were most acute among central bankers. The United States had never experienced the hyperinflation that had so impoverished Germany and parts of Europe earlier in the twentieth century. Despite this, Americans didn't fade as a potential danger in the minds. Indicators for measuring inflation are essential and had the origins in the work of Irving Fisher. Fisher gravitated toward that view of economies as mechanical systems, and he believed economic patterns could be observed and quantified (see Footnote 8).

Similar limitations existed for how any one individual or government approached economic affairs. They needed tools to understand the workings of a system, and the tools for economies were better data, more rigorous math, and coherent statistics.

Fisher grappled with the mathematical challenges of index construction to invent formulas to inch closer to his personal Holy Grail, "the ideal index" (see Footnote 8).

In describing his approach, Fisher used the analogy of many different weights being placed on a bar was balanced, and that would be the average price of a basket of differently weighted goods. After testing hundreds of different formulations, Fisher concluded that the "ideal" index would be a combination of constant weights and "chained" weight. Constant weights meant that the basket of goods was set, and then amended periodically as new goods entered and old ones dropped out. A "chained" index was more fluid and adjusted more frequently to account for the simple fact that as prices fluctuate, so does consumption.

Later observers remarked Fisher, Kuznets, and others that they suffered from "physics envy," because they wanted to capture the inner workings of the world of economic affairs with the same level of precision they thought that physicists possessed.

This historical period highlighted that economic indicators had become the symbol of economic performance. The representatives were the titans of the economics profession, Irving Fisher included. But they also received criticisms. This is because; few leading indicators were invented to help the "average" person navigate the world. Instead, they were created to empower officials and give policymakers clarity. The consumer price index (CPI), as a leading indicator that can measure people, disconnects with the inflation that people experienced or understood. The indicators may have measured the economy, but they could not determine whether the economic system was meeting people's needs and expectations. The leading indicators were meant to measure the health of the system, but are deviating from the intention. These metrics began to justify trillions of dollars of government spending and business investment. According to Karabell's summarization (see Footnote 8),

> *In fact, it is now almost impossible for government to spend money, for businesses to plan ahead, and for individuals to make decisions about investing and retiring without reference to these numbers. Rather than taken as guides, they are used as inputs into a system that is assumed to operate like a machine: calibrate the inputs correctly and you can, with reasonable accuracy, predict the outputs. It all seems very neat, and it is. It is also, with disturbing frequency, wrong.*

To sum up: the development process of economic indicators—from mathematics, to statistics, and to physics envy provides us a glimpse of Thomas Piketty's dissatisfaction with "childish" practices. That is,[14]

> *the discipline of economics has yet to get over its childish passion for mathematics and for purely theoretical and often highly ideological speculation, at the expense of historical research and collaboration with the others.*

In the next section, I'm going to explore the replacement of indicators that may require us to have data-informed insights.

Avatars of Leading Indicators

The prior section of tracking history shows that none of these indicators were designed to carry the weight they now do. They were invented to offer clarity where there had been none and to help policymakers navigate the many and different policy experiments. But today's indicators are not used following these original intentions. Instead, these statistics often act as barriers to innovative approaches rather than facilitating them.

What, then, is to be done? Whether it exists any changes or alternatives, I call avatars for these leading indicators? Faced by massively available data, what should we do? I'm going to talk about their existing avatars, and then to glance at new methods or new thoughts. At last, we can look forward to leading indicators' future avatars.

[14] ***See:*** Piketty T. Translated by Goldhammer A. Capital and Ideology. Cambridge, MA: Belknap Press: An Imprint of Harvard University Prss (March 10, 2020). 2020: 1104 pages.

Existing Avatar 1: Economic Indicator Dashboard

Economic indicator dashboard is a rather stunning yet straightforward invention: a visualization tool that allows anyone to present global indicators not as inert numbers, but as fluid and dynamic tables or graphs. Bloomberg's dashboard of key economic indicators, for example, has the 12 statistics to watch, such as global PMI (i.e., purchasing managers index), US employment, US consumer spending, Brazil GDP, Chile copper exports, and so forth.

Someone may say this system is similar to the epidemiologic dashboard like the John Hopkins dashboard that tracks COVID-19 cases in different geographic areas across the globe. It is correct. Actually, one apostle of visual economic indicators is an epidemiologist named Hans Rosling. He found that the eye of his passionate students grew heavy with boredom when numbers appeared. Rosling wanted to do something to struck what his students did not know or what they assumed they did. How then to communicate useful information to the student using numbers without their attention waning and the lesson being lost? He developed a visual system that put the numbers in motion along a graph, shaping deformation and movement. On a screen, circles representing economic indicators of multiple countries, or wealth, or employment rates, or any set of variables, start pulsing and migrating, like the trajectory of disease spread. The effect is quirky, funny, and the point is that anyone can appreciate it fully if seeing it.

But indicator dashboard inevitably has the limitations. It is only a data visualization tool, where raw data is taken into consideration to then convert into graphs, charts, and infographics to present the data, and has no data analysis. A famous spatial visualization is "Napoleon's March to Moscow" Map.[15]

> *The map depicts the advance and retreat of Napoleon's army in Russia, showing the changing size of the army due to death and desertion represented by the thickness of the line. Beginning at the Polish-Russian border, the thick band shows the size of the army at each position. The path of Napoleon's retreat from Moscow in the bitterly cold winter is depicted by the dark lower band, which is tied to temperature and time scales. The map displays multiple sets of data: the number of Napoleon's troops; distance traveled; temperature; latitude and longitude; path of travel; and location relative to specific dates.*

It is easy to find the map on the Internet, so I don't want to cite it in this book. Still, I provide a reference link for the interested reader. While a visual image is easy to understand, even for people with little background information on the topic or quantitative skills, visualization still has limitations. A display may be biased to make focused people to be wrong or to skip core messages because its challenge is to be transparent, accurate, and rich in information. For instance, if

[15] ***See:*** Advantages and Limitations of Data Visualizations. ESRI. Retrieved from https://www.arcgis.com/apps/MapJournal/index.html?appid=eaf2999fe4a04f50b1051fedcd438ab5.

the audience is not focusing on the core message for which the visualization presented, it may not seem satisfactory. Then such people may remain untouched to the core purpose of them doing any analysis. It is worth noting that the benefit of today's indicators is possible to sacrifice when only using an indicator dashboard. The advantage is that these indicators may not have been around for very long, but there is now just enough history of providing useful comparisons and of allowing people to gauge how things have evolved either for the better or the worse. One reason why the indicators are central and potent over time is that useful patterns and points of reference can be found after having analyzed. GDP growth of 4% could be good or bad; for example, it is excellent if giving an average growth rate of 3% over the past two decades.

Existing Avatar 2: CPI vs. Billion Prices

Tracking inflation is pretty likely the prior task of the economy. When hearing about inflation at the dinner table, it's often a conversation about how everything seemed to cost so much less in the past; for example, how we used to be able to buy three gallons of gasoline for a dollar and so forth. Key points relating to inflation are (1) price level that is measured by constructing a hypothetical basket of goods or services and calculating how the total cost of buying that basket increases over time, (2) the rate of inflation that is measured as the percentage change between price levels over time, and (3) an index number that is a unit-free number derived from the price level over several years that makes computing inflation rates easier; and (4) inflation is that the general and going rise in the level of prices in an economy.

In early 2013 Argentina, as the first country ever, received a threat of expulsion from the International Monetary Fund (IMF) due to its failure to maintain the integrity of its official economic data. The IMF executive board called on Argentina to align "these indicators with the international statistical understandings and guidelines that ensure accurate measurement," or else face sanctions culminating in suspension from the fund. What happened? In 2007, the National Institute of Statistics and Census (INDEC) of Argentina reported the first month of inflation was 1.5%, suggesting an annual rate of about 20%. Then president Néstor C. Kirchner was unpleased. High inflation would undermine his claim that the Argentine economy had begun to recover. In response, he fired the head of INDEC, Graciela Bevacqua, on the grounds that her approaches were inaccurate and replaced her with someone who, unsurprisingly, reported a newly inflation was less than the initial one, as Kirchner had supposed. The firing triggered an international uproar. The political intervention was seen as a threat not only to Argentina's ability that gauges its economic life independent of political pressures but also as a dangerous precedent for other leaders worldwide who might follow what Kirchner did when the indicators are not favorable. In the following years, some private groups began to assemble their independent reports of inflation to offer more accurate data. But they were warned by the economy ministry to be revoked the nonprofit status if reporting an inflation level that is not the same as what INDEC announced. After years of such shenanigans and increased unreliable crisis of its official data, Argentina got a severe warning from one of the world's most signification financial institutions.

Apart from a reminder of how vital independent indicators are to the smooth functioning of international diplomatic and economic affairs, the Argentine story spurred for academics. At MIT, Roberto Rigobon and Alberto Cavallo had a new idea. Cavallo, who born and raised in Argentina, was particularly interested in creating statistics that the government could not interfere with. They believe that a government may fire official statisticians or harass domestic private institutions that provide independent analysis, but cannot prevent so easily is a web crawler from scanning and scraping through all price online.

Rigobon and Cavallo's idea resulted in the Billion Price Project. They coordinated with the programmers to develop a real-time index of consumer prices. While being small compared with brick-and-mortar stores, online sales offer prices for much of their inventory. That means a well-designed program can find and record all of those prices with automatic updates. Plugged into the same formulas used by the US Bureau of Labor Statistics (BLS) and other national statistical agencies, the prices can be calculated precisely in the same way as prices obtained through the laborious process of sending out questionnaires and then getting the responses. And where the BLS and most national agencies spend thousands of staff-hours obtaining the information (that costs about US$250 million to collect prices for nearly 80,000 types of commodities), adjusting the basket of goods, and then producing monthly indices, a crawler designed to sour the Web acts far faster. This project once detected inflation of the US domestic trends about several months ahead of the official data release. Karabell commented on this project that (see Footnote 8),

> *The work of Rigobon and Cavallo is vital in several ways. First, it makes it much harder for inflation conspiracy theorists to argue that government statisticians are purposely and deviously understating actual inflation. The electronic bots that scour the Internet for retail prices have no political agenda. They move at inhuman speed, dispassionately assembling information. Of course, a good conspiracy never lets facts get in the way. One could credibly argue (in the absence of any information to the contrary) that the professors at MIT have a secret agenda and are receiving hidden financing from powers-that-be who wish to perpetuate the optic of low inflation in order to rob the masses of their hard-earned income and undermine the value (what little there is) of the work of the Billion Price Project has the virtue of confirming official statistics using a different methodology and technology.*
>
> *It also has the advantage of creating a key indicator in real time at far less cost. The budgets of statistical agencies may be pennies compared with other programs such as defense and health care, but such budgets still amount to billion in the United States and substantial sums in other countries. Calculating the CPI alone costs Americans*

more than US$200 million each year, and even then, it is released with a time lag and subject to revisions and statistical massaging to compensate for erratic but nonrecurring factors such as extreme weather. Both chained CPI (which adjusts the basket of goods based on consumer spending patterns) and traditional CPI generate intense debate and political controversy, as the proposal in 2013 to shift Social Security payment to chained CPI (thus lowering the increase of future payments) demonstrates. The Billion Price Project avoids those problems because it is almost entirely electronic, does not involve much human mediation, makes no distinction between different products, and attempts no value judgments about what people buy and in what quantities. And it is much less expensive.

The automatic means and much lower cost are two reasons why I advocated in the previous chapter about "government roles in data as innovative currency" (see Chap. 8). I'll be back at this point, especially when talking about the government that reduces public debt. Note that the Billion Price Project is different from traditional or chained CPI. The former is on a big data-based full sample, and the latter use a typical sampling despite about 80,000 commodities were selected. Besides, this project did not optimize the CPI in essence.

Existing Avatar 3: GDP and Consumer Surplus

Using the concept of consumer surplus as a way of determining what consumers consider to be the cost of free goods or services, is a methodology that attempts to make up for GDP (gross domestic product) flaws in the information economy. The methodology was proposed by Erik Brynjolfsson (also at MIT), who is one of the most critics of contemporary GDP. Consumer surplus is the difference between the actual price that consumers pay and the maximum price that they are willing to pay.[16]

An example of a good with generally high consumer surplus is drinking water. People would pay very high prices for drinking water, as they need it to survive. The difference in the price that they would pay if they had to, and the amount that they pay now is their consumer surplus. The utility of the first few liters of drinking water is very high (as it prevents death), so the first few liters would likely have more consumer surplus than subsequent liters.

Since our measurement framework for economic growth and well-being has not fundamentally changed since the 1930s, there exist some problems with economic indicators as we saw with the way the trade statistics have not kept pace with the globalization. The primus inter pares of these indicators is GDP. Even though it is now enshrined as an all-encompassing indicator of national success, it came into prominence in the US only in 1991, where the

[16]The primary reference here is the entry of "Economic surplus" on Wikipedia. ***See:*** Retrieved from https://en.wikipedia.org/wiki/Economic_surplus.

Bureau of Economic Analysis made it, rather than the gross national product (GNP), the primary gauge. Yet, neither GDP nor GNP has kept pace with technological change. The economic system that GDP and national accounts measure have evolved more rapidly than the methods used to measure them.

Brynjolfsson has been investigating how GDP hasn't entirely kept pace with technology. He found it is not due to the lack of awareness or attention from the Bureau of Economic Analysis. Instead, official statistics, as the product of years of testing and calibrating, have investigated the effect of these new digital tools, but the fact that they didn't create an untenable divide between the economy as measured and economic life as lived. With a boost from the proliferation of the Internet, the economy receives a driver from a host of "free" goods or services available online, such as Google's gratuitous search, Wikipedia's free retrieval, online and mobile banking's unrestricted use, online booking sites' no-fee reservation, etc. Tens of thousands of applications that allow individuals and institutions to perform tasks that used to consume considerably more time and money, are free. The consumer price index will eventually register a new product as long as it bought or sold, because a basket of goods changes over time to reflect purchasing behavior. But free goods won't show up due to no price. Because of this, the productivity promoted and enhanced by information technology is not a part of the market exchanges that constitute "the economy." Just as housework and the black market (about 96% of the Internet is the dark web) and volunteer work are invisible statistically, so are these free stuffs on the Internet. Free goods or services may not have a price, but still require time. Brynjolfsson, thus, argues that GDP undercounted a large part of the overall output. In a society with capital, time is money, and formulas to translate time spent into money spent must be created. To measure this part output, Brynjolfsson and his colleagues spent many months of hard work to craft the formulas based on consumer surplus, come up with a credible calculation, to refine the time-survey of consumer behavior. Their preliminary findings are that the Internet added US$34 billion each year to consumer surplus between 2002 and 2011. Note that consumer surplus is a term of art and cannot capture that the goods or service contribute to overall economic activity. It attempts to put a specific number on the value through the time that consumers were assigning to these goods or services.

The figure calculated by Brynjolfsson and his colleagues might not be accurate. For example, no one knows how much online social connectivity facilitates commerce or enables the business to be better operated internally. Many advertisers have embraced targeted advertising. Even if they told economists that half of the original cost they expected to spend saved, should we add this half to the output? This question may trigger a lot of academic arguments. Given no accurately counting the information industry and facing the emerging data economy, we should consider inventing new methods rather than to patch old economic statistics.

Start Being Data-Informed: Economists' Preparation

The above avatars of leading indicators, such as simple visualization, change data capture, and update old statistics, didn't aim to use new approaches. This is what Karabell's thought (see Footnote 8),

> ***... enough to demonstrate that the world is not without extremely adept individuals tackling the limitations of economic numbers, refining them and deriving alternatives.***

All indicators, however, are single numbers. That is the problem. Someone may argue there exists no evidence to support "simplification is bad." The words are quite right. I would say the simplification we need is traceable—sufficiently backed by the data and data results. These indicators show what we do with the data they assemble, and constructs meant to make sense of information. But the raw data is itself equally valuable to give more details according to those indicators. Every inflation reading comes with changes in demand for products, and the correlation among these products (e.g., beer and diapers). Each trade report not only shows the vagaries of geography but allows insights into dozens of industries and sectors. And also, every GDP report has hundreds of charts and tables for tracking various national incomes literally. People may gravitate toward answers that simplify reality yet embrace the truth as it is, and customize it tailored to the specific needs.

So, again, what should be done? What do we need to prepare? The answer varies, but I want to emphasize that it is time to start being data-informed.

Based on my knowledge, there exist roughly four types of economists. First is de facto economists due to their job title. We can see them in government agencies, engineering firms, international organizations, and think tanks (especially in the bank). They mostly make graphs, tell what these graphs say using economics words or concepts with generalities and jargon to a specific industry, or work for the media relating to the article, blog, podcast, and something else. The second is trained economists. They often have a doctoral degree; they can do the regression and statistical work for drafting more authoritative briefs and internal memos, or for advising colleagues. They usually do less graph making to focus their time on econometric software (e.g., SPSS, statistical product and service solutions). But they might not spend much time on writing academic papers. The third is academic economists. Most of them spend time engaged in conferences, debate equations, and publications. Of course, some try their hand at business (e.g., consultants and advisors), or leave academia for more policy or commercially oriented work. The last type is guru economists, who are a rare breed and generally known as the economic thinker. They often give fancy speeches in somewhere like Davos.

Whichever type of economist devises methods and procedures for obtaining the data they need, depending on the topic under study. In essence, economists are similar to epidemiologists, who receive substantial training in study design, measurement, and the art of causal inference, and also in the theory and application of methodological techniques, in particular mathematical models. I think what economists needed is not data-driven but data-informed. One might question the two are both buzzwords. But they are different: data-driven means letting the data guide during the decision-making process, and data-informed refers to let data act as a check on intuition. I mentioned in Chap. 3 when talking about the stock market is that data and data technology can act as intuition. According to Stephen Ross, the intuition is the combination of experience based guesswork and imagination. When data guides our decision making, we need to collect more of the data to have the most accurate models and find trends. But here's the catch—data is never going to tell us the full story, and it's never going to tell us exactly what we should do next. Data could not generate the idea or guide for study design in the first place. Human guidance was needed for that. Data was there merely to help. Being data-informed is to let economists formulate a research question using both intuition and data to ensure that it is answerable methodologically. And then, economists can relegate all data-related tasks to a single or more data scientists.

For a data scientist, the only thing is to simply observe what produces naturally with multiple models over time. While this processing is called the data experiment, data scientists often do not engage in any categories of study design. These advanced methods include, but not limited to, data collection, database system management, novel visualization techniques, sophisticated system modeling, the software development lifecycle, data security, data privacy, and algorithm ethics. And it also involves various algorithms that can represent a data-adaptive approach in the estimation of causal inference parameters, such as targeted minimum loss-based evaluation and double/debiased machine learning. At this point, we want to draw a clear distinction between sophisticated methods used by data scientists and predictive models used by economists. In a predictive model, one seeks candidate factors from among a broader set that is statistically associated with an outcome. This is useful for hypothesis generation but may lead to the identification of spurious associations. Occasionally a correlation between covariates in the data may be present but not intervenable and possibly irrelevant from an economic perspective. In causal modeling, a specific exposure is examined to test its causal relationship with the outcome, potentially to intervene upon (if harmful) or promote (if protective).

The data scientist may employ computational, and visualization approaches for high-dimensional data. And the hypothesis the data scientist uses is not one or more only; instead, are thousands of hypotheses generated automatically from, for example, a predictive Super Learner model. So, the collaboration among data scientists and economists t is an iterative process. The data scientist, armed with the data, can begin the process in partnership with the research team, with careful consideration of testing the algorithm, the appropriate type of implementation, evaluating the ongoing operation of the algorithm. Before the processing, the economist should provide expertise with data-informed insights in deploying the intervention and can design evaluation of its effectiveness that should, in turn, result in further refinement by the team. The confluence of challenges compels to support and encourage the convergence of these disciplines to obtain the new methods tailored to the specific needs or applications.

Future Avatar 1: The New Reality Without a Statistical Matrix

The future avatar of "no statistical matrix" was envisioned by Karabell in his book *The Leading Indicators*. He uses the 1999 popular movie *The Matrix* as a metaphor that a forest of indicators has formed by numbers, row upon row. He said (see Footnote 8),

> *While a forest of indicators doesn't create the same risks as too many weapons, as far as our ability to gauge the world around us, more is less. ... Our suite of leading indicators is also a construct, a statistical matrix. They create a virtual world defined by a set of numbers that may reflect aspects of reality but also obscure and distort.*

When commenting on various updates for the indicator forest, he wrote that (see Footnote 8),

> *As we saw, the commission convened by former French president Nicolas Sarkozy has already suggested a dashboard approach to national economic indicators. That, in effect, means expanding the array of statistics that are utilized in determining policy. ... But simply broadening the array isn't sufficient.*

This future avatar Karabell expected is driven by a group of "Neo," the hero of *The Matrix*, who "are measuring 'the economy' in novel and eclectic ways" (see Footnote 8).

> *They build on the efforts of the past, but they aren't beholden to them. They understand the importance of states, but they don't necessarily see them as the most important unit. And they recognize above all that the world see through the lens of current indicators is not the world. It is the world: a construct built on numbers derived from surveys and statistical formulas.*

The indicator forest described by Karabell looks like a statistical matrix that is obviously sparse. In data science, sparse data refers to the data where most values are missing or equals to zero. Sparse data is not useless, but its information is not complete. In some cases, the sparseness can reach more than 95%. Traditional statistical methods are not suitable for processing this data type. There exists a bias in sparse data, sometimes called "small sample bias," but in fact can occur in quite large datasets and thus is termed sparse data bias. Just as Karabell pointed out, "simply broadening the array isn't sufficient" and cannot reduce sparsity. One way to not further increase such a sparseness lies at indicator correlation analysis; for example, finding association rules among the indicators to add dimensions.

Also, analyzing the raw material to make the corresponding indicator traceable is quite good. It is equivalent to increasing the information flows. A simple example is a comparison of traditional money-based and future data-rich markets. The former is only with price and money as information, and the latter can consider more information to identify the transaction in better matches. The example applied in indicators is the same.

Future Avatar 2: Unbundling Work, Look to Symbiosis

Employment was perceived as connected to many economic problems. The first is labor income (or wage) inequalities involving the issues of a race between education and technology, the role of institutions the minimum wage, the super-manager phenomenon, and marginal productivity. Secondly, employment is related directly to the expansion or contraction of what the economy was producing calculated by GDP. Also, un-/ or employment may be about the conditions where people are able to pursue well-being. In terms of the above content, measuring labor utilization belongs to critical economic considerations. As a result, both Piketty and Karabell included one or more chapters in their book to talk about the topic; and Viktor Mayer-Schönberger did as well. A newly initiative suggested by *Nature*, aiming to "ensure economists and natural scientists working together,"[17] unfolds from an example point of seeing in "wage stagnation and widening inequality."

The most widely accepted theory to explain the inequality of income from labor, especially wage inequality, is that of a race between education and technology. Although suggesting interesting and essential clues for solving certain historical evolutions, this theory is too native to offer a satisfactory explanation for the others. The theory rests on two hypotheses. One is, a worker's wage is equal to his individual contribution that corresponds to the output of the collective he works; namely, his marginal productivity. The other is, the worker's productivity depends on his skill and on supply and demand for the skill in a given society. For instance, in a society in which few people are qualified data scientists (i.e., low supply) and the prevailing technology requires many data scientists (i.e., high demand), then it is likely that the combination of low supply and high demand results in high pay for data scientists relative to the others. This theory is in some respects limited. It has the virtue of emphasizing two fundamental social and economic forces relating to wage inequality: the supply

[17] ***See:*** Economists and scientists: solve big societal problems by working together. February 26, 2020. Retrieved from https://www.nature.com/articles/d41586-020-00532-4.

and demand for skills. In practice, the supply of skills depends on the educational (or training) system: the number of people having access to these tracks, the quality of the training, the degree of integration on teaching and professional experience, and so forth. The demand for skills lay on the technologies available to produce the goods or services that society consumes. Of course, the two factors themselves depend on many other forces. Taking about the role of institutions contains expectations of the ideal society. Because the theory of marginal productivity fails to explain the diversity of the wage distributions we observe in different countries at different times, we must introduce other factors, such as the institutions and rules that govern the operation of the labor market in each society. When considering the institutions, there exists inevitably an issue of how the government supervises the labor markets. And labor market regulations depend on each society's perceptions and norms of social justice, which are intimately related to each country's social, political, and cultural history. There is no doubt that the minimum wage plays an essential role in the issue (see Footnote 14).

> *The United States used the minimum wage to increase lower-end wages in the 1950s and 1960s but abandoned this tool in the 1970s. In France, it was exactly the opposite: the minimum wage was frozen in the 1950s and 1960s but was used much more often in the 1970s.*

It would be easy to multiply examples from other countries. So, the institutional comparisons among countries can be derived, especially between developing and developed countries. In some developed countries, it exposes the phenomenon of very high salaries due to a central role of the institutions. That is, "in all the English-speaking countries, the primary reason for increased income in e quality in recent decades is the rise of the super-manager in both the financial and nonfinancial sectors" (see Footnote 14).

The above theories of marginal productivity and a race between education and technology are not enough to explain employment. Let's glance at explanations according to economic indicators. Karabell gave an example of a stimulus billing totaling US$787 billion—the American Recovery and Reinvestment Act during the Obama administration. The US$787 billion figure emerged from a relatively simple calculus. Government spending is believed to have a "multiplier" effect meaning that a dollar of stimulus spending can have more than a dollar of consequence, laying on means the government takes. These means include tax breaks, grants to states, and direct spending. Closing the gap between what the economy was producing and what it could produce, was supposed to reverse the decline trend in employment. The former figure is from GDP, and the latter is related to the "natural rate of employment." Specifically (see Footnote 8),

> *The equation goes something like this: if the economy is at or near full employment, it will generate its full potential output. And when economic activity is heated, the "output"*

> *gap can even be positive; that is, the economy can, statistically speaking, generate more activity than its "potential." That, in turn, is supposed to lead directly to inflation.*

But employment is not static. In the United States, truck driving is one of the few remaining jobs that don't require a college degree but provides a relatively decent median income (an annual income of more than US$40,000), according to the US Bureau of Labor Statistics. Despite no concern that a factory worker in Shenzhen will replace a trucker in Minnesota, truck driving has gotten reasonably sophisticated. Today's trucker is less a manual laborer than a controller of digital equipment, and even has to handle a fair amount of administrative work. Trucks now come equipped with power steering and braking, satellite navigation, and modern flee management system. In addition to labor's connotation, the percentage of people participating in the workforce and the amount of the nation's income allocated to worker or employee compensation both fall.[18]

> *In the United States, the percentage of people participating in the labor force has declined from its peak in 2000 to be a rate that was last seen in the 1970s.*
>
> *... the "labor share" has declined considerably since the 1980s, from 67 percent of value added to 47 percent, much of the drop happening after 2000. The United States is not alone. In most advanced economies, labor share has declined. It has also fallen in the large economies of India and China. In fact, economists estimate that labor share globally has been dropping since the 1980s.*

But the new fact is that technology and globalization are unbundling work. In other words,[19]

> *Work is becoming "unbundled," fueled by the combination of automation and globalization, and driven by the rapid pace of change.*

The word "unbundling" is a concept advanced in a 1999 *Harvard Business Review* article[20] by John Hagel of Deloitte's Center for the Edge, with Marc Singer. What does the means for unbundling jobs? Unbundling an entire job allows different people to finish different layers. Take, for example, geography. With the combination of automation and globalization, a single work can be performed in various places separately (e.g., in the Western and Eastern Hemispheres), to uninterrupted 24-h operations to improve efficiency. As work becomes unbundled, the likelihood increases that hirers will look for ways to distribute tasks wherever it's cheapest. Besides, more examples of the unbundling of work can be seen in the dynamics of what's frequently alternative

---

[18] ***See:*** Mayer-Schönberger V, Ramge T. Reinventing Capitalism in the Age of Big Data. New York, NY: Basic Books (February 27, 2018). 2018; 288 pages.

[19] ***See:*** Bolles GA. Unbundling Work: Learning to Thrive in Disruptive Times. February 8, 2017. Medium.Com. Retrieved from https://medium.com/@gbolles/unbundling-work-learning-to-thrive-in-disruptive-times-427b172b1470.

[20] ***See:*** Hagel III J, Singer M. Unbundling the Corporation. The March-April 1999 Issue. Harvard Business Review. Retrieved from https://hbr.org/1999/03/unbundling-the-corporation.

work—where a worker is working either part-time or temporary with low-paying—called the gig economy. Today, it's driving for Uber and fetching for TaskRabbit; tomorrow it will be sweeping a street, or filling a shift. With the adoption of data-rich markets, the unbundling phenomenon will get even more complicated by encountering more digital errand runners and data partners.

It has a somewhat positive idea, suggesting that we must be relentless in a flight against other competitors to be an adaptive worker so that we can thrive in disruptive times. That means "survival of the fittest" or "eat or be eaten," which is a derivative of the law of the jungle. As we mentioned before, data productivity belongs to ecosystem productive forces. In that case, let's first take a look at ecology, and then make a conclusion.

Japanese ecologist Kinji Imanishi has a rather un-Darwinian view. He saw nature as inherently harmonious. Species fit together in an ecological whole, adjusting to each other, each finding its own niche. This is true for mayflies (i.e., crawlers, clingers, burrowers, and swimmers are staying different areas in a river), and so are North American warblers (e.g., five kinds of living at different heights on a tree). A Zuckerkandl Prize awarded paper that describes an experimental molecular evolution for the Escherichia coli population also supported this view.[21] As shown in figure 9.2a, the simplified experimental evolution is to place two Escherichia coli populations (i.e., a transformed with mutated genes and the normal one) in an unstructured environment of the chemostat. Figure 9.2b shows that (1) the mutant population grew fast to reach a three-fourth niche in the chemostat; (2) the mutant population stopped expand-

Figure 9.2 The simplified experimental evolution of Escherichia coli. Note: the red one is the mutant population, the green one is the normal population. (Data source: Author drew the two figures according to the literature (see Footnote 21))

[21] ***See:*** Kashiwagi A, Noumachi W, Katsuno M, *et al.* Plasticity of fitness and diversification process during an experimental molecular evolution. J. Mol. Biol. 2001;52(6):502-509. DOI: 10.1007/s002390010180. PMID: 11443353.

ing itself after 1 month, so as to remain a quarter niche for the normal ones. American famous microbiologist Lynn Margulis' ideas similarly focused on symbiosis—a living arrangement of two different organisms in an association that can be either beneficial or unfavorable.[22]

> *Among her most important work was the development of the serial endosymbiotic theory of the origin of cells, which posits that eukaryotic cells (cells with nuclei) evolved from the symbiotic merger of nonnucleated bacteria that had previously existed independently. In this theory, mitochondria and chloroplasts, two major organelles of eukaryotic cells, are descendants of once free-living bacterial species. She explained the concept in her first book, Origin of Eukaryotic Cells (1970). At the time, her theory was regarded as far-fetched, but it has since been widely accepted. She elaborated in her 1981 classic, Symbiosis in Cell Evolution, proposing that another symbiotic merger of cells with bacteria – this time spirochetes, a type of bacterium that undulates rapidly—developed into the internal transportation system of the nucleated cell. Margulis further postulated that eukaryotic cilia were also originally spirochetes and that cytoplasm evolved from a symbiotic relationship between eubacteria and archaebacteria.*

From this perspective, we can feel life itself is not selfish. And even, flora or fauna sometimes will sacrifice itself for the survival of other populations. Only food chains and food webs exist; an entire ecosystem can survive.

While remaining a dispute between the two research thoughts, symbiosis can give enlightenments for our human society, especially for the economy. Given a data economy that is with a property of "increasing profits of other industries," a future avatar in favor of symbiosis is what we needed. The concept "symbiosis" stretches to the society can be used to explore the issues of social interaction among actors from innovation and complementarity. But this requires a complete study rather than some remedies, which will be the content of Part 4 in this book.

---

[22]***See:*** A topic relating to Lynn Margulis at Encyclopedia Britannica online. See: Retrieved from https://www.britannica.com/biography/Lynn-Margulis.

Part IV
Regulating Data Capital in the Twenty-First Century

Abstract The data confirmed to be capital may reinvent the current global economy. This part presents a feasibility study with three steps. The first is to imagine a future society with data capital. Regarding this global society, next is to give a set of schemes relating to worldwide taxation on data capital for the regulation purpose. Lastly, this plan is discussed based on the latest technique of data self-governing openness.

In the first three parts of this book, I analyzed the evolution of capital forms from tangibles to intangibles and the rise of data capital. I offered multiple examples (and a brainstorm) of the hopes brought from data capital, and provided potential responses to the challenges science, industry, and government need to solve alone or together. From the analysis, I must now try to conduct a feasibility study for regulating this emerging capital. The new data economy has brought with it both immense hopes (such as data-rich markets, better coordinate human activity, and the appearance of more free will) and equally tremendous inequities (such as the individuals' privacy problem, and inequitable data profit distribution). It is impossible for this regulation to no involves political or democratic issues. The ideal plan for avoiding the two points would be from a perspective on taxation. After all, the dream of world harmony is somewhat through the complementarity that depends on the difference. Centralization aims to efficiency, while decentralization is equity and resilience. Capitalism has many flaws in itself, and socialism cannot be achieved overnight. Seeking common ground while reserving differences and meet each other halfway to develop is the theme of globalization.

Chapter 10
A Global Society with Data Capital

Capital and Ideology is a newly book written by Thomas Piketty following his 2013 book *Capital in the Twenty-First Century*. Piketty believes the worth of "a society is measured by its Gini coefficient alone."[1] And his core argument is that "rising inequality is fundamentally due to the political choices" because;[2]

> *great driver of human progress over the centuries has been the struggle for equality and education and not, as often argued, the assertion of property rights or the pursuit of stability. The new era of extreme inequality that has derailed that progress since the 1980s ... is partly a reaction against communism, but it is also the fruit of ignorance, intellectual specialization, and our drift toward the dead-end politics of identity.*

These are related to "education,"[3] he concludes, educational institutions cannot foster social mobility due to unequal opportunity in higher education—caused by the payment of extremely high tuition fees and the inequality of university access.

The most prestigious private universities are not merely with the high cost of tuitions, and their admissions depend in significant ways on the partners' financial capacity to make donations to the universities. For example, Chinese real estate tycoon couple Shiyi Pan and Xin Zhang gave Harvard University US$15 million as a gift in 2014, and their son was admitted to Harvard after 5 years (in 2019). By comparing various sources of data, one study used in Piketty's book has shown that (see Footnote 3) the average income of the parents of Harvard students is estimated at US$450,000 annually, which is in line with the top 2%

[1] ***See:*** A bestselling economist sets out the case for socialism. Economist.Com. March 6, 2020. Retrieved from https://www.economist.com/books-and-arts/2020/03/05/a-bestselling-economist-sets-out-the-case-for-socialism.

[2] ***See:*** Piketty T. Translated by Goldhammer A. Capital and Ideology. Cambridge, MA: Belknap Press: An Imprint of Harvard University Prss (March 10, 2020). 2020: 1104 pages.

[3] ***See:*** PiKetty T. translated by Goldhammer A. Capital in the Twenty-First Century. Cambridge, MA: the Belknap Press of Harvard University Press; Reprint edition (August 14, 2017), 2017: 816 pages.

© Springer Nature Switzerland AG 2021

C. Tang, *Data Capital*, https://doi.org/10.1007/978-3-030-60192-8_10

of the US income hierarchy. This finding also indicates that the admission seems not to be entirely compatible with the idea of selection based solely on merit. Just like explained by Jason England, a former dean of admissions at an elite liberal arts university;[4]

> *As a society, we've created a system of credentials that keeps the wealthy in place. The wealthy define "deserve" to ensure their self-interest is disguised as the greater good. The wealthy teach the middle class the checklist for college admissions success and who to blame if things don't go their way, engendering a suspicion that their birthright is being curtailed by an undeserving "other."*

The total absence of transparency regarding selection procedures, moreover, often companies with some scandals. In March 2019, the largest college admission corruption case[5] in the history of the Department of Justice made public. Fifty people have been arrested nationwide of the US, including 33 parents and a California businessman who allegedly collected money in bribes to get students in some of the nation's top colleges.

> *William "Rick" Singer leaves Boston Federal Court after being charged with racketeering conspiracy, money laundering conspiracy, conspiracy to defraud the United States, and obstruction of justice on March 12, 2019 in Boston, Massachusetts. Singer is among several charged in an alleged college admissions scam involving parents, ACT [American college testing] and SAT [scholastic assessment test] administrators and coaches at universities including Stanford, Georgetown, Yale, and the University of Southern California.*

People were very keen to talk about the case itself and famous guys involved in this scandal. But it is one of the essential problems that most countries worldwide must face in the twenty-first century, not just a problem solely in the US. To date, no country has come up with a genuinely satisfactory response, including Germany, France, Italy, Spain, and even some Nordic states with much lower university tuitions like Sweden. An example given in *Capital in the Twenty-First Century* is below (see Footnote 3).

> *In practice, the French system of "grande sécoles" leads to spending more public money on students from more advantaged social backgrounds, while less money is spent on university students who come from more modest backgrounds. Again, the contrast between the official discourse of "republican meritocracy" and the reality (in which social spending amplifies inequalities of social origin) is extreme. According to the available data, it seems that the average income of parents of students at Sciences*

[4] ***See:*** England J. The mess that is elite college admissions, explained by a former dean: Eight things I wish people understood about my old job. Vox.Com. May 8, 2019. Retrieved from https://www.vox.com/the-highlight/2019/5/1/18311548/college-admissions-secrets-myths.

[5] ***See:*** Goggin B. Meet the accused ringleader of the massive college-admissions scandal, William "Rick" Singer, the owner of Edge College & Career Network. Insider.Com. March 12, 2019. Retrieved from https://www.insider.com/william-rick-singer-the-accused-ringleader-college-admissions-scandal-2019-3.

Po is currently around 90,000 euros, which roughly corresponds to the top 10 percent of the French income hierarchy. Recruitment is thus 5 times broader than at Harvard but still relatively limited. We lack the data to do a similar calculation for students at the other grandes écoles, but the results would likely be similar.

Piketty's solution for this, first and foremost, is to begin by working toward greater transparency regarding admission procedures than exists today. Second is to imitate what public health insurance systems do: universities are granted a certain independence, but the cost of education is a collective responsibility, thus ensuring that students have equal access to higher education. One could do with universities and students excluding substantial public financing, which is to "range from charging tuition fees that vary with parents' income to offering loans that are to be paid back by a surtax added to the recipient's income tax" (see Footnote 3).

This solution is an obvious way of treating the symptom, not the cause. As Pierre Bourdieu analyzed in his book *The Inheritors: French Students and Their Relations to Culture*, "more subtle mechanisms of social and cultural selection, which often do the same work as a financial selection" (see Footnote 3). The root cause lies in the financial capital—using price and money is too naive to be a solely criterion for partitioning income, identity, behavior, and others. It might not be the same in a society having in terms of data capital.

Political Returns on Data Capital

Each form of capital, in essence, is to pursue any political returns. It is an inevitable outcome for capitalists who participated in politics in many ways. Take financial capital as an example; financial capitalists start with influencing local economic governance, next integrate with other forms of capital (e.g., industrial capital) to get a monopoly, and then dominate social or political policies nationally, transnationally, and even globally. A typical case is the financialization of housing. In other words, housing and real estate are transformed by financing tools to be financial assets, which are used to obtain new profits on capital markets and financial excess. The subprime mortgage-backed securities that triggered the 2008 financial crisis is the one.

Someone may wonder why I tend to interpret data capital (that is an alternative form of capital and even covers financial capital) in a somewhat favorable perspective. My point is possibly affected by a characteristic of the data economy—improves or enhances the advantages of other industries.

Less Financial Crisis

The financial crisis of 2007–2008 was a severe worldwide crisis caused by financial capital. The crisis also covers a variety of commercial, economic, political, social, and ideological aspects. Its fuse is that BNP Paribas suspended the redemption request for three funds because of problems in the US subprime mortgage sector on August 9, 2007. Later that day, the European Central Bank (ECB) was forced to inject additional liquidity in the euro area banking sector to prevent a liquidity crisis and bank runs. One day later, on August 10, 2007, the ECB, the US Federal Reserve, the Bank of Japan, the Reserve Bank of Australia, and the Bank of Canada Reserve made the first joint action to inject large-scale international liquidity in the global financial system. One year later, in September 2008, when many people believed the subprime mortgage crisis was alleviated substantially, Wall Street had begun to suffer its biggest disaster since 1929. Lehman Brothers declared bankruptcy early on the morning of September 15 with US$639 billion in assets and US$613 billion in debts, making it the largest bankruptcy filing in U.S. history. Washington Mutual Bank, the US' largest savings and loan institution, declared bankruptcy. The US Office of Thrift Supervision was forced to accept and arrange for the acquisition of JP Morgan urgently. American International Group, powerhouse in the business of selling insurance, and in September 2008, was on the brink of collapse. It was rescued by the Federal Reserve courtesy of an $85 billion emergency loan, and in exchange, the federal government acquired a 79.9% equity stake. Global investment bank giant Merrill Lynch was in deep crisis and had to cheap sell it to Bank of America. A month later, on October 3, 2008, the US Congress authorized the US Department of the Treasury to use the funds of crisis relief, forcing the nine most substantial financial companies, including Goldman Sachs and Morgan Stanley, to sell their equities to the government. Gerald F. Davis lamented in his book that won the 2010 George R. Terry Book Award *Managed by the Markets: How Finance Reshaped America*:

> *In a period of just over a month, a conservative presidential administration devoted to free markets and financial deregulation had engineered a level of direct government control over the nation's financial institutions undreamed of by V. I. Lenin a century before.*

The global stock market value evaporated by US$16 trillion from the outbreak of the subprime mortgage crisis to the bankruptcy of Lehman Brothers but has lost a half around US$35 trillion in several weeks since then. At the same time, the value of housing and real estate markets fell precipitously, and the losses in the US alone were as high as US$7 trillion. The total equity capital loss of listed companies, listed non-companies, and other commercial institutions caused by the crisis is close to US$5 trillion, which is equivalent to a loss of 80% of the total GDP globally in 2008. Other casualties of economic recessions, unemployment, and a stage of low-speed growth caused by this crisis cannot be estimated accurately.

The financial crisis should blame for asset price bubbles. Given the money is oversupplied, according to the Walras' law (that asserts an examined market must be in equilibrium if all other markets are in equilibrium), underdeveloped financial markets will be excessive demand for goods and services with the risk of a sustained rise in inflation, whereas developed ones will cause asset price to rise sharply and even to form asset bubbles. After decades of evolution, various countries of central banks' policy response to financial markets relating to the nearly unlimited global expansion of base currencies and liquidity caused by floating exchange rate systems, are very similar. Central banks are often by lowing interest rates, and as rates hit rock bottom, banks in some nations even changed negative interest for deposits. While in practice, interest rates on deposits cannot be lowered much beyond zero, a real low-interest-rate often means a meager discount rate of the entire socio-economic system. Assuming that the expected profits or future income flow is constant, a lower social discount rate represents a higher present value of investments and assets. It is easy to interpret or intervene in low-interest rates in an individual country or a solely period. But the current low-interest rates are long-standing globally. It is difficult to do intervention due to too many hypotheses, including the excess savings of emerging market, the Triffin dilemma (that is the conflict of economic interests that arises between short-term domestic and long-term international objectives for countries whose currencies serve as global reserve currencies), Albert Hirschman's unbalanced growth hypothesis, endogenous wealth changes, and so forth. The crux of all the problem lies at price and money.

We explained earlier (see Chap. 7) that money as the market's preeminent conveyor will prompt a decline in the future data-rich markets. So, it is an inevitable trend that data capital will replace financial capital. Viktor Mayer-Schönberger described "the end of finance capitalism"[6] as follows.

> *If capital is abundant, but fewer companies are looking for capital, supply outstrips demand on capital markets, and this means returns on investment are plummeting. This spells the end of finance capitalism as we know it—the mental association of working markets with huge investor returns. And it's unlikely the good days of finance capital will ever come back: as markets turn data rich, there is less need to signal with the money. The economy will thrive, but finance capital will not thrive with it on account of the shift from money-based markets to data-rich ones. With the market economy advancing with the help of data, we may no longer label the future "capitalist" in the sense of power concentrated by the holders of money. Ironically perhaps, as data-driven markets devalue the role of money, they prove Karl Marx wrong, not Adam Smith.*

[6] ***See:*** Mayer-Schönberger V, Ramge T. Reinventing Capitalism in the Age of Big Data. New York, NY: Basic Books (February 27, 2018). 2018; 288 pages.

In sum, forecasting future sales will be more accurate as the price is no longer the sole indicator of goods or services (financial products included), thereby no much asset bubbles and further fewer financial crises.

Identity Politics No Longer Run Amok

When summarizing Piketty's work on Capital in the Twenty-First Century (see Chap. 2), I highlighted his interests in classification. According to the sum of income from labor and income from capital, Piketty divided a society into five groups. They are upper class (defined as the top 10%, in which top 1% refers to dominant class and the rest is well-to-do class), middle class (the middle 40%), and lower class (the bottom 50%). He interpreted the designations he gave that seems to correspond more closely to common usage; for example (see Footnote 3),

> *the expression "middle class" is generally used to refer to people who are doing distinctly better than the bulk of the population yet still a long way from the true "elite." ..., in other words, the point is to say that such individuals are not privileged and fully deserve the indulgence of the government, particularly in regard to taxes.*

He also said that, to register more precisely, he breaks down the groups using centiles or even thousandths on dividing income.

Contrasting Mayer-Schonberg's remarks about "no longer label the future capitalist in the sense of power concentrated by the holders of money," we must admit the vast differences between economists and data scientists. Economists use a variety of economic markers as identity labels, such as race, gender, religion, education, health, profession, geographic location, etc., to divide up the population. While dividing, it usually reflects an implicit or explicit position concerning the justice and legitimacy claimed by a particular group. This is consistent with identity politics. Identity politics aim to form exclusive socio-political alliances, manifesting in various dialogues with multiple nationalist, ideological, and postcolonial movements. Data scientists may oppose any representation of objects that only relies on a small number of characters. For example, in the retail that is a field of more business intelligence applications, customer profiling and targeted marking of many commodities (e.g., Martin boots) are no longer to separate males and females. The approaches of using a single identity label or few labels compounded together to resonate with most people to identify with groups they are familiar with can cause social unrest, such as the Occupy Wall Street and Black Lives Matters. Data scientists might not try to determine whether "the 1%" had more power under Louis XVI or under Barack Obama. They are more concerned about determining a particular group effectively through features as more as possible and then discovering valuable relationships, breaking the limit of communications, and expanding personal networks on strong or weak ties.

In terms of the solution for equal opportunity in higher education, thus, the core issue data scientists believe is to satisfy the needs and wants of vulnerable groups rather than economists' thoughts of transferring income from the rich to the poor directly (Piketty's solution is to range from charging tuition fees that vary with parents' income). Also, any given level of group always varies following the shape of the distribution in society. There is never fixed between social classes or between "people" and "elite," particularly in the ear of big data.

There exists an online survey called "the future of social relations" with over a thousand answers collected in 2010. Its question is as follows.[7]

> *In 2020, when I look at the big picture and consider my personal friendships, marriage, and other relationships, I see that the Internet has mostly been a positive/a negative/no response force on my social world. And this will only grow more true in the future.*

While the question is so simple, many of them had long, free-text comments. They believe that,

> *The internet has been embraced globally at an amazing speed because of its capabilities for communication and connection – for creating, cultivating, and continuing social relationships. One big advantage is that email, social networking tools and other apps allow people to maintain bigger social networks and allow people to learn more about those in their networks. Richer social relations emerge from this greater awareness. Another advantage that it is much easier now to build communities on the fly via personal broadcasts. All these trends will continue to hold strong over the next decade.*
>
> *"The enemies of social connectivity are silence, disengagement, distance, and abandonment. In the past, how many individuals and families have suffered from these degenerative influences? ... On the Internet, social alienation remains a factual force. But never before has a person had more opportunity for social integration. More than ever, being inside or outside now is a matter of personal choice." – Eric James, president of the James Preservation Trust and publisher of Stray Leaves, author and lecturer.*
>
> *"The internet is communications gold mine. We can already find people with whom we've lost contact, communicate with people independent of time zones, ... The trend I see only improves with time!" – David Moskowitz, principal consultant at Productivity Solutions, Inc., and lead editor of OS/2 Warp Unleashed.*
>
> *"It's now easy for me to find people who share characteristics or interests, whereas for much of my pre-internet life I mainly felt like I didn't fit in anywhere. Also, it's made it easier for me to find and interact with many types of people who are very different from me, giving me a wider range of experience. ... I've gone to many real-world get-togethers coordinated through that service, and have made many friends that way. That, to me, blends the best of the net and the real world." – Amy Gahran, contributing writer at eMeter Corporation, senior editor at Oakland Local, co-creator and community manager at Reynolds Journalism Institute.*

---

[7] ***See:*** Anderson J, Rainie L. The future of social relations. Pew Research Center. July 2, 2010. Retrieved from https://www.pewresearch.org/internet/2010/07/02/the-future-of-social-relations-2.

"Despite the media narrative, cultural relations and social engagement mediated by virtual spaces is a plus. At a minimum, it is an opportunity for complex dialogue with no opportunity for physical violence." – Joshua Fouts, leader of Dancing Ink, digital diplomacy expert, senior fellow at the Center for the Study of the Presidency and Congress, and founding editor of Online Journalism Review.

"As humans, we can have both strong and weak tie relationships in our normal networks. We can multiplex these relationships. We are weak ties to some of our connections and strong ties to others. Just like a network multiplexer our weak ties can carry both types of signals around our network. The weak tie between Ego and his acquaintance, therefore, becomes not merely a trivial acquaintance tie but rather a crucial bridge between the two densely knit clumps of close friends ...individuals with few weak ties will be deprived of information from distant parts of the social system and will be confined to the provincial news and views of their close friends. This deprivation will not only insulate them from the latest ideas and fashions but may put them in a disadvantaged position... Weak ties might bring you the crucial information about a new job opportunity, a new start-up business, or new connections into other areas of your peripheral business. Your relationship with your weak ties should be maintained and cultivated, knitting your networks together to encourage information free flow between the different parts of your networks." Mark Granovetter, an American sociologist and professor at Sandford, refers the words in his paper "Notes on the strength of weak ties."

It is obviously what William Gibson said, "The future is already here; it is just not evenly distributed." Let's get rid of ingrained thinking to believe in disruptive innovation, and to make innovation as the standard response to any problem we meet. Data capital and advanced analytics may help us social mobility in both reality and cyberspace without any fixed identity labels.

International Assistance: Not Just An Obligation

A seemingly much broader range of political returns is international assistance. From an economic perspective, the poor and emerging countries cannot fulfill much more than its traditional regalian functions and responsibilities, due to the relatively lower government revenues. If paying every sector—the police force, judicial system, education, and healthcare, it is unlikely that any of these public services will work will. It may further lead to a vicious circle: poorly functioning public services undermine confidence in government, which makes it hard to increase taxes, resulting in less government revenue. Therefore, these governments usually choose to support the teams of police and judges fully. As a result, education and health are without insufficient funds and even no money; and they require international assistance from the rich and developed countries.

While having the practice of the UN's organs, some international conventions, and many significant arguments concerning foreign assistance in recent years, the issue is often beyond economics. It usually falls into the debates that attempt to give more visibility to social and cultural rights in human rights, such as social or distributive justice at the international level, extraterritorial human rights violations involved in obligations, and international humanitarian law and the emerging rules on the protection of persons in disasters. The provider and recipient countries typically wear down a lot each other during the break-in period. The provider is accused of meddling too much about what the recipient ought to be doing. The recipient trends to be used by the provider as a field of experimentation that aims to resolve chaotic conflicts and problems. This illustrates a more general phenomenon: some of education and healthcare issues the developing countries face are specific to their own situation or their own historical development; and in many cases, the previous experience of developed countries is not applicable.

Data capital can offer greater convenience to such a field of experimentation. Although having no more data found in a poor nation, we can conduct data experiments. Suppose that a researcher, who wants to study consumer behavior of shopping in a supermarket, can neither collect the data from a specific group of consumers, nor obtain enough point-of-sale data from the retailer to verify the algorithm. In this case, a data experiment can be used to simulate a data scene or an experimental environment to generate more data with a small amount of sampling data. In a medical scene, for instance, big data now can support the entire health informatics life cycle. This is to connect medical treatment research and clinical practice; namely, from medical issue understanding, data understanding, data preparation, exploratory analysis or modeling, validation, visualization & presentation, to final decision-making. Few pieces that are challenging to analyze can make up of other parts that are easy to do. Assume to use machine learning to study the role of respiratory biomarkers in managing COVID-19 severity, whether the patient will require intensive care in the next days and even hours can be predicted. The results are extremely useful to other purposes; for example, making decisions about intensive care unit rooms, where to build temporary hospitals, or where to boost supplies of personal protection equipment, ventilators, or diagnostic tests.

Note, also, that international assistance does not take the form of "one-to-one" helping as we imagine. According to China's Second White Paper on Foreign Aid (on July 10, 2014),[8] China provided aid to 121 countries from 2010 to 2012, 51 in Africa, 30 in Asia, 19 in Latin America and the Caribbean, 12 in Europe and 9 in Oceania. International assistance can quickly form a complex structure among countries (or other stakeholders; e.g., non-profit international organizations) in which data experiments may support cooperation and competition. Such cooperation and competition are similar to how scientists pieced together the new-famous image of a black hole. Besides, role swaps may exist

[8] *See:* Zhang D. China's Second White Paper on Foreign Aid: In Brief. Australian National University. December 2015. Retrieved from http://bellschool.anu.edu.au/sites/default/files/publications/attachments/2015-12/IB-2014-26-Zhang-ONLINE.pdf.

in this complex structure that consists of countries or other stakeholders. In other words, the recipient can role a provider; and vice versa.

In game theory, a win-win game is a voluntary transaction where all parties can gain wealth in some way. Data capital enables international assistance to shift from the nature of obligations to profit in both the short- and long-term (fact that the profits are always sought, but no one states clearly). Truly, business interests and societal interests can be one and the same. Profitable cooperation (the profits here are not limited to money) can achieve a win-win situation, which is a real healthier future for globalization.

The Change of State in Twentieth Century

The advanced economies of the world are enmeshed in a seemingly unending debt crisis. Especially in the wake of the Great Recession, a multi-year sovereign debt crisis broke out in the Eurozone. In the US, some would say Obama's accumulated budget deficits that added more than nine trillion dollar during his two terms. (When Obama was sworn in on January 20, 2009, the debt was US$10.626 trillion. When he left office on January 20, 2017, it was US$19.947 trillion.) Because fiscal expenditure is not only facing a large amount of financial aid to save the banking sector, of budgetary stimulus for markets' recovery from the economic recession, but also maintaining not a small amount of series expenses on the military, social welfare, and unemployment relief. Japan has a high debt to GDP with a ratio of exceeding 220%, which is "almost always … the result of overspending relative to tax revenues." The UK looks like other regions. The crisis came first; significant deficits were a response to the downturn, not precursors to it; the rising debt ratios came last, driven mainly by falling GDP. As things now stand, governments must pay massive interest on their outstanding public debt, and these interest payments weigh just as heavily on the public exchequer. As a result in economic circles, scholars brought to light: can we imagine political institutions that might harmonize today's systems on electoral democracy versus social welfare with prudent monetary and fiscal policies for a stability purpose?

How can a public debt be significantly reduced? Piketty summarized three main methods that can be combined in various proportions: taxes on capital, inflation, and austerity. He believes (see Footnote 3):

> *An exceptional tax on private capital is the most just and efficient solution. Failing that, inflation can play a useful role: historically, that is how most large public debts have been dealt with. The worst solution in terms of both justice and efficiency is a prolonged dose of austerity – yet that is the course Europe is currently following.*

Other developed countries outside Europe like the US also do the same; for example, the US recently withdrew from multiple the UN's organizations to save money (e.g., UNHRC: the UN Human Rights Council in October 2017; UNESCO: The UN Educational, Scientific and Cultural Organization in June 2018). In this section, I'll introduce how data capital help the worst solution of austerity.

Emerging Facts on Social Spending

In case of a strange paradox: Europe "has both the highest level of private wealth in the world and the greatest difficulty in resolving its public debt crisis" (see Footnote 3). One direct approach to reduce public debt to zero would be to privatize all public assets, namely:

> *According to the national accounts of the various European countries, the proceeds from selling all public buildings, schools, universities, hospitals, police stations, infrastructure, and so on would be roughly sufficient to pay off all outstanding public debt. Instead of holding public debt via their financial investments, the wealthiest European households would become the direct owners of schools, hospitals, police stations, and so on. Everyone else would then have to pay rent to use these assets and continue to produce the associated public services.*

While this approach is easy to implement and some solemn people advocate, Piketty thought he "should to mind be dismissed out of hand." Countries have been existing and will exist for a long time; new public debt may be generated. What, without any public assets, do? The expropriation of these assets without compensation by the government did first, then sell them again? From the standpoint of the general interest, it is impossible. If a state is to fulfill its mission adequately and durably, especially in the areas of education, health, and security, it must continue to own the related public assets.

The longer-term goal of reining in the public sector I envision is on public goods or services, instead of public assets. The variety of countries' approaches to delivering public goods and services should include outsourcing. As mentioned previously (see Chap. 8), the public sector is with the largest and most customer segments in the data economy. From demographics to investigate people's achievement, governments' social spending involves data of all kinds with their analysis. And now that the commercial market offers more accurate data in near-real-time, and technology to perform more sophisticated data analytics. It is time for the public sector to explore the benefits of outsourcing public data goods or services. Working with reliable third-party data producers, government agencies may increasingly look to reduce the costs as well as security and liability-related risks. It might not be an innovative idea. As firms would be permitted to obtain a tax credit to pay a fraction of their tax liability (not least because the government needs money to fulfill its role), both parties can benefit from this idea. Put, in other words, the government saves this part of expenditure without cash, and the data producer gets tax savings and is possible to profit from its derivative products. Because of data products' characteristics that are incremental updates and replication, besides, the subsequent maintenance costs of government agencies are not very high. In other words, in the following years, the product will get better, whereas social spending may decline.

Data Capitalized for Public Pension Systems

As positive tax expenditures due to tax relief associated with welfare, pensions account for the lion's share of total government transfer and is analytically equivalent to negative taxes. Public pension systems worldwide are pay-as-you-go (PAYGO) generally. That is, "contributions deducted from the wages of active workers are directly paid out as benefits to retirees" (see Footnote 3). The primary justification of the PYGO is to best guarantee a reliable and predictable payment. The PAYGO schemes nothing is invested, and based on the principle of intergenerational solidarity, the rate of return is, by definition, equal to the growth rate of the economy. In theory, if today's active workers pursue tomorrow's retirees rise, they should promote a higher birth rate and invest in their children's education. When introducing the PAYGO in the middle of last century, conditions were idea for such a virtuous series of events to occur. Demographic growth and productivity growth were both high. According to an example given by Piketty (see Footnote 3),

> *The growth rate was close to 5 percent in the countries of continental Europe, so this was the rate of return on the PAYGO system. Concretely, workers who contributed to state retirement funds between the end of World War II and 1980 were repaid (or are still being repaid) out of much larger wage pools than those from which their contributions were drawn. The situation today is different. The falling growth rate (now around 1.5 percent in the rich countries and perhaps ultimately in all countries) reduces the return on the pool of shared contributions.*

With acceleration peaks in the elderly population, incoming funds, and many others like it, will run out of money. The 2019 Social Security and Medical Insurance Trustee Report released by the US Treasury Department in April that year, for example, the balance of the US Social Security Fund will be exhausted by 2035. That means this fund won't be able to pay all the benefits to the citizens on schedule. Apart from revising the retirement bill to extend the retirement age with a broader collection, economists have been thinking to conclude that the PYGO system should be replaced as quickly as possible by a capitalized system. Put it simply, contributions by active workers are invested rather than paid out immediately to retirees. It seems to boost total returns 4% a year in financing the pensions of today's workers, and the investments can pay them adequately after several decades. There are several major flaws in actual practices, however. First must bear in mind that the return on capital is volatile. It would be risky to invest all retirement contributions in global financial markets. No one can ensure an average annual return on investment that is more than 4% in the following decades. Once encountering the financial market crash, all of the investment will be lost.

Second, even if assuming that a capitalized system is indeed preferable to a PAYGO system, the transition between the two is possible to raise a fundamental problem: "an entire generation of retirees is left with nothing" (see Footnote 3). This means trouble for the generation that is about to retire, who paid for the pensions of the previous generation with their contributions, and they would take a steady monthly income in retirement (e.g., that expected to pay their rent and buy their food) from the contributions of today's workers. which would in fact be invested in assets worldwide. This may lead to the widespread dissatisfaction among them; they will feel cheated out of their dream of a cozy retirement.

The third is another difficulty of pension reform concerning the complexity of systems. For a person who didn't experience different types of jobs (e.g., civil servants, private sector workers, and nonworkers), which is increasingly common in the younger generations, it is sometimes challenging to have a clear understanding of what they are entitled to. According to Piketty (see Footnote 3),

> *Some even think that they will get nothing even though they are paying a substantial amount into the system (something like 25 percent of gross pay).*

So, Piketty's solution is (see Footnote 3):

> *to establish a unified retirement scheme based on individual accounts with equal rights for everyone, no matter how complex one's career path. Such a system would allow each person to anticipate exactly what to expect from the PAYGO public plan, thus allowing for more intelligent decisions about private savings, which will inevitably play a more important supplementary role in a low-growth environment.*

But a recent research from the OECD (Organization for Economic Cooperation and Development; in February 2020) showed that one additional dollar of pension wealth reduces financial wealth by 53 cents.[9] The reason that this occurs is because mandatory pension schemes induce households to reduce their precautionary savings. In order to explore the heterogeneity of this effect, the study used cross-country household-level data on wealth, income, sociodemographics, and pension entitlements for seven European countries.

A more radical measure uses the form of universal basic income (UBI), coving not only pensions but unemployment relief. The measure has garnered

---

[9] ***See:*** d'Addio A, Roger M, Savignac F. Pensions and household savings: Cross-country heterogeneity in Europe. European Central Bank, Working Paper Series. February 2020. Retrieved from https://www.ecb.europa.eu/pub/pdf/scpwps/ecb.wp2372~2565732234.en.pdf?47af863d68ddc6ff0d940f1d4b92177e.

surprising support, particularly Elon Musk who is working to revolutionize transportation both on Earth and in space. SpaceX, Musk's rocket company, is now valued at more than US$20 billion. There are many variations of this measure, but the core idea is very similar (see Footnote 6).

> *Everyone receive a monthly check for a fixed amount that would be sufficient to pay for food, clothing, basic education, a warm, dry home, perhaps even some form of health insurance.*

Universal basic income is an idea that has been circulating among economists and politicians since the late eighteenth. Thomas Paine proposed one for everyone above 50. Nobel laureate Milton Friedman made it somewhat more complex to administer by a negative income tax. Libertarians claim its twofold benefits. On the one hand, they see in the UBI a comprehensive welfare program without stigma. On the other hand, it eliminates the need for a vast government bureaucracy to reduce a public debt significantly. Also, libertarians believe that "by giving everyone just enough to get by and no more, a UBI would not necessarily eliminate the incentive to work" (see Footnote 6).

Both universal basic income and Piketty's unified retirement scheme are not necessarily a bad idea. But the biggest challenge of the two is finding the money for it. If data enables us to go beyond money, why does the social innovation designed to solve the problems emphasize fiscal funds? My suggestion is that a unified plan for coving universally to make everyone receive a monthly check is acceptable, but needs to exchange the data. We are generating truly mind-boggling amounts of data daily, and they have been collecting. While one half is we made our private life public deliberately, the other half is collected unknowingly by digital providers from the Internet, mobile apps, and the Internet of Things. A fixed individual amount would allow each person to authorize precisely; for example, which data has what kinds of access permissions for what types of data user of how long to purpose keeping. Data profits from these data can be estimated and settled automatically, and then to distribute monthly in every single account. With growing amount of data with its production, income in each account may increase. The data and unused money in the account can be capitalized at the same time based on the user's authorization.

Just like Mayer-Schönberger's statement, "If we accept that a general shift in the economy from reliance on money to reliance on rich data, we ought to think more creatively" (see Footnote 6). Consider using data-rich profits to pay social welfare is another innovation on open-data access: not only data itself but the purpose of producing the data. One needed to reiterate is that the value of data lies in its use.

Government and Capital Accumulation in the Twenty-First Century

Regarding governmental debt, many people may ask: In an ideal society, what level of public debate is desirable? Piketty thought it is particularly challenging because of uncertainty about the answer. And he suggested to take a step back; that is, to relax this optimization problem into: What level of public capital is desirable, and what is the idea level of total national capital? Piketty quoted the equality $r=g$ the "golden rule of capital accumulation" baptized by Edmund Phelps in 1961 to give a criterion for an ideal government capital accumulation in his book (see Footnote 3), where r is the capital/income ratio and g is the economic growth rate.

> *Under certain hypotheses, ... one can establish a ceiling on the quantity of capital that one can envision accumulating a priori. The maximal level of capital is attained when so much has been accumulated that the return on capital, r, supposed to be equal to its marginal productivity, falls to be equal to the growth rate g.*

Piketty also believes that the answer given by the golden rule relating to a "capital saturation" strategy approaches too abstractly, and is not very useful in practice. Because any human society is likely ever accumulate that much capital. He argued that (see Footnote 3),

> *If one takes it literally, the golden rule implies much higher capital/income ratios than have been observed historically, since, as I have shown, the return on capital has always been significantly higher than the growth rate. Indeed, r was much greater than g before the nineteenth century (with a return on capital of 4–5 percent and a growth rate below 1 percent), and it will probably be so again in the twenty-first century (with a return of 4–5 percent once again and long-term growth not much above 1.5 percent). It is very difficult to say what quantity of capital would have to be accumulated for the rate of return to fall to 1 or 1.5 percent. It is surely far more than the six to seven years of national income currently observed in the most capital-intensive countries. Perhaps it would take ten to fifteen years of national income, maybe even more. It is even harder to imagine what it would take for the return on capital to fall to the low growth levels observed before the eighteenth century (less than 0.2 percent). One might need to accumulate capital equivalent to twenty to thirty years of national income: everyone would then own so much real estate, machinery, tools, and so on that an additional unit of capital would add less than 0.2 percent to each year's output.*
>
> *In practice, ... accumulate capital worth dozens of years of national income ... might require several generations to forgo consumption.*

Piketty went on say that if betting on the other side of the mathematical formula (i.e., focusing on economic growth), the only source of growth is "demographic" (see Footnote 3).

> *if one assumes that the population will grow forever at 1 percent a year and that people are infinitely patient and altruistic toward future generations, then the right way to maximize per capita consumption in the long run is to accumulate so much capital that the rate of return falls to 1 percent. But the limits of this argument are obvious. In the first place, it is rather odd to assume that demographic growth is eternal, since it*

> *depends on the reproductive choices of future generations, for which the present generation is not responsible (unless we imagine a world with a particularly underdeveloped contraceptive technology). Furthermore, if demographic growth is also zero, one would have to accumulate an infinite quantity of capital: as long as the return on capital is even slightly positive, it will be in the interest of future generations for the present generation to consume nothing and accumulate as much as possible. According to Marx, who implicitly assumes zero demographic and productivity growth, this is the ultimate consequence of the capitalist's unlimited desire to accumulate more and more capital, and in the end it leads to the downfall of capitalism ...*
>
> *If productivity growth is even slightly positive, ... the social optimum then becomes more difficult to resolve. If one knows in advance that productivity will increase forever by 1 percent a year, it follows that future generations will be more productive and prosperous than present ones. That being the case, is it reasonable to sacrifice present consumption to the accumulation of vast amounts of capital? Depending on how one chooses to compare and weigh the well-being of different generations ... : that it is wiser to leave nothing at all for future generations ...*

Economists are too entangled in the analysis based on assumptions. In my opinion, we can directly solve the optimization problem instead of its relaxation. In our society, lower government debt is a form of capital accumulation. The approach is through data production to avoid unnecessary waste in the process when the government pursues to achieve common goals, and to carry out valuable assistance and cooperation globally.

Will Data Sovereign Own the World?

When discussing a danger that the forces of financial globalization lead to a higher concentration of financial capital, Piketty wrote a section called "Will Sovereign Wealth Funds Own the World?" Now that I image the era of data capital, I took a corresponding section title. I suspect that data sovereigns may own the world. After all, the evidence on data of every minute in 2019 shows that[10] their businesses nearly cover all human life.

> *Google conducts 4,497,420 searches; Skype makes 231,840 calls; Netflix streams 694,444 hours of video; YouTube is watched 4,500,000 videos; Airbnb books 1,389 reservations; Uber takes 9,772 rides; Venmo helps transfer 162,037 payments.*

The main reasons that economists do not have much respect for the wealth rankings (e.g., published by magazines such as Forbes in the US and other weeklies in other countries) are as follows. One is that such rankings suffer from some biases and methodological problems, so in their way, they cannot offer reliable information about the global dynamics of wealth. The other is too few wealthier people (i.e., only a few hundred or at most a few thousand individuals at the present time represent a little over 1% of global wealth) to be genuinely

[10] ***See:*** Data Never Sleeps 7.0. Domo.Com. Retrieved from https://www.domo.com/learn/data-never-sleeps-7.

significant in macroeconomic terms. That is to say, this leaves out nearly 99% of global capital, in which the sovereign wealth fund is the largest one. Piketty's estimation is as follows (see Footnote 3).

According to available (and notoriously imperfect) estimates, sovereign wealth funds in 2013 had total investments worth a little over $5.3 trillion, of which about $3.2 trillion belongs to the funds of petroleum exporting states (including, in addition to ... [the Norwegian fund and the Middle East funds such as e Abu Dhabi, Saudi Arabia, Kuwait, and Qatar], the smaller funds of [Russia,] Dubai, Libya, Kazakhstan, Algeria, Iran, Azerbaijan, Brunei, Oman, and many others), and approximately $2.1 trillion to funds of nonpetroleum states (primarily China, Hong Kong, Singapore, and many smaller funds). For reference, note that this is almost exactly the same total wealth as that represented by the Forbes billionaires (around $5.4 trillion in 2013). In other words, billionaires today own roughly 1.5 percent of the world's total private wealth, and sovereign wealth funds own another 1.5 percent. It is perhaps reassuring to note that this leaves 97 percent of global capital for the rest. One can also do projections for the sovereign wealth funds just as I [Piketty] did for billionaires, from which it follows that they will not achieve decisive importance – 10~20 percent of global capital – before the second half of the twenty-first century, and we are still a long way from having to pay our monthly rent to the emir of Qatar (or the taxpayers of Norway).

Sovereign wealth funds, thus, often refers to petroleum exporting countries. Take the case of Saudi Arabia's sovereign wealth fund as an example; according to Piketty (see Footnote 3), it has chosen to keep a very low profile, while ranking third after Abu Dhabi and Norway among sovereign wealth funds of petroleum exporting states and ahead of Kuwait, Qatar, and Russia.

The Saudi reports are ... sober and provide information not only about oil reserves but also about national accounts and the government bud get. These are clearly addressed to the people of the Kingdom of Saudi Arabia, whose population was close to 20 million in 2010 – still small compared to the large countries in the region (Iran, 80 million; Egypt, 85 million; Iraq, 35 million) but far larger than the microstates of the Gulf. ... Saudi funds seem to be invested ... aggressively. According to official documents, the average return on the Saudi sovereign wealth fund was no more than 2~3 percent, mainly because much of the money was invested in US Treasury bonds.

Why would Saudi Arabia choose to invest in US Treasury bonds when it is possible to get far better returns elsewhere? Piketty explained that (see Footnote 3),

the political and military aspects of the choice must also be taken into account: even though it is never stated explicitly, it is not illogical for Saudi Arabia to lend at low interest to the country that protects it militarily. ... To my [Piketty] knowledge, no one has ever attempted to calculate precisely the return on such an investment, but it seems clear that it is rather high. If the United States, backed by other Western powers, had not driven the Iraqi army out of Kuwait in 1991, Iraq would probably have threatened Saudi Arabia's oil fields next, and it is possible that other countries in the region, such as Iran, would have joined the fray to redistribute the region's petroleum rents. The dynamics of the global distribution of capital are at once economic, political, and military. This was already the case in the colonial era, when the great powers of the day,

Britain and France foremost among them, were quick to roll out the cannon to protect their investments. Clearly, the same will be true in the twenty-first century, …

Compared with sovereign wealth funds, it is similarly for data sovereigns who involve data resources with their protection as the force from data technologies. This is why I often emphasize that the so-called "attack" faced by data sovereigns is very elastic, and in case of infinite attack may be in line with the same level of offensive operations from US cyber forces. In this section, I'm going to talk about data resources and data force, respectively.

Energy Resources vs. Data Resources

An energy resource relating to sovereign wealth funds is oil that belongs to fossil fuels. The first petroleum reservoirs exploited by humans were found as a result of seeps on the surface. The first oil well was drilled in Titusville, PA., in 1859. Oil wells eased petroleum recovery and initiated an oil boom. Fossil energy now are our primary source of energy, as about 80% of world energy consumption (39% oil, 23% coal, and 24% natural gas).[11] In 2011, the world's energy expenditure on energy totaled over US$6 trillion,[12] in which Europe spends close to one-quarter, North America is about 20%, and Japan 6%. But fossils fuels have a disadvantage in that they are non-renewable on a human time scale, and cause potentially harmful effects on the environment. The current known oil reserves will likely run out sometime between 2050 and 2150.

Solving the problem of energy is equivalent to no issues related to war, hunger, and poverty. This is a fundamental core problem for human society. So, a new concept has emerged on this problem in economics—free energy as a free renewable energy.

Renewable energy is the energy that is naturally replenished on a human time scale, such as sunlight, wind, waves, geothermal heat. Solar Energy arrives from the Sun by electromagnetic radiation. It is a nearly unlimited source, it is renewable, and primarily, non-polluting. But solar energy requires other resources (Li, Rare Earth Elements) to exploit, and its efficiency is very low (the cost of the solar support on electricity generation is 5–10 times that of thermal power generation). The gravitational pull of the Moon on the Earth causes tides. Tidal flow can be harnessed to drive turbines. This is also a nearly unlimited non-polluting source of energy but only locally. Wind power is limited to areas with consistently high winds, and so is limited to geographical conditions. The windmills

[11] ***See:*** Nelson SA. Energy Resources in Physical Geology. Tulane University. October 19, 2015. Retrieved from https://www.tulane.edu/~sanelson/eens1110/energy.htm.

[12] ***See:*** The entry of "World energy consumption" on Wikipedia. Retrieved from https://en.wikipedia.org/wiki/World_energy_consumption.

are not aesthetically pleasing to look at, make a lot of noise and kill large numbers of birds, all problems that would need to be overcome to expand this resource. As for hydroelectric resources, they will not likely increase, since most rivers are already dammed, and there are few places left where new hydroelectric facilities could be built. Geothermal energy is limited to areas of known thermal activity (mainly recently active volcanic regions). It is a great local resource, but will never play a significant role as an energy resource.

The other three fundamental sources of energy are from the reactions of nuclear fission, nuclear fusion, and antimatter. The latter two require high costs. That means collecting energy from the last two would spend more (that needs to use energy generated from other resources). The former one is to use radioactive uranium that is concentrated and made into fuel rods that create large amounts of heat as a result of radioactive decay. But uranium is also non-renewable on a human time scale. The known uranium reserves are currently less than five million tons. According to global nuclear power plants' annual consumption (around 70 thousand tons), an estimation was to use 60–70 years. But the yearly consumption has exponentially increased so that it might only be used for 20–30 years. Once proposed as a cheap, clean, and safe way to generate energy, nuclear fission has come under some disfavor. Costs of making sure nuclear power plants are clean and safe and the problem of disposing of radioactive wastes, which are unsafe, as well as questions about the safety of the plants under human care, have contributed to this disfavor.

Note that none of those energies (mentioned in the above two paragraphs) can serve in daily transportation, distribution & logistics despite having the nuclear-powered submarine; namely, the devices using these energies cannot be installed on vehicles, aircraft, and ships. This is why our society relies mostly on fossil fuels for energy. Thus, we need to ask how much longer the society can rely on oil. Further, what are the options for the future?

My question, nevertheless, is whether we need to think differently? Given US$1 billion, for example, what do you think on this amount is unlimited or not? If spending US$1 million per day, the fund is not inexhaustible, but how about only spend US$1 per day? We always complain about the lack of energy resources. Do we consume them too much? Then, can we get global energy consumption to be reduced? Or, if that is impractical, could the increase in demand at least be slowed? Some environmental scientists believe that people need to change their lifestyles to live more sustainably. If they are right, then people will have to live simpler lives, scaling back their expectations of everything in life. But I don't think we should change a lot to make economy down because the answer to reduce energy consumption lies at data resources.

First, traffic flow analysis can shed light on the operations of transportation, distribution & logistics to help do optimization. Our industrial structure now is mass production. That means the output of a large number of standardized products in a constant manufacturing process, especially on assembly lines. After mass production, logistics and warehousing are needed to deliver products as commodities to the retailers (e.g., supermarkets), or direct to the customer's home. Many energy (e.g., gasoline) that would consume in logistics (and warehousing) can reduce based on traffic flow optimization. Also, on-demand manufacturing as a new model based on consumer needs and wants can keep costs down for everyone (both business and its customer) and reduce waste. Japan throws out 6.21 million tons of perfectly edible food each year while, also in Japan, an estimated 3 million kids don't have enough to eat. The 6.21 million ton figure is from the Japan agricultural ministry for the year 2017.

Second, better utilizing available data resources may further help reshape the current industrial structure. Today, for example, data technologies in construction are being developed at a breakneck pace. Data connected equipment and tools, telematics, mobile apps, autonomous heavy equipment, drones, robots, augmented and virtual reality, and 3D printed building are being deployed and used on jobsites across the world to reduce the costs and energy.

I've mentioned in the prior section, in every single minute across the world, places are collecting data at a speed that would shock anyone. The above examples are numerous in other industries and sectors. As we trend to better utilize data resources, the increase in demand for energy resources will be slowed. The value of data resources is reflected in the compensation of energy that would be wasted in large quantities. Thus, I believe data resources are free energy.

The Use and Abuse of Data Force

For the sake of clarity in expression, the distinction between data force and cyber force makes sense. A data force like a mercenary belongs to a specific data sovereign or an alliance of data sovereigns. A data force could be constituted by personnel to conduct data activities. Exaggeratedly, the activities include an attack on other data sovereigns to obtain the data, an defense to protect own data resources from other data forces, and the exploitation to do the routine operation on data and data products. But in essence, data force aims to enhance data and data products by repairing or mitigating vulnerabilities using data innovations.

A cyber force is part of a state military branch or a combined command of a country, under the framework of "self-defense" prescribed in international legal (e.g., Article 51 of the UN Charter). The framework defines how to use cyber force or a force of an armed nature in the technical sense during a cyber conflict. There are now existing cyber forces in several countries. Some well-known activities involve an attack made by the US Cyber Command's attack that uses a virus "stwxnet" to destroy Iran's nuclear facilities and a response from the Norwegian Cyber Defense Force that was against Russian's cyber-attack carried out by "Fancy Bear."

Data force can also be abused, and one way to avoid the abuse of data force is wargaming. This is precisely what Person of Interest season 4's If-Then-Else demonstrates. You've seen the episode, but I'll sum it up quickly: faced with a shitstorm of Samaritan operatives and a Twinkie's chance at fat camp, Team Machine (which consisted of Finch, Reese, Shaw, and Root; Fuso included) fought to battle the economic crash started by Samaritan in the last episode, and needed the assistance of The Machine to hightail it out of a basement of Wall Street's brain in one piece. The Machine ran tens of thousands simulations top find an optimal exit strategy, which played out in real-time to us, showing the often undesirable outcomes of attempting to escape.

Wargaming is a simulation of different confrontations. Early wargaming is the Warring States period of China about 4500 years ago when the ancients used pebbles and twigs to demonstrate maneuvers and evaded the need for actual hostilities. In 1811, Baron Leopold von Reisswitz, the Prussian war councilor at Breslau, invented a wargame that finally broke with the grid and alternating-move conventions of chess. Wargaming has been changing by contemporary computers but is still with an impasse of intensely depending on manual operations. One way to solve the impasse is to have more data resources. So, wargaming caters to needs from data forces. With developing in data technologies, future military counterwork can also be decided by virtual wargaming.

Chapter 11
Worldwide Taxation on Data Capital

There exist two ways for a government to finance its expenses (such income is called revenue): debt and taxes. In the prior chapter (i.e., Chap. 10), I introduced the issue of how data capital that helps reduce public debt significantly. But in general, taxation is by far preferable to debt from a perspective of justice and efficiency. The problem with debt is that it has to be repaid. In terms of the standpoint of the general interest, it is normally preferable to tax some people rather than borrow from them. Economists think taxation that is preeminently a political and philosophical issue, instead of a technical matter. The core of taxation is "how sovereign citizens can democratically decide how much of their resources they wish to devote to common goals,"[1] such as education, health, retirement, inequality reduction, employment, sustainable development, and so on. Taxation is the main source of a country's fiscal income. Using Piketty's words (see Footnote 1),

> *At the heart of every major political upheaval lies a fiscal revolution. The Ancient Régime [that was from the Late Middle Ages (circa 15th century) until the French Revolution of 1789] was swept away when the revolutionary assemblies voted to abolish the fiscal privileges of the nobility and clergy and establish a modern system of universal taxation. The American Revolution was born when subjects of the British colonies decided to take their destiny in hand and set their own taxes. ("No taxation without representation").*

Actually what concrete form taxes take is the crux of political conflict in any society.

According to *Paying Taxes 2020*[2] that is a joint annual publication by the World Bank Group and PWC looks at tax regimes in 190 economies with a

---

[1] ***See:*** PiKetty T. translated by Goldhammer A. Capital in the Twenty-First Century. Cambridge, MA: the Belknap Press of Harvard University Press; Reprint edition (August 14, 2017), 2017: 816 pages.

[2] ***See:*** World Bank Group and PwC. Paying Taxes 2020. November 26, 2019. Retrieved from https://www.doingbusiness.org/en/reports/thematic-reports/paying-taxes-2020.

© Springer Nature Switzerland AG 2021

C. Tang, *Data Capital*, https://doi.org/10.1007/978-3-030-60192-8_11

database, tax reforms have always been a top priority for government worldwide, especially during the economic downturn.

China is the most frequent tax reform country ever. Since economic reform and opening up in 1978, China has sought to set up a streamlined tax system geared to its unique market with multiple changes over 40 years. In 1980, China allowed the introduction of both enterprise and individual income tax. In 1983 and 1984, China carried out two steps of a "profit for tax" reform. Tax-sharing reform in 1994 was a comprehensive tax reform by the division of national and local taxes to mobilize the enthusiasm of local governments to participate in fiscal and tax. China announced the abolition of agricultural taxes at the end of 2005. By adopting a new law in 2007, China ended the dual corporate-income-tax system by unifying domestic and foreign enterprise income tax at a rate of 25%. In 2009, China's refined oil price and tax reform took an important step, to promote energy conservation and emission reduction and to reform the fee system in China. Since the start of 2012, China has been undergoing a vast reform of its indirect tax regime including a replacement of the sales tax (or business tax) with a Value Added Tax (VAT) regime based on OECD principles. Apart from the loosened corporate income tax, China renewed individual income tax law in 2019, to provide a big relief on the tax burden of personal income. China passed two new Laws—the Tobacco Leaf Tax and Vessel Tonnage Tax Law to take effect on July 1, 2018. Also, drafting of draft tax laws such as deed tax law, resource tax law, consumption tax law, stamp tax law, and real estate are also in the pipeline.

I'll briefly interpret the economics of taxation here. Taxes as governmental revenue are often obtained by three economic bases: income, consumption, and wealth. In the US, for example, the federal government relies mainly on income taxes (or taxes on income), state governments depend on both taxes on income and consumption, and local governments use wealth taxes (or taxes on capital) to raise their revenue. Of course, these three categories are not exempt from ambiguity, and the dividing lines are sometimes not clear. For instance, the income tax applies in principle to capital income (as well as earned income) but is also from wealth. The two taxes I mentioned above in China's VAT reform (i.e., value-added tax and business tax; since 2012) are both on consumption. Also, there exists a tax called a consumption tax. Ideally, a properly designed consumption tax system would reward savers and penalize spenders (one thinks the salt tax). The three are often called "indirect" taxes: they are paid indirectly, as part of the selling price of a purchased good, and do not depend on an individual taxpayer's income or wealth. The word "fee" in the above expression of "reforming the fee to tax" is a kind of tax in nature that appeared in the twentieth century: contributions to government-sponsored social insurance programs. But when being collected, it may have different correspondences (e.g., an exchequer).

Beyond the above classification, another criterion for characterizing different types of tax is the degree to which each type is proportional or progressive.

A tax called "proportional" refers to the term "flat tax," with which everyone enjoys the same rate. For example, the rate of 25% stipulated by China when ending the dual corporate-income-tax system unified domestic and foreign enterprise income tax. China's individual income tax, whatever being renewed or not (i.e., before or after 2019), is progressive. The differences are only on the tax threshold and tax bracket. The tax-free threshold is a basic allowance at which a person begins paying taxes. A tax bracket is progressive when its rate is higher for some than for others, whether it be those who earn more, those who consume more, or those who own more. A tax bracket can also be regressive when its rate decreases for wealthier individuals.

Rethink a Global Tax on Capital

Piketty believes that the "utopian" idea of a global tax on capital (see Footnote 1) he proposed is an ideal tool for innovating tax systems.

> *Such a tax would provide a way to avoid an endless in egalitarian spiral and to control the worrisome dynamics of global capital concentration.*

Also, he listed some reasons to proof it is useful (see Footnote 1).

> *First, even if nothing resembling this ideal is put into practice in the foreseeable future, it can serve as a worthwhile reference point, a standard against which alternative proposals can be measured. Admittedly, a global tax on capital would require a very high and no doubt unrealistic level of international cooperation. But countries wishing to move in this direction could very well do so incrementally, starting at the regional level (in Europe, for instance). Unless something like this happens, a defensive reaction of a nationalist stripe would very likely occur. For example, one might see a return to various forms of protectionism coupled with imposition of capital controls. Because such policies are seldom effective, however, they would very likely lead to frustration and increase international tensions. Protectionism and capital controls are actually unsatisfactory substitutes for the ideal form of regulation, which is a global tax on capital – a solution that has the merit of preserving economic openness while effectively regulating the global economy and justly distributing the benefits among and within nations. Many people will reject the global tax on capital as a dangerous illusion, just as the income tax was rejected in its time, a little more than a century ago. When looked at closely, however, this solution turns out to be far less dangerous than the alternatives.*

Specifically, Piketty's global capital tax is (see Footnote 1):

> *the capital tax I am proposing is a progressive annual tax on global wealth. The largest fortunes are to be taxed more heavily, and all types of assets are to be included: real estate, financial assets, and business assets – no exceptions. This is one clear difference between my proposed capital tax and the taxes on capital that currently exist in one country or another, ... One drawback of these taxes is that they are based solely on real property. (Financial assets are ignored, and property is taxed at its market value regardless of debt, so that a heavily indebted person is taxed in the same way as*

a person with no debt.) Furthermore, real estate is generally taxed at a flat rate, or close to it. ... in some countries (such as the United States) rely on fairly sophisticated assessment procedures with automatic adjustment to changing market values, ... In some Europe an countries (including France, Switzerland, Spain, and until recently Germany and Sweden), there are also progressive taxes on total wealth. Superficially, these taxes are closer in spirit to the ideal capital tax I am proposing.

... It would make sense to tax net wealth below 200,000 euros at 0.1 percent and net wealth between 200,000 and 1 million euros at 0.5 percent. This would replace the property tax, which in most countries is tantamount to a wealth tax on the propertied middle class. The new system would be both more just and more efficient, because it targets all assets (not only real estate) and relies on transparent data and market values net of mortgage debt. ... there is no reason why the tax rate on fortunes above 5 million euros should be limited to 2 percent. ... To avoid divergence of the wealth distribution (that is, a steadily increasing share belonging to the top centiles and thousandths), ... one might envision rates of 10 percent or higher on billionaires.

Why Reinvent the Wheel?

Piketty does have some wheels. His "utopian" idea at least innovates at two points. The first is on a global concept. Piketty envisions moving forward this ideal solution step by step, "first at the continental or regional level and then by arranging for closer cooperation among regions" (see Footnote 1). His proposal can indirectly resolve the issues relating to both tax evasion and exemptions. Overall without considering globally, the merits of a wealth tax depend in part on how a country taxes capital gains—the income accrued from financial capital—and estates. On a conventional basis, designing of wealth taxes requires that "tax exemption should be high to ensure that the net wealth tax is only levied on the very wealthy," and that "tax rates should be low and take into account tax rates on capital income to avoid imposing excessively high tax burdens on capital so as to prevent capital flight." One of the difficulties countries face when they try to impose a capital tax on their own is to have some mechanisms to combat tax evasion. For example, a mandatory audit rate requiring a certain number of people who pay the wealth tax to be subject to an audit every year, or a one-time tax penalty for wealth-taxpayers who try to renounce their citizenship. A global concept enables that each member state of a continent (or at least a region) could generate similar revenues by applying such a wealth tax on its own. In practice, however, the taxation of wealth is often riddled with exemptions. This is why, in recent decades, such a tax has fallen out of favor in several countries. In 1990, 12 OECD number countries imposed some form of wealth tax. By 2017, that number had dropped to just four: France, Norway, Spain, and Switzerland.

Secondly, Piketty claims to use an automatic solution for asset valuation before the taxpayer fills the tax statement, and financial assets held in foreign banks included. He understood very well that there is no obstacle to have an automatic solution relying on techniques and some legal attempts (e.g., the practice from the Foreign Account Tax Compliance Act adopted in the US in 2010 and phased in by stages in 2014 and 2015). He also cited a negative example to illustrate the reason for taking financial assets as a top priority in tax on capital.

> *In 2012, the Italian government, faced with one of the largest public debts in Europe and also with an exceptionally high level of private wealth (also one of the highest in Europe, along with Spain), decided to introduce a new tax on wealth. But for fear that financial assets would flee the country in search of refuge in Swiss, Austrian, and French ..., the rate was set at 0.8 percent on real estate and only 0.1 percent on bank deposits and other financial assets (except stocks, which were totally exempt), with no progressivity. ... the system also had the unfortunate consequence of imposing a regressive tax on wealth, since the largest fortunes consist mainly of financial assets and especially stocks. This design probably did little to earn social acceptance for the new tax, which became a major issue in the 2013 Italian elections; the candidate who had proposed the tax – with the compliments of European and international authorities – was roundly defeated at the polls.*

Still, as he explained, this is the difficulty for a country acting on its own to impose a progressive tax on capital.

Piketty's supporters are generally activists. Emmanuel Saez and Gabriel Zucman, two economists at the University of California, Berkeley, advised Elizabeth A. Warren (she withdrew from the race in the 2020 US presidential election on March 5, 2020) on her campaign proposal. Warren's proposal is to levy a 2% wealth tax on Americans with assets above $50 million, as well as a 3% wealth tax on those who have more than $1 billion. "The wealthiest 1% of families currently face a total tax burden, including state and local taxes, of about 3.2% relative to wealth,"[3] Saez and Zucman highlighted, "but the bottom 99 percent of families currently has a tax burden of 7.2 percent relative to their wealth." One of the key motivations for introducing a progressive wealth tax, the two encomiasts believe, is to curb the growing concentration of wealth. They estimated that the wealth tax would raise $2.75 trillion over a 10-year period from about 75,000 families, or less than 0.1% of U.S. households. Warren's proposal caused some controversy in early 2019. Daniel Hemel, at the University of Chicago, specifically wrote an article in Time Magazine to criticize it as "the wrong solution to the right problem."[4] One of Hemel's criticisms is that "a wealth tax would be an administrative nightmare" (see Footnote 4).

---

[3] ***See:*** Stein J, Ingraham C. Elizabeth Warren to propose new 'wealth tax' on very rich Americans, economist says. Washington Post. January 24, 2019. Retrieved from https://www.washingtonpost.com/business/2019/01/24/elizabeth-warren-propose-new-wealth-tax-very-rich-americans-economist-says.

[4] ***See:*** Hemel D. Elizabeth Warren's Wealth Tax on the Super-Rich Is the Wrong Solution to the Right Problem. Time.Com. January 30, 2019. Retrieved from https://time.com/5516903/elizabeth-warren-wealth-tax-income-assets.

> *It would require us to calculate the value of a household's assets year in and year out – which may be easy for publicly traded stocks with daily price quotes, but is much harder for illiquid assets like real estate or stakes in privately held companies ... – a rather wide range. With other assets like fine art (remember, we're talking about the super-rich here), valuations are even less certain: works sometimes sell at auction for less than their pre-sale estimates and sometimes for fourfold more.*
>
> *Warren says that she would ensure proper enforcement of her wealth tax by hiring more IRS employees and conducting more audits. But valuation problems won't be solved by more manpower. In many instances, there is no one "right" answer to the question of how much an asset is worth, no matter how many people we throw at the problem. Wealthy taxpayers will likely pick a low-end figure, and it will be impossible for the IRS to prove that they are wrong.*

Warren's proposal designed by Saez and Zucman failed to learn the two innovations of Piketty's global tax on capital. Regarding "harder for calculating the value of a household's assets year in and year out" mentioned by Hemel, it is also the point where being criticized for *Capital in Twenty-First Century* most frequently. But financial assets or business assets, "which may be easy for publicly traded ... daily price quotes," most people think, are what Piketty wants at all.

> *... especially for ["financial assets" or] "business assets" and, in practice, for nearly all large stakes in listed and unlisted companies. To do this is to drain much of the content from the progressive tax on capital, and that is why existing taxes have generated revenues so much smaller than ...*

The Problems with Piketty's Solution

Instead of exploring the approach of distributing the wealth tax among member states, Piketty's solution would place its hopes of redistribution through immigration while layering on an automatic transmission of banking information. They are the reasons why this idea is unwise.

First, one point that needs to be emphasized is that a significant fraction of global financial assets is already hidden away in tax havens such as the Cayman Islands. According to official statistics (e.g., relying on national data collected by the IMF and so forth), it seems that the net asset position of the wealthy countries (e.g., the US) compared with the others is negative. That means American households, firms, and even governments own much more foreign assets than other countries that hold US domestic assets. In principle, the negative official position should be counterbalanced by an equivalent positive position for the rest of the world. But this is not the real case: a finding is that the world as a whole is in a substantially negative situation if one adds up the financial statistics globally. This is a so-called "statistical anomaly," and it has gotten worse in recent years according to various international organizations. Recent research shows that (see Footnote 1)

The most plausible reason for the discrepancy is that large amounts of unreported financial assets are held in tax havens. ... Now, all the evidence indicates that the vast majority (at least three quarters) of the financial assets held in tax havens belongs to residents of the rich countries.

Since the issue of financial transparency and information sharing is closely related to the ideal tax on capital, it is not appropriate to merely design a progressive annual tax without considering tax distribution among member states. In particular, these tax havens implement a flat tax system generally.

Secondly, relying on traditional banking that has been in precarious situations is a big failure. It is useless to condemn people to keep their money out of the country where they pay taxes. One has the right to choose the proper administration of wealth. Of course, the tax havens can invoke some excuses like maintaining bank secrecy. But the problem is, so far, the results from banks joining forces with financial technology firms have been mixed. Although in theory, banks should be comfortable working with the data due to the fact that they have gathered a variety of financial data for decades by operating large data stores that contain detailed customer information. In fact, they haven't done much with the data they have. "In this context, they are rich in data but poor in insight,"[5] Viktor Mayer-Schönberger summarized. In more detailed (see Footnote 5),

This may be a remnant of earlier times, when analyzing data was difficult and costly. And banks are traditional. Their ethos generally is to preserve rather than to risk, and their staffs reflect that ethos. Perhaps it also reflects a strong customer preference; maybe people don't trust their banks enough to let them use the rich personal data banks have collected to create new products or services. But it is likely structural as well: if an organization has been focused on helping money lubricate the economy, it will have a hard time thinking fare beyond what has made it successful.

Learned from Tax Paid in Data

In addition to sharp comments on traditional banking, Mayer-Schönberger is with unique insights on regulating data-rich markets. Denying "a robo tax" and envisioning "tax paid in data" are both impressive.

Mayer-Schönberger denied using a robo tax (in this book, I called a digital errand-runner) to solve the problem of which workers are made redundant by data-driven automation. This opinion stands on the opposite side of Piketty's solution, to address "the shrinking role of labor and the shift in income distribution," by using indirect taxes (e.g., consumption tax) rather than direct taxes (i.e., tax on capital). He said that (see Footnote 5),

[5] ***See:*** Mayer-Schönberger V, Ramge T. Reinventing Capitalism in the Age of Big Data. Now York, NY: Basic Books (February 27, 2018). 2018; 288 pages.

> *As income seems to shift from labor to capital, the proposed measures on the distributive side aim at taxing the source of this automation-driven income. This can be achieved, for instance, by taxing machines such as computers running machine learning systems (or more precisely taxing the value that machines add) that make this new wave of automation possible. Colloquially called a "robo tax," this mechanism would supplant payroll taxes as workers are being replaced by machines. The goal is not necessarily to tax the economy more, but differently. ... the European Parliament discussed and ultimately rejected a proposal to apply a robo tax throughout the continent. ... it was dismissed because of fears it might stifle innovation.*
>
> *A somewhat different distributive strategy is to impose a tax not on the value generated by machines but rather on the capital that supposedly claims a rising share of the economy's income. This could be achieved, for example, through an increase in the capital gains tax or by instituting a new wealth tax. The latter is the more popular option and the one that is being advocated in the several European nations. In most proposals, these distributive measures would keep overall tax revenues flat or increase them only incrementally, just enough to offset the decrease in tax revenues resulting from a decline in labor share.*
>
> *These distributive measures seek to establish a more just tax system that focuses on where the value add is generated, and to offset some of the negative tax consequences of this upheaval in labor markets. But in practice, their effects may be limited. ... the robo tax is [not] a solution to the social problems this contraction will bring about. Similarly, a wealth tax may feel fair, but experts predict it will yield only a modest volume of tax revenue and do little to remedy social disruption.*

The second of "tax paid in data" is what we all should learn. It is on a great idea of "channeling the large profits back into the economy as investments" with abandoning "money" altogether. Mayer-Schönberger's design is as follows (see Footnote 5).

> *... we ought to think more creatively about how to get companies to pay the taxes they owe. Governments might consider a partial payment of taxes in data rather than money. ...*
>
> *Together with the data-sharing mandate we propose, this would make data available to small firms, especially start-ups, so that they can compete against the big players. It may also be a good way to jump-start innovation. The data could also be used by government to improve it services. And it might be offered to nonprofits, researchers, and society at large so that everyone can benefit from the profits of superstar firms.*
>
> *Unlike a conventional tax, a tax in data may be less onerous for those who must "pay" it. In contrast to money, data is non-rivalrous. IT isn't used up, and a firm can make use of its own data even if it allows others to use it as well. ...*
>
> *... even a modest amount of a firm's annual tax bill paid in data could make a sizable different. A resulting boost in innovation would also lead to new economic growth, taxes on which would mostly be paid the traditional way: with money.*

Of which, a so-called "data-sharing mandate" is defined as below.

> Building on this idea, we suggest what we term a progressive data-sharing mandate. It would kick in once a company's market share reaches an initial threshold – say, 10 percent. It would then have to share a randomly chosen portion of it ... data with every other player in the same market that requests it. How much data it must make available would depend on the market share captured by the company. The closer a company is to domination, the more data it would have to share with its competitors.

> *Our plan for progressive data sharing is targeted at the growing number of companies that use ... data and adaptive machine learning to improve their service offerings, from Google, Facebook, Apple, and Microsoft all the way to Tesla. ...*

Corporate Profits and Investment

The differences between Mayer-Schönberger and Piketty are that the former looks at superstar firms' profits to channel them back into the economy as investments. At the same time, the latter sees the top rich people's capital income to transform the structure of inequality. Of course, their focuses are not the same: data and money, respectively.

The so-called "superstar firms," Mayer-Schönberger cited MIT economist David Autor's definition, is "a particular type of firm is reaping unprecedented profits far exceeding what would be expected in an environment of efficient competition."[6] And the profits these superstar firms earn aren't fully reflected in their profit-and-loss statement. The characteristic has a sign that is to "invest huge amounts every year in research and the development of new products and lines of business." These superstar firms, listed by Mayer-Schönberger as examples, are Amazon, Google, Apple, Facebook, and Netflix. Mayer-Schönberger's view is nuanced; that is, finance capitalism is resurging through a data-fueled labor-destroying, and finance capitalism's resurgence lies on massive skyrocketing profits. Mayer-Schönberger believes (see Footnote 5), therefore, consumption taxes like the VAT and retail sales taxes are possibly better than taxes on income or taxes on capital.

> *On the distributive side, new and increased taxes on capital (including a tax on machines) would essentially accelerate and deepen the decrease of capital share while leaving untouched the huge pile of corporate profits that lies at the root of the problem. That would be close to a textbook example of a bad idea.*
>
> *A far more appropriate distributive policy response to skyrocketing profits accruing to certain firms is to tax there firms. ... That is what's needs as both labor and capital shares decline.*
>
> *... large profits may be less worrisome if they are channeled back into the economy as investments. There has been a move to implement a so-called progressive consumption (PCT) to replace the progressive income tax. With it, personal income would be taxed only to the extent that income isn't reinvested. ... And remember that in general, investments don't pay high returns anymore, so don't expect a raft of new investments.*

In fact, labor-income (wage) taxes can be thought of as pre-paid consumption taxes, while consumption taxes are viewed as post-paid consumption taxes. The public finance literature demonstrates the equivalence between consumption and wage taxes, despite the question of whether to tax consumption or income

---

[6] ***See:*** Autor D, Dorn D, Katz LF, *et al.* The fall of the labor share and the rise of superstar firms. The Quarterly Journal of Economics, 2020;135(2), 645-709.

repeatedly arises in tax policy over the decades. A real-effort experiment to compare tax behavior showed that,[7] subjects tend to prefer consumption tax to income tax when the nominal tax rates are identical. It's no wonder Mayer-Schönberger chose consumption taxes. Also, I've mentioned in the early of this chapter; the income tax can apply to capital income from wealth. As shown in Case 11.1, either way, the guy and the government end up in the same place.

Case 11.1: A comparison of income tax and tax on capital

Imagine that an individual has $100, which he invests at a 5% annual rate of return. If he is subject to a 2% tax on her wealth calculated at the beginning of the year, he will owe the government $2. If he is subject to a 40% tax on her investment income calculated over the course of the year, he will also owe the government $2.

The figures in the example above—a 40% tax rate and a 5% rate of return—were chosen to make the math easy, but they turn out to be reasonably realistic. The average rate of return on 10-year Treasury bonds from 1928 to 2018 was approximately 5%, and the top federal income tax rate under current law is 40.8%. If high-net-worth individuals are paying income taxes at the top rate and if their portfolios are growing at 5% annually, they already are giving approximately 2% of their wealth to the government each year. As the game is quiet, you decide to play in the library.

Another point I would like to emphasize is that an economic conclusion of the inverse relationship in incomes between labor share and capital share cannot apply to data capital. The revenue generated from data capital is both the capital income and labor income. Particularities determine this in "data as capital" (see Chap. 5). Also, I argued in this book repeatedly, that is, no value but high costs without data on production. So, it is unnecessary to struggle with taxes of type. We, data scientists, have to leave something for economists to resolve.

A Data Sharing Based Coordination and Collaboration

Someone may be curious about Mayer-Schönberger's purpose on "tax paid in data." The answer is obvious: to reach coordination and collaboration. Just like he and his coauthor Stian Westlake wrote in their book ***Reinventing Capitalism in the Age of Big Data*** (see Footnote 5):

[7] ***See:*** Kurokawa H, Mori T, Ohtake F. A Choice Experiment on Taxes: Are Income and Consumption Taxes Equivalent? (April 6, 2016). ISER Discussion Paper No. 966, Available at SSRN: https://ssrn.com/abstract=2755864 or https://doi.org/10.2139/ssrn.2755864.

Despite the importance of the moments when humans first tamed fire, invented the wheel, or developed the steam engine, these discoveries and inventions pale compared to our human ability to coordinate. Without coordination, a flame would not warm more than one human being; the wheel could not transport but a single individual; and the steam engine would have no tracks to roll on and no factory to operate in. If there is a single crucial thread that has persisted through human history, it is the importance of coordination, whether our aim is to build a castell or a country.

The future data sharing based coordination and collaboration, I envisage, will center around a data sovereign or a data sovereign alliance rather than the government. In general, being a data sovereign is possibly one or more existing superstar firms and a unicorn league which both can produce data using advanced analytics. The government also can be a data sovereign in specific fields but shows typically as a provider of converging data and purpose. Concerning a data sovereign, there exist more partnerships such as nonprofits and researchers. Because of injecting a piece of market DNA, a data sovereign's interior is decentralization aiming to make internal information flows faster than data-rich markets. This is a "small" market with market-based competition and cooperation, internal talent marketplaces included. I think it may be the most natural place to pilot with "tax paid in data," due to no external data exchange involved. Before having outsourced a particular public data product like a census, the government still needs to do data analytics, such as estimating the costs of traditional data gathering activities (e.g., salaries paid to temporary workers), and comparing the prices of commercial data for verification, and so forth. The government can bid for the public product according to analytical results. Of course, some data sovereigns may lobby. This is because; while coming from different sectors, the data can also be replaced. In the case of the census, survey providers, mobile vendors, and online retailers may offer similar information to the Census Bureau needs, such as demographic, lifestyle, and other data on individual households. As a result, there will be more and more "small but mighty" enterprises that can provide people-oriented products or services; meanwhile, data sovereigns have been possibly contributing to enhancing their data forces—creating social value is much beyond business interests. Note that sharing data-rich profits with individual data owners is included.

A Global Tax on Data Capital

Given the two ideal tools, a global tax on (financial) capital and a progressive data-sharing mandate based tax (paid in data), designed by Piketty and Mayer-Schönberger, respectively, each cannot be directly applied to data capital but may be measured against these ideals. In this section, I'm going to design a tax on capital to regulate the globalized data capital of the twenty-first century for adapting it to today's society. I will look forward to having the feedbacks of all parties.

Suppose that the global system of individual data accounts is established. An individual account enables a person data self-sharing through an authorization mechanism to obtain data profits. A batch (that is a collection of data points grouped within a time interval) corresponding to the specific data products with their producers, is transparent. Even if the person does not know exactly, they can learn by performing queries. Data profits shared by each data product can be automatically estimated and settled to distribute to the individual account monthly. The individuals are fully informed about the market share of each data product globally. The first assumption clarifies that the subject of a global tax on data capital is the individuals. Explicitly, such a tax aims to capital incomes; that is, personal data capital self-accumulated by the individual authorized the others to produce and then to obtain profits in proportion.

Suppose next that all data producers globally have data accounts as well. In terms of the data producer, it is not limited to a data sovereign or some groups within a data sovereign. A data producer account may open in the name of one person or enterprise. And government specifics, nonprofits, and researchers can also have a data producer account although their products or services might not be profitable. Compared with an individual account, a data producer account has one more function relating to data requests. Put, in other words, a data producer account can request the data corresponding to a data product that needs to design. For example, a pharmaceutical manufacturer asks for some patients' genetic data concerning specific cancers via the account. If all available authorized data is used timely, data profits from the relevant data product or service can be obtained to settle automatically via the data producer account. The second assumption clarifies that the subject of a data capital tax also includes a variety of data producers. The incomes generated by these data producers are the combination of capital income and labor income, and their proportion can be learned from the technical level. Explicitly, such a tax aims to capital incomes; that is, personal data capital self-accumulated by the individual authorized the others to produce and then to obtain profits in proportion.

Someone may ask how should the profits from data resources these data producers formerly owned or controlled be shared? The problem could be solved by setting a time period of mandatory disclosure. During the period, the individuals or other data owners can self-claim their data. The idea of some mandated disclosure of the information is not a new thing. The patent system is on the idea: the privilege of a patent is only granted to those who reveal the working of their invention in their application (at least that is the principle) so that their creativity and novelty can be proved. Data ownership self-claim allows profit sharing to the data owners, and the data was formerly collected through the unfair format clause (see Chap. 5). Of course, such an ownership claim can be made as a data product for paid services. For instance, after having obtained authorization, using the anti-search engine technology deeply buried negative results, in response to someone's request. When having received confirmation from one patient, another example, collecting all EHRs via multiple hospitals' open APIs integrated into the patient's individual data account.

It is essential to recognize that there is no technical obstacle to doing so. I will give a complete solution for each point based on the latest self-governing openness of data in the next chapter.

The global tax on data capital I envisaged possibly overlaps with taxes on incomes in terms of the tax burden. Because the current inheritance tax in the rich countries is against landed wealth, the two charges (i.e., the inheritance data capital taxes) won't across functions to have tax burden overlapping. Note that the digital part of landed wealth and real estate belongs to data capital and can also participate in data production to have profitability.

Based on the above statements and assumptions, the data capital tax has two distinct justifications: a contributive and an incentive justification. The contributive logic is quite simple: income is often tight for the more needy individuals who can generate data but lack money, the profit-sharing behavior of data producers is not only direct for paying taxes but also indirect for making up the possible loss of function of government public relief. The tax may attenuate this dynamic of wealth inequality partially. Regarding the incentive justification, it likes what Piketty's statement (see Footnote 1).

> *It relies on a logic of incentives. The basic idea is that a tax on capital is an incentive to seek the best possible return on one's capital stock. ... According to this logic, the purpose of the tax on capital is thus to force people who use their wealth inefficiently to sell assets in order to pay their taxes, thus ensuring that those assets wind up in the hands of more dynamic investors.*

That is to say; the data capital tax encourages data capital accumulation and data production. Again, the data is without the value but with the costs, if no production.

There exist a premise in the tax schedule for my proposed data capital tax; namely, the proposed tax is not intended to replace all existing taxes. The primary purpose of the data capital tax is to regulate data capital. It represents a newly modest supplement to the government revenue streams derived from technological progress. So, I'll clarify three aspects of tax system design, tax sharing among governments, and data resource rents.

Flat Tax on Data Capital Globally

Unlike Piketty's capital tax, my data capital tax is the flat tax. Capital in the form of land has been taxed since time immemorial. And property is generally taxed at a low flat rate. This is because; the property tax aims to "guarantee property rights by requiring registration of titles; it is certainly not to redistribute wealth" (see Footnote 1). Given the principle that "each type of capital would be taxed in the same way, with no discrimination a priori" (see Footnote 1) and the four reasons as follows, I thus didn't choose the progressive tax. Of course, whether choosing the flat or progressive tax system, the machine runs in the same.

The first reason is from a historical perspective. It is undeniable that some countries, such as Germany and Sweden, established an annual progressive tax on capital as early as the late nineteenth century or early twentieth. But a fundamental flaw in assess valuation was fatal to a progressive tax on capital: the value of the assets subject to taxation depended on the infrequently revised assessments by the tax authorities, rather than on the annually revised market value. More than this, the threshold for each tax bracket was also on arbitrary factors. Challenges to such arbitrary taxation became increasingly common, in a period of rapidly rising real estate prices. Often the courts became involved to rule on violations of the principle of equal taxation. Germany and Sweden abolished their annual taxes on capital in 1990–2010.

One needs to point out is that the value of the current data products is high but volatile. I'm not sure whether it can be changed by more advanced data technologies, user preferences, or the others. Therefore, the profits from a specific data product might not lie in short-term traffic. It should balance from its wealth growth and speed in a relatively long period. The progressive tax is not a possibility right choice. Imagine that a big M or W shape (with twin peaks and tall sides) shown in paying a fee in consecutive months (even if years), the majority of taxpayers will reject it. And they don't think it is caused by a volatile payable. Still, they may believe that the new tax is unreasonable in asset valuation, which is a similar situation as Germany and Sweden experienced historically, then to let the courts involve intervening in the tax's implementation.

The second reason is on tax burden overlapping. The data capital tax is obvious to conflict with income taxes for both individuals and enterprises. Generally speaking, taxes on income mostly adopt the progressive tax. A realistic fact is the rise of tax competition in a world of free-flowing capital has led many governments to exempt capital income from the progressive income tax. This is particularly true in Europe, whose relatively small states have proved incapable of achieving a coordinated tax policy. Because of the overlapping in both taxes on capital and income, it is natural to deduct the repeated payment from taxes on income when processing the final annual settlement; for example, to the exemption of data product profit-sharing as dividends from the taxes to which capital income is subject. Suppose that the data capital tax is progressive as well; it is no conducive to manual verification. While calculating by the machine automatically, spot checks by tax officials and self-inspection by taxpayers are required. This concern is for convenience only.

Thirdly, from the perspective of revenue sharing, redistribution through immigration is not reasonable. Piketty's solution (that places its hopes of redistribution through immigration) cannot apply to digital, intangible data capital in cyberspace. Future citizens can choose their nationality without immigration in another country. And even their identity can be a virtual one (see Chap. 5). Even though it is via immigration, people cannot carry out data capital in person. I wonder if the future Amazon Snowmobile business will support moving massive volumes of data across borders. In addition, the tax rate in tat havens is mostly flat. If designing a new progressive tax, I think it is not enough to counter them. Furthermore, it may do not help implement the data capital tax in the counties with a negative net asset position.

Lastly, conducting progressive taxation may play only a limited role in regulating data capital income. The fact that tax overall is relatively close to proportional for the majority of the population crowding together. This part of income is possible not to have a strong influence on the structure of inequality. The evidence suggests that it does not require progressive taxation to adjust and increase government revenue. Still, one consequence of progressive taxation is that in most countries, taxes have (or will soon) become regressive.

Tax Sharing Through Progressive Tax Treaties Among Governments

The critical step toward a global tax on data capital should be to extend the international level of visualizing tax sharing and tracking its evolution. The United States has tax treaties with many foreign countries. Under these treaties, residents (not necessarily citizens) of foreign countries may be eligible to be taxed at a reduced rate or exempt from US taxes on certain items of income they receive from sources within the United States. These reduced rates and exemptions vary among countries and specific details of income. With certain exceptions, the treaties do not lessen the US taxes of US citizens or treaty residents (who are subject to US tax on their worldwide income). Treaty provisions generally are reciprocal (apply to both treaty countries). Thus, a US citizen or treaty resident who receives income from a treaty country and who is subject to taxes imposed by foreign countries may be entitled to certain credits, deductions, exemptions, and reductions in the rate of taxes of those foreign countries. US citizens residing in a foreign country may also be entitled to benefits under tax treaties of that country with third countries. These treaties can serve as a basis for tax sharing among governments. I recommend considering a progressive plan for tax sharing. The next step is to make the sharing process of a global tax on data capital systematically and transparently, such as entrusting digital errand runners to handle and visualize the process automatically.

The data capital tax would generate massive data about distribution, collection, and tax sharing. These include, but not limited to, specific data resource utilization data (e.g., the geographic information with each country rescaled in proportion to relevant products), several similar data products' comparable data of global market share, and a particular product's profit distribution data based on its life cycle. National governments, international organizations, and statistical offices around the world would, at last, be able to obtain reliable data about data capital. That means it offers a shortcut to make regulation. The stakes for data capital regulation are also considerable. Take the current financial capital as an example (see Footnote 1),

> *The international organizations currently responsible for overseeing and regulating the global financial system, starting with the IMF [i.e., the International Monetary Fund], have only a very rough idea of the global distribution of financial assets, and in particular of the amount of assets hidden in tax havens. ... Navigating our way through a global financial crisis blanketed in such a thick statistical fog is fraught with peril.*

We may understand the Internet via some volunteers like Ruslan Enikeev's map[8] (that is a bi-dimensional presentation of the Internet at the end of 2011, encompassing over 350 thousand websites from 196 countries). But that still wasn't far enough away for data capital. Because data capital exists in data nature that we know very little.

Another use is to provide auxiliary decision-making on future energy resources for energy saving, emission reductions, and health co-benefits. This is also of great benefit to the government. While having an independent environmental protection agency (with people, fiscal and material resources), these data allow the government thinking from an economic perspective, which will be conducive to the future of nationally institutional arrangements, the cost of the transition to new sources of energy, and state-building, and so on.

Note, in passing, that this visualization can also open access to the taxpayer with appropriate evidence. In practice, such as question with sharing evidence would force governments to clarify and broaden international agreements concerning tax treaties and other implementation rules.

[8] ***See:*** Enikeev R. The Internet Map. Retrieved from https://internet-map.net.

Data Resource Rents vs. Tax Paid in Data

Rent in economics is a surplus-value after all costs, and regular returns have been accounted for (see Chap. 8). When referring to rent in natural resources such as coastal space or minerals, it can be conceptualized as abnormal or supernormal profit. For example, petroleum rents constitute a unique problem to make the Middle East region, where locates the most petroleum exporting countries in the world, to be a militarily sensitive area. Piketty once envisaged (see Footnote 1):

> *If the world were a single global democratic community, an ideal capital tax would redistribute petroleum rents in an equitable manner. National laws sometimes do this by declaring natural resources to be common property. Such laws of course vary from country to country. It is to be hoped that democratic deliberation will point in the right direction.*

But the world is not a single democratic community, therefore, the redistribution of natural resources is often decided in far less peaceful ways.

Is it easier to resolve the problem within a country? Maybe does no yes. China's land ownership and land-use rights are separate. In the era of China's planned economy, there was no the concept of land-use rent that is a part of state-owned enterprise profits. But China's government and economists ignored the rent after having transited to a socialist market economy. It said that at least 20 trillion RMB of land-use rant has been lost since 1990. Data as a human-created resource can have rent as well. At the very beginning of the data capital era, we should ask whether the government can be a rent-seeker to obtain data resource-use rents, apart from being an open-access data provider. I envision the difference depends on the purpose of data use. If providing the data for non-commercial use, the role of government should be an open-access data provider; otherwise, it can be a data rent-seeker. Mayer-Schönberger's "tax paid in data" happens to help the government to be a data rent-seeker.

Mayer-Schönberger's "tax paid in data" is a great idea. It can revitalize some infrequently used data. But I feel it might not be efficient. The first is a data storage issue. Once having decided to use the idea, the government should find a place to store the data paid as a tax. Randomly designating a data center of any data sovereign is certainly not realistic. Using the existing data centers owned by the government might not have enough capacities. The only possibility is to build a new one by allocating fiscal resources. Building a new one may be easily solved in China due to "China Speed." But in developed countries such as the US and Australia, even if it enjoys support across the political spectrum, it is no guarantee it will be approved into existence anytime soon. Also, constructing will take a long period that is often for several years.

Secondly, it is difficult for the government to determine the availability and value of the data paid as a tax. The government may find it futile because they spent more public expenditure and workforce on data identification. The data payer may feel at a loss because they paid the unique valuable data they think. And the subsequent user might not have excellent feedback on the former two.

Thirdly, the basis of "tax paid in data" is a "progressive data-sharing mandate" that aims to share a randomly chosen portion of data with competitors in the same market. Such kind of data sharing is mandatory without any basis for cooperation. It can belong to a disguised attack from a data sovereign perspective. It is not very practical.

If applying in the issue of data resource-use rent, however, this idea may be useful. First, the data belongs to the government so that there exists no problem with data storage. Next, the data can be cleaned and identified incidentally, when producing an exclusive public data product or service. Lastly, at this time, no competitors are requesting a mandated data sharing. I mentioned in the previous chapter (see Chap. 10); if the government can open both the data and purposes, it would reduce public debt and promote market growth. For the data producer, their additional costs in the public data product can be a tax credit after having negotiated with the government. The extra cost, for example, is about adding alternative data to obtain a complete analytical dimension.

In this sense, we believe that "tax paid in tax" can be renamed as "data as fiscal policy."[9]

[9] ***See:*** Zhu Y. The governments need to change "land finance" to "data as fiscal." China XiaoKang. 2015;23:24-25. In Chinese. Retrieved from http://www.chinaxiaokang.com/renwu/2016/0215/4346.html.

Chapter 12
A Blueprint Via the Latest Self-Governing Openness of Data

In the first chapter of this part (i.e., Chap. 10), I had a prospect of the global society with data capital with a focus on the comparison of the society dominated by financial capital. In the next chapter (i.e., Chap. 11), I purposed a set of schemes for the regulation use on data capital (i.e., a global tax on data capital supplemented by "data as fiscal"), by rethinking Thomas Piketty's "global tax on capital" and Viktor Mayer-Schönberger's "tax paid in data." The regulation can visualize to track the global data-rich profit flows, and then achieve the data capital tax sharing among the government worldwide. So, in this chapter, I will examine more closely based on the latest self-governing openness of data to give a blueprint.

Piketty designed a blueprint for his capital tax concerning an automatic transmission of banking information within Europe. He believed that the plan could be achieved via democratic deliberation if bank information is automatically shared and the tax authorities have accurate assessments of who owns what. However, the data flow through such a transmission may be too small to reach the democratic and financial transparency he expected. I think the latest self-governing openness of data can help. Still, some additional considerations are requested.

In the beginning of this chapter, I'll briefly describe the core principle of this technology.[1] Then, I am going to elaborate problems relating to our blueprint. And they are: establishing a people-oriented mechanism, identifying the transaction in the best matches, and making system transparency via a feedback mechanism.

[1] ***See:*** Zhu Y, Xiong Y, Liao Z, *et al*. Self-governing openness of data. Big Data Research. 2018;4(2):3-14, in Chinese.

© Springer Nature Switzerland AG 2021

C. Tang, *Data Capital*, https://doi.org/10.1007/978-3-030-60192-8_12

Introduction to Self-Governing Openness of Data

The self-governing openness of data is a new data tool beyond traditional (closed) data governance. The problem of the self-governing closeness mode is that only data owners can use data resources for apps design featured as follows. There are: known in advance, basically internal, quantity limited, safety controllable, and privacy controllable. Correspondingly, the products or services produced under the mode of self-governing openness can be: unknown in advance, basically external, quantity unlimited, safety uncontrollable, and privacy uncontrollable. As shown in figure 12.1, the self-governing openness of data can solve open-related requirements that cannot be handled by existing database management systems.

In fact, the self-governing closeness mode has largely hindered open government data worldwide. Some studies reveal that:[2]

> *(Peled, 2011 at data.gov): although the amount of the total datasets on data.gov dramatically climbed from 47 to 389,681 in 2 years, five out of 169 participating agencies contributed 99.37% of all datasets (as of 8 May 2011). It was evident that most participating agencies were increasingly reluctant to update or add new datasets, and only a few agencies remained active.*

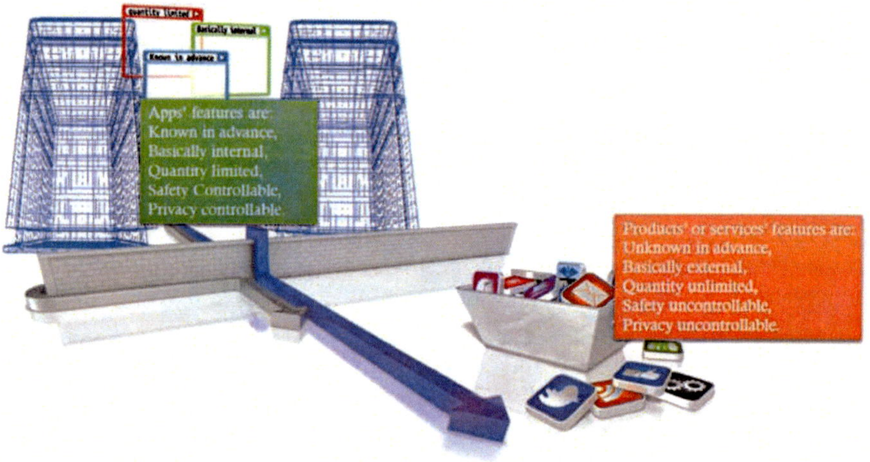

Figure 12.1 The difference between self-governing closeness and openness of data. Author redrew this figure according to literature (see Footnote 1)

[2] ***See:*** (1) Peled A. When transparency and collaboration collide: The USA open data program. Journal of the American Society for Information Science and Technology 2011;62: 2085-2094; (2) Yang T, Lo J, Shiang J. To open or not to open? Determinants of open government data. Journal of Information Science. 2015;41(5): 596-612; (3) Stevens H. Open data, closed government: Unpacking data. gov.sg. First Monday, 2019, Retrieved from https://firstmonday.org/ojs/index.php/fm/article/view/9851/7746.

(Yang, 2015 based on the site logs of open data portals): only 10 percent of the datasets account for 90 percent of network traffic load on our open data platform. Some agencies' open data are not attractive and receive no feedback or usage from the public. The intention and the willingness of those agencies to open more datasets may be negatively influenced and significantly reduced.

(Stevens, 2019 at data.gov.sg): most of the time series data have only been recorded for relatively short time periods. Over one quarter of the dataset span one year or less and significantly more than half (57 percent) span ten years or less. The vast majority of the data is either reported annually (45 percent) or in an ad hoc fashion (25 percent). There are very few datasets that provide fine-grained data (less than monthly).

These scholars finally concluded that (see Footnote 2),

(Peled, 2011): This research highlights the contradiction between Open Data's transparency goal ("All data must be freed") and federal agencies' goal of collaborating with each other through data trade. The research also suggests that agencies comprehended that Open Data is likely to exacerbate three critical, back-office data-integration problems: inclusion, confusion, and diffusion.

(Yang, 2015): ... the complexity of open data initiatives from four perspectives: technology, organization, legislation and policy, and environmental context. Responsible central government agencies should also take the lead to develop unified standards of metadata in specific policy areas. In addition, the identified factors are discussed with the theory of planned behavior to perceive how the factors act as determinants to influence agencies' intentions to participate in open data initiatives.

(Stevens, 2019): ... cannot just put data out there and hope for the best: "sharing data" will not prove to be some sort of magic bullet for solving economic, political or social problems. Innovative uses of data will require not just access to data but more attention to who and what are represented, the contexts within which it can be used and who can work with the data.

Let's go to the point. Based on the self-governing openness mode, a prototype system could be designed for supporting data resources to open access, sharing, and transaction, we call "self-governing openness of data." This prototype provides a DataBox (see Footnote 1) to organize and manage the data, serving both data owners or data providers (i.e., the seller) and data users (i.e., the buyer) at the same time. Figure 12.2 indicates the minimal flow of this prototype. First, the prototype allows a seller to get one or more virtual DataBoxes for their data and helps a specific data description after one DataBox is filled. The prototype next packages the DataBox and has been real-time monitoring to wait for any requests from the seller and other prospective buyers. Then, the prototype supports any paid or unpaid transactions for one or more data users who meet the restrictions given by the data provider. Finally, after having reached the transaction conditions, the prototype opened the DataBox to deliver the data to the buyer with a series of necessary toolkits under the development framework additionally. The core principle is simple and easy. Regarding the meaning of "transaction" here, it is not to pay by money only. I suggest adopting distribution or redistribution if having data profits. Because the data does not mean data product or service; still, a data product doesn't mean profits. Later in this chapter, I'm going to explain how to apply the prototype to address practical issues.

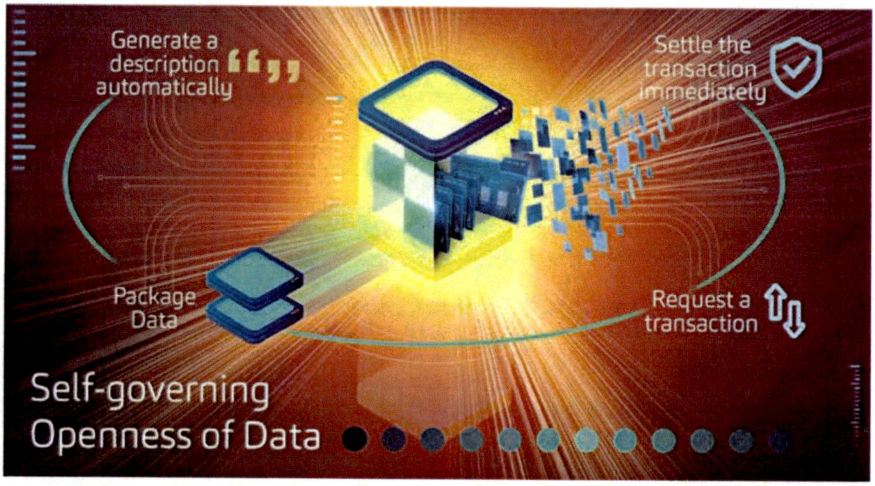

Figure 12.2 A diagram of interpreting self-governing openness of data: DataBox

Authorization and Distribution: A People-Oriented Logic

Authorization and distribution are at the core of the system of collecting a global tax on data capital (hereinafter referred to as "the system"), aiming to control the dissemination and abuse of the data. While having built individual data accounts for most of the global citizens, each one can have multiple DataBoxes corresponding to a variety of data. In other words, the tax system should enable data owners or data providers (i.e., the seller) to organize and to manage data open access in the form of DataBox. Concerning the purpose (i.e., to control the dissemination and abuse of the data), there exist two corresponding technical issues here. First is how the system makes the data both "self-governing" and "openness." The word "self-governing" means that the ownership of data is always in handling by data owners (unless the data owner is willing to give up). The word "openness" is that the data can be opened access to data users (i.e., the buyer), and the data user cannot disseminate the data. Therefore, the system requires a new data model to organize and manage data resources. DataBox is the latest data model. The outsider can see any available data resources through various DataBoxes to determine whether they use or not. The DataBox enables that the system supports data users to carry the data link to the apps and to use it with guidelines given by the DataBox.

Second is how the system enables data security and privacy, especially on protecting data scarcity. Data security and privacy are well to understand. There exist a large quality of international- and national level legal or technical measures. We just need to follow these instructions. Data scarcity, as one of the major bottlenecks for open data access, involves the value and benefits closely. More detailly, open data access is possible to mean the loss of the original profit for the data. Even though a data resource is with more meaningful content, no one is willing to buy if everyone has a copy or its copy is available at any time. Since the data is easy to be replicated to disseminate at a slight cost, it is a contradiction. The value is limited if the data is not open and only used by a single data owner itself, while the scarcity may be lost if the data is opened. In reality, data resource owners are not ready to risk the unusual, so they are unwilling and dare not to settle for open. The system should guarantee that, a limited quantity of data users, who meet the restrictions given by the data owner, can only use the data within a specified time period, so that the data won't lose its scarcity.

Someone may be curious about where the data is from, and they are mind-boggling to think that we are creating massive data. Every interaction with one computer or phone has data. Every interaction on social media creates data. Every time we walk down the street with a phone in our pocket, it's tracking our location through GPS (global positioning system) sensors: more data. Every time we buy something with our contactless debit card: data. Every time we read an article online: data. Every time we stream a song or movie: also, data. Other statistics include: 1.5 billion people are active on Facebook every day that is one-fifth of the world's population, and two-thirds of the world's population now own a mobile phone. In fact, all of the world's data is the result of our increasingly digitized way of life. The connectivity of modern smart devices—not just phones, but cars, TVs, and fridges, and so on—also plays a huge role. Cars are now often with automatic sensors that produce massive trajectory data when moving. These trajectory data, for instance, can be used to understand urban hot zones caused by a spatiotemporal aggregation effect and further facilitates researching the stop-and-wait behavior of people in specific regions during daily driving. Smart TVs can be a user-centric wellness platform that syncs across phones, pedometers, and other wearables to integrate the data. These workout data is researchers are seeking to quantify human on aspects of a person's daily life in terms of physiological variables, environmental exposures, and psychological mood states. Smart fridges have several cameras inside that take an image or a video every time the door closes, and people can then check the levels of milk from a smartphone app. They have advanced humidity and temperature sensors that can work either as a fridge or a freezer, and they will, of course, recommend recipes according to the food inside. Also, they can order fresh direct groceries automatically when the food is gone.

All the above and other devices are constantly gathering and transmitting data in ways we know or don't know. In the future, the individual data account enables merge these data as data resources belong to our individuals or families. So, each people needs to do authorization on these data. You might choose to whether allow your government to use it or not. You may approve a foreign enterprise to produce. Also, You can set up various restrictions on the data. The most important thing is that the authorization should follow your intentions and purposes of freedom.

Any data to be opened access is required to apply, test, and use a DataBox. Thus, DataBox should meet some technical standards. As a basic unit of providing open data to data users, first and foremost, DataBox requires a formal specification of conceptualization on: (1) data description as general knowledge such as which is the data type or application domain; (2) data operations, such as optimize cross, void merge, and other operations; (3) data restrictions as data owners' wants and needs, such as reporting some findings when using the data; and (4) measures that are based on data users' purposes and requirements such as demand modeling, matching, and pricing.

Secondly, because loaded data that are unknown in advance, basically external, quantity unlimited, safety uncontrollable, and privacy uncontrollable, DataBox needs some functions and mechanisms as follows. They are: (1) the ability of prevent data leakage and benefit protection, (2) anti-theft and anti-tamper technologies, and (3) anti-debugging and anti-reverse engineering techniques. That means the above functions and mechanisms should be encapsulated within the DataBox.

The third involves DataBox splicing. In general, data users won't apply a single DataBox. Contrary, they need to have DataBoxes in a batch. Thus, what we call "anti-data jigsaw technology" is required. Now consider the jigsaw itself. The pieces are randomly scattered. Some are face up, some aren't. Imagine that each DataBox is a piece. While nor do the data user know how many pieces there are, how big it is supposed to be, or whether any pieces are missing, the finished picture can be made after a data user obtained multiple pieces, or several users work together. We defined the "data jigsaw" as below.

Definition 12.1 *Data Jigsaw* Data jigsaw is a behavior that is to splice multiple data fragments or numerous times to restore the data to be integral.

Through data jigsaw, the data user is possible to obtain a protected data object without authorization to privatize it. This behavior will bring substantial harm to the system, such as the period or purpose of data use. Moreover, the data can be re-disseminated to other unauthorized people, resulting in a secondary infringement for the original data ownership. Apart from setting on the DataBoxes, it also requires discussing on the system level. I will come back to this in the next section when I turn to the question of the system.

To sum up: authorization and distribution are built around a people-oriented logic and a principle of equal access to personal data ownership and data-rich profits. This is the first step to walk away from surveillance capitalism pointed out by Shoshana Zuboff. Purely disputes on the means of effecting real improvement in the living conditions of the least advantaged is pointless. The long-word privacy format clause that forces the user to sign (no signature means no service) is no longer needed. The step consists of two components: constructing the global system of individual data accounts is a starting point; and personal authorization at the core of this system.

Identifying the Transaction in the Best Matches: With Malice Toward None

I want to caution the reader; in Chap. 7, I cited a Viktor Mayer-Schönberger's hypothetical example about "a frying pan," which may be one of the fastest cases we can enjoy the transaction in the best matches. While existing in some sectors, such as travel and ride-sharing, better matching in transactions is still rare. This is because, conventional markets often focus on price, and the preferences of the buyer or the seller are trivial. Specifically,[3]

> *All preferences are condensed into the price a buyer is willing to pay and a seller is willing to accept. Bringing the two together is supposed to happen pretty much by itself, as long as buyer and sellers state their various preference in terms of price and as long as there are sufficient (and sufficiently diverse) market participants. In practice, valuable preference information gets lost, perhaps because market participants fail to reflect correctly all their preference in price, but also because others erroneously deduce preferences from a price. Under those conditions, something that looks like a good match, in fact, is not. We may think the market works, but in truth it doesn't, and it leaves everyone worse off.*

Over the decades, some algorithms were developed to evaluate sets of multiple preferences and their relative weights and to identify best matches. But the algorithms are not perfect, in particular, most shallow applications require more and more private information. As shown in Case 12.1,[4] Google rewrote the same privacy policy into a sprawling 4000-word explanation of the company's data practices.

Just like the following comments given by the New York Times (see Footnote 4),

> *This evolution, across two decades and 30 versions, is the story of the internet's transformation through the eyes of one of its most crucial entities. ... Google has collected much of that information to help shape world-changing products, while also serving as the fuel for a money-printing advertising machine.*

In addition to the excessive dependence on private information, algorithmic bias is often in the news. It has become a bit of a hot button in ethical AI as algorithms are increasingly deployed in sensitive applications to notice the selection of job candidates, to inform bail and parole decisions, and to filter loan application, among many others.

---

[3] ***See:*** Mayer-Schönberger V, Ramge T. Reinventing Capitalism in the Age of Big Data. Now York, NY: Basic Books (February 27, 2018). 2018; 288 pages.

[4] ***See:*** Warzel C, Ngu A. Google's 4,000-Word Privacy Policy Is a Secret History of the Internet. NY Time. July 10, 2019. Retrieved from https://www.nytimes.com/interactive/2019/07/10/opinion/google-privacy-policy.html.

Case 12.1: Google's Privacy Policy across two decades and 30 versions

Case V.2 Google's Privacy Policy across two decades and 30 versions

Year	Cause	What Google collect
1999	Words: 609 Important Content: Google may share information about users with advertisers, business partners, sponsors, and other third parties. However, we only talk about our users in aggregate, not as individuals.	1. aggregated search activity 2. personal information you provide 3. clickthrough information 4. cookies
2009	Words: 3,999 Important Content: If you have an account, we may share the information submitted under your account among all of our services in order to provide you with a seamless experience and to improve the quality of our services. We will share personal information with companies, organizations or individuals outside of Google when we have your consent to do so. We require opt-in consent for the sharing of any sensitive personal information. When you upload or otherwise submit content to our Services, you give Google (and those we work with) a worldwide license to use, host, store, reproduce, modify, create derivative works (such as those resulting from translations, adaptations or other changes we make so that your content works better with our Services), communicate, publish, publicly perform, publicly display and distribute such content.	Things you create or provide to Google 1. your name 2. password 3. phone number 4. payment information 5. content you create, upload, or receive from others when using our services If you use Google services for calls or messages 1. telephone information 2. phone number 3. calling-party number 4. receiving-party number 5. forwarding numbers 6. time and date of calls and messages 7. duration of calls 8. routing information 9. types of calls Your activity 1. terms you search for 2. videos you watch 3. views and interactions with content and ads 4. people with whom you communicate or share content 5. Chrome browsing history 6. activity on 3rd party sites and apps that use our services 7. voice and audio information when you use audio features 8. purchase activity Apps, browsers, and device data 1. unique identifiers 2. browser type and settings 3. device type and settings 4. operating system 5. mobile network information including carrier name and phone number 6. application version number 7. IP address 8. crash reports 9. system activity 10. date, time, referrer URL of your request Data from publicly accessible sources Data from partners 1. trusted partners 2. marketing partners 3. security partners 4. advertisers Location data 1. from GPS, IP address, device sensor data, wifi access points, cell-towers, Bluetooth-enabled devices From Android devices with Google apps, collected periodically 1. device type 2. carrier name 3. crash reports 4. which apps are installed

Definition 12.2 *Algorithmic Bias* Algorithmic bias describes repeatable errors caused by algorithms and models that create unfair outcomes. Bias can emerge due to many factors, including but not limited to the design of the algorithm or the unintended or unanticipated use or decisions relating to the way data is collected, selected, or used to train the algorithm. Algorithmic bias can have impacts ranging from inadvertent privacy violations to reinforcing social prejudices of race, gender, sexuality, and ethnicity.

Some recommendation algorithms, for example, treat members of different classes differently, perpetrating biases, providing varying degrees of utilities, and inducing disparities. The cases that have emerged are quite similar. In predicting whether a student will drop out from a massive open online course, by using features such as high school GPA, SAT scores, AP courses taken and scores, intended major, and levels of physical activity, Gardner *et al.* reported that,[5] in some cases there are noticeable differences between the models' predictive performance on male students compared to female students. In detecting people with false identities, by comparing driver's license or ID photos with other departments of motor vehicles images on file, Buolamwini and Gebru reported that,[6] the overall classification accuracy on male individuals was higher than female individuals. They also found that the predictive performance on lighter skinned individuals was higher than darker skinned individuals.

Fortunately, in recent years data scientists and statisticians have been hard at work exploring algorithmic fairness. In the prestigious International Conference on Machine Learning 2018, 2 of 5 best papers selected from more than 600 accepted papers focus on algorithmic fairness. Among the two, a paper called *Delayed Impact of Fair Machine Learning* led by Liu *et al.* from a UC Berkeley research group shared one of two top honors.[7] This article investigates whether fairness would hurt the advantaged groups. In another one, Tatsunori *et al.* developed an approach based on distributionally robust optimization, to solve a problem caused by traditional models that is to only consider the majority groups.[8] Besides, Romano *et al.* presented an operational methodology that achieves the goal of constructing prediction intervals by offering rigorous distribution-free coverage guarantees holding in finite sample, to support equi-

[5] ***See:*** Gardner J, Brooks C, Baker R. Evaluating the fairness of predictive student models through slicing analysis. In Proceedings of the 9th International Conference on Learning Analytics and Knowledge. 2019;225-234. Association for Computing Machinery.

[6] ***See:*** Buolamwini J, Gebru T. Gender shades: Intersectional accuracy disparities in commercial gender classification. In Friedler, S. A. and Wilson, C., editors, In Proceedings of the 1st Conference on Fairness, Accountability and Transparency. PMLR 81:77-91, 2019.

[7] ***See:*** Liu LT, Dean S, Rolf E, Simchowitz M, Hardt M. Delayed impact of fair machine learning. In Proceedings of the 35th International Conference on Machine Learning, PMLR 80:3150-3158, 2018; arXiv preprint arXiv:1803.04383.

[8] ***See:*** Hashimoto TB, Srivastava M, Namkoong H, Liang P. Fairness without demographics in repeated loss minimization. In Proceedings of the 35th International Conference on Machine Learning, PMLR 1929-1938, 2018; arXiv preprint arXiv:1806.08010.

table treatment on all protected groups.[9] This is why I always think the human is a driving force for the algorithms to become perfect. As Zuboff wrote in her book's introduction as:[10]

> *Our effort to confront the unprecedented begins with the recognition that we hunt the puppet master, not the puppet. A first challenge to comprehension is the confusion between surveillance capitalism and the technologies it employs. Surveillance capitalism is not technology; it is a logic that imbues technology and commands it into action.*

It is quite right that we need to develop more advanced algorithms to get rid of the increasing dependence on privacy information, and to design more fair ones as well.

Let's back to the system designed for our blueprint. Because lying on the self-governing openness mode to allow the user accessing DataBoxes through external apps (and software), it is necessary to track, monitor, and analyze all behaviors of the user (in particular the data user, the data provider also included). As a result, here involves how the system chooses fair algorithms to utilize the user's private information and evaluate their behaviors as well. Take data jigsaw as an example; the purposes these algorithms designed for include, but not limited, dynamic analysis (e.g., for modeling transaction dynamics for the majority group), trace analysis (e.g., for running the traces affected by data loss to produce a coverage report for minority groups), change detection analysis (e.g., for identifying, describing, and quantifying differences between behaviors of the similar scenes to mine abnormal groups).

In addition to considering the effects of algorithmic bias, the system should focus more on best matching that is an essential service and will be optimized continuously. If a customer repeatedly buys DataBoxes from a particular field (e.g., healthcare) from data owners regardless of other fields' providers, that buyer relevels a preference for this domain; the system does not need to know why the buyer is relatively other fields insensitive. When that customer starts buying DataBoxes of another area (e.g., transportation), it not only signals that preferences may have changed but also implies that the buyer becomes interested in the intersection of two realms (e.g., medical transportation). The system needs to adjust automatically and adaptively. The matching function is on understanding the user's behaviors, then to get their preferences. The content of matching involves between user demand and the sample of data resources (or the data descriptions of DataBoxes), or and internal data technologies (e.g., text

[9] ***See:*** Romano Y, Barber RF, Sabatti C, Candès E. With Malice Toward None: Assessing Uncertainty via Equalized Coverage. Harvard Data Science Review. 2020. https://doi.org/10.1162/99608f92.03f00592.

[10] ***See:*** Zuboff S. The Age of Surveillance Capitalism: The Fight for a Human Future at the New Frontier of Power. New York, NY: PublicAffairs; 1 edition (January 15, 2019). 2019; 704 pages.

mining), or and external techniques (e.g., NoSQL/Open language, SDK as software development kit), which are to use DataBoxes easily.

System Transparency: Feedback Effects

But there is yet another element that's needed for the adaptive system to optimize. Rich feedback data streams and improved matching abilities are like a machine without the fuel if they're not paired with a robust, rational way to help system participants express three preferences. While matching abilities rely on algorithms, algorithms alone are not enough to improve matching abilities because algorithms are not the raw material we need for our adaptive system to learn. That is (see Footnote 3),

> *Algorithms are both recipes for learning and its outcome. As recipes, often they are already widely accessible and in the public domain. As concrete outcomes of data fed into adaptive systems, they may change each time a system learns from new data, so having access to them is but getting a condensed glimpse into the past.*

Considering how much we depend on getting reviews in now shopping online, it is easy to understand why a feedback mechanism is vital for the system. Suppose an Amazon shopper; he first logged in his Amazon account to search a product by entering the keywords. Next, he was browsing the relevant products from top to bottom according to the list given by the Amazon. This is a screening process of information. He then clicked into one of the product pages to look through the buyer reviews and was linked to another brand similar product mentioned in those comments. And he read comments to check another product page or learn to feedback on the current one. And he spent several hours shopping a product he needed, but the most time is on reading the comments. This is unsolicited feedback, for example, when the buyers open up their packages, they compare what they get with their expectations to give ratings and comments. The same foundation allows scholars to touch new research focuses. A new researcher often needs to read one or more review articles to understand what has been studied, how these studies are processed, and what is expected in future research. So, a feedback mechanism is required everywhere, in particular, after having turned into the data.

Our system's strategy is not unique; it is representative of big data, an approach to data analysis from user behavior and feedback. The proposal is to capture and identify particular (both expected and unexpected) "patterns" embedded in the data, to optimize and update the system itself. It is often characterized as an "artificial intelligence" approach, even though the term originally referred mostly to systems that had been fed general rules rather than having them learn from the data.

In sum, a global tax on data capital is a new idea aiming to regulate data capital. This idea partially referred to Piketty's capital tax. The blueprint I described above is not the only one, and can be every-changing based on the latest techniques. The intention of the blueprint is to give a recommendation; that is, the designers of the tax must consider to track and control the tax source from data-rich profit sharing, and the tax authorities need not rely too much on taxpayers to declare their own asset holdings. Our goal in this century is possible to offer more opportunities to get rich, rather than to eliminate wealth inequality. The blueprint is the one that may be can reach the goal. Once again, new technologies and a global tax on data capital are the right answers.

Chapter 13
Conclusion

I have presented an emerging type of capital—data capital, and I have attempted to do some original work with "letting data speak for itself." While being the first scholarly writing, there exist multiple early related work on this topic of "data as capital." I put forward some of my own opinions by summarizing their well-done research. The sources on which this book uses are much broader than any previous author has assembled. All my conclusions, however, are inherently optimistic and deserve to be questioned and debated since I'm a data optimist.

In the conclusion part of this book, I would like to talk about what data capital's role is in the future society. Of course, this topic is also related to the monetary aspects: industry usually pays more attention to it than academia and offers more contributions and active participation. I'm going to start with some ideas of data from the industry, such as data as "a currency," "governance," and "a service." Nevertheless, I do not want to make any so-called "definitive conclusion" for this issue. I believe this is a practical process of trial and error. So, I open this topic to the public and hope all shades of opinion through democratic debate, specifically from a variety of industries.

The Data-as-Currency World

A quotation given by Mayer A. Rothschild, the first Rothschild of Rothschild family, is that, "give me control of a nation's money, and I care not who makes the laws." It still makes sense today. Yet, I would like to say the word "money" in this quotation may be changed to "data" now. I think Viktor Mayer-Schönberger, a data scientist who wrote the first significant book to identify big data trends, may agree with me, although he didn't expect money and price prevailed. Another person includes Jane Barratt, chief advocacy officer at MX Technologies, a Utah-based data firm. In a conversation with the Knowledge@

© Springer Nature Switzerland AG 2021

C. Tang, *Data Capital*, https://doi.org/10.1007/978-3-030-60192-8_13

Wharton radio show on SiriusXM, Barratt said "it is still very early days in this data-as-currency world, but it is an absolutely one-sided trade."[1] Some financial data come out of the exhaust pipes of financial services, and financial institutions made money off these data while consumers had no idea who was using it or how valuable it was to them. "There's a long way to go," Barratt said, "getting to the point where data really acts as currency requires technological and structural changes, a whole new regulatory standard and body included." Barratt also believed the implications from data-driven innovation for financial institutions are very positive, within open banking especially, for instance—a loan through a startup lender. It should be noted that Barratt mentioned "data exhaust" here, which is the most collected data from individuals excluding personal data. The data is made up into the sovereign view of who we. That is captured, analyzed, and sold via each click when visiting the online world, or via location tracking when using mobile phones, and then targeted right back at us to sell us more things. This process is without our approvals and was everything portrayed in Shoshana Zuboff's *The Age of Surveillance Capitalism: The Fight for a Human Future at the New Frontier of Power*. Obviously, as shown in figure 13.1, it happens now rather than will occur in the future.

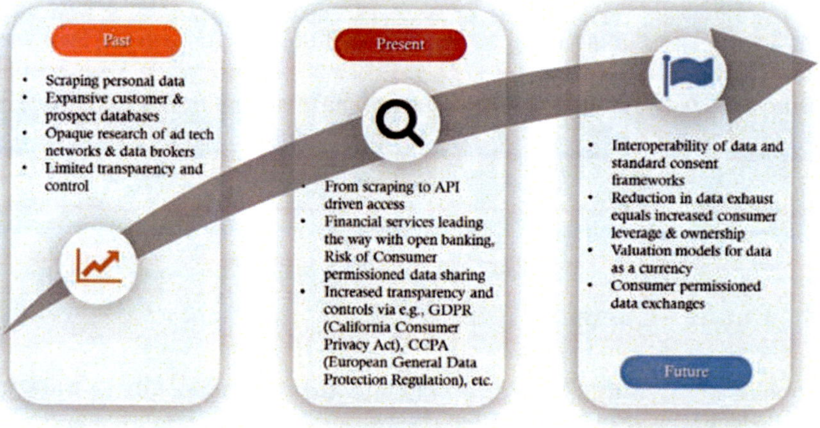

Figure 13.1 A possible path to data as currency within the financial sector. (Source: Author drew this figure according to some materials found on the Internet)

[1] *See:* Buff, A., Wixom B.H., Tallon P.P. Foundations for Data Monetization. SAS White Paper, SAS Institute and MIT CISR, August 2015. Retrieved from https://www.sas.com/en_us/whitepapers/foundations-for-data-monetization-107980.html.

In data science, data monetization is not about exchanging into dollars, euros, or something similar, but rather turning data into "irrefutable value." For example, one often finds that multiple values of the same data entity reside in a dataset or different datasets. With this comes that need for identifying the current value of the entity and to answer queries with the current values. The value here cannot be turned into a legal tender directly but has perceived equivalent value.

Someone may question that data as both the capital and currency against Karl Marx's logical construction is obvious. The abstract formulation of Marx's theory of money in Capital, Volume I, in fact, is just the first step of a materialist analysis of concrete monetary phenomena. His concrete analysis of monetary phenomena in Capital, Volume III can be used here. The fundamental insight provided by Marx is that money is "a social relation"[2] that offers absolute command over the capital and property of others. Put in another way, the absolute value of money the most human desires may be found in the data ultimately.

I envision that data as both the capital and currency may globally result in increasing transparency on the value exchange, which helps people move their lives forward, loans and persons included. While seeing massive potential social issues with income and wealth inequality, it is huge and transformational if data and analytics can provide technology solutions on some systemic issues within the micro and the individual level, as well as up to the macroeconomic landscape.

Sovereignty-Based Data Governance as the Enabler of the Data Economy

The overall conclusion of this study in the book is that, the approaches of deriving profits from data remain unclear, if data producers cannot be confirmed as data sovereigns; so that the inequity of data-rich profit distribution continues to exist, which are potentially threatening to the global society and to the values of social justice on which it is based.

Data sovereign identities, to some extent, are incentives for data producers (e.g., organizations, corporations, industry bodies, or government specifies) to spontaneously engage in management of data as resources. Before proceeding, I want to clarify the goals of resource management. According to Anany V. Levitin and Thomas C. Redman:[3]

[2] ***See:*** Vasudevan R. The significance of Marx's theory on money. Mronline.Org. September 18, 2017. Retrieved from https://mronline.org/2017/09/18/the-significance-of-marxs-theory-on-money.

[3] The reference may be an early one in this research topic. ***See:*** Levitin A.V., Redman T.C. Data as a resource: Properties, implications, and prescriptions. MIT Sloan Management Review. October 15, 1998. Retrieved from https://sloanreview.mit.edu/article/data-as-a-resource-properties-implications-and-prescriptions.

> *Management of any resource seeks to achieve the following goals: to possess a sufficient but not excessive supply of the resource, based on user needs and usage patterns; to provide legitimate users with timely and efficient access to the resource; to protect the resource from unplanned destruction and unauthorized access and use; to maintain and improve the quality of the resource; and to promote effective and efficient usage of the resource for the maximum benefit.*

Traditional resources originate typically outside the organization; but data can be generated both externally and internally, involving every transaction with suppliers and customers, some internal operations, and managerial and professional work all supply data. Data resources come from a tremendous number of sources, and its sheer diversity adds enormous complexity. Simple solutions with data management or data operations in the general sense cannot meet the requirements.

Governance can be described as the processes, structures and organizational traditions that determine how power is exercised, how stakeholders have their say, how decisions are taken and how decision-makers are held to account.[4] When employed properly, governance is an excellent tool organization use to create strategies, establish goals, and make decisions. But too much governance may derail even the best of intentions.

Data governance goes beyond data management and data operations; and its essential function is "to ensure data availability and data quality (accuracy, completeness, reliability) for all possible management tasks and operations that contribute to the vision, mission, and strategy"[5] of the organization, see figure 13.2. This opens access to the data that stakeholders need to maximize business value. Data governance can be seen when looking at the system, process, and strategy level of a data sovereign and its data ecosystem.

As I explained earlier, I tend to pin the hopes on the enterprises and entrepreneurs who owned or controlled the data because of the interesting butterfly effect that is often from the bottom of the pyramid. Despite this, as an "unaffectionate" technician I never think about ideology if there exists any possible technology solution to one problem. Data governance can also serve governments and the global market. In January 2019, Japanese Prime Minister Shinzo Abe called for a summit that starts worldwide data governance, which aims to figure out how to govern the complex data ecosystem, both enabling its potential and managing its risks. The proposed issues are to fill key gaps in the existing global architecture of data governance frameworks; for example, the need for consistency, interoperability, and coordination of the myriad international,

[4] ***See:*** Gill M. Building effective approaches to governance. Non Profit Quarterly. 2020.

[5] I referenced Figure 2 of which Engels B revised Aliu Q *et al's*. ***See:*** (1) Engels B. Data governance as the enabler of the data economy. Intereconomics, 2019;4:216-222. DOI: 10.1007/s10272-019-0827-y; (2) Aliu Q., Azkan C., Bresser P., *et al.* Data economics and management of data driven business. Demand White Paper. 2019. Retrieved from https://demand-projekt.de/paper/DEMAND-DataEconomicsAndManagementOfDataDrivenBusiness(WhitePaper).pdf.

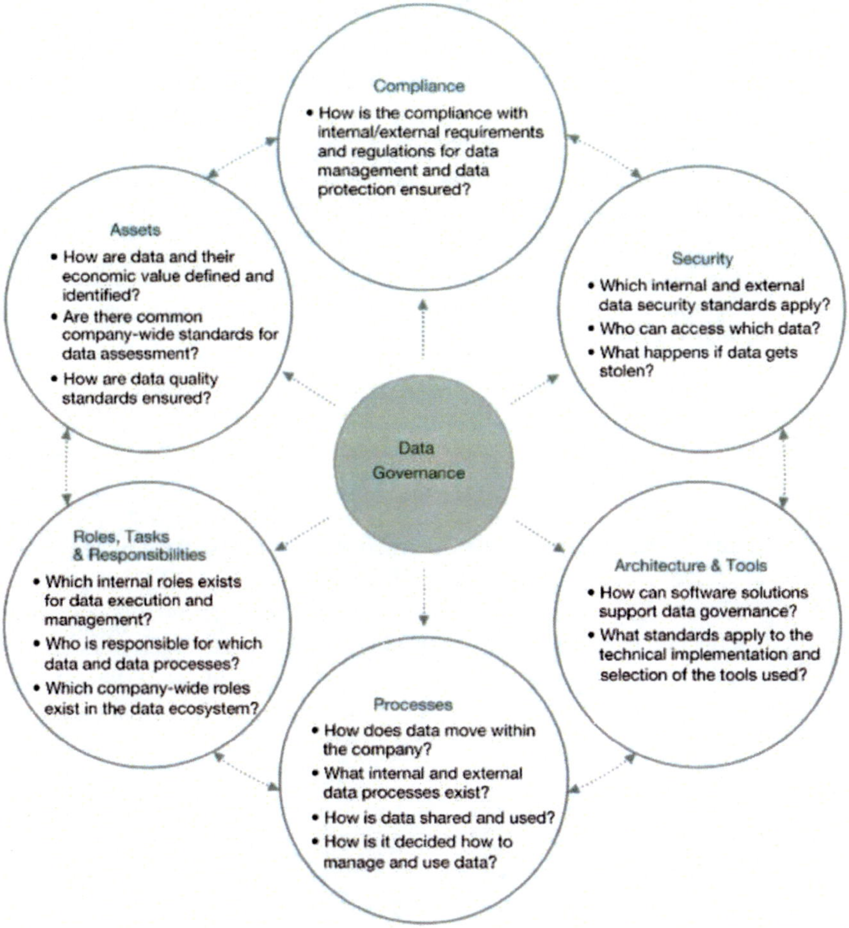

Figure 13.2 Dimensions of data governance within a single enterprise. Note: Data governance consists of several correlating dimensions with sub-dimensions that partially overlap. (Source: Author got the approval of permission from Engels B at Footnote 5)

regional, national, and local laws and regulations that impact data. The others include but not limited to, (1) governance mechanisms around privacy, human rights law, and digital trade or e-commerce, and (2) the terms and conditions specified must include the allowed use of the data when one organizational unit accesses to another, such as a fair transfer price, the technical means of access, quality assurance, and a means of resolving conflicts. Of course, all of them require everyone's efforts, ranging from the most mundane operation to the most far-reaching strategic decision.

For a Future Economic Science

I would like to conclude with a few words about economic science. As indicated in his *Capital in the Twenty-First Century*, Thomas Piketty dislike the expression "economic science," which is terribly arrogant because it suggests that economics has attained a higher scientific status than the other social sciences. He criticized economists' grand but over-abstract principles as:[6]

> *For far too long economists have sought to define themselves in terms of their supposedly scientific methods. In fact, those methods rely on an immoderate use of mathematical models, which are frequently no more than an excuse for occupying the terrain and masking the vacuity of the content.*

His words belong to internal criticism, which is actually "love well, whip well." Like my specialized field of study as data mining, my blame is; for example, the most is on optimization algorithms or models rather than the pioneering work.

The fact that abstraction is not bad. The objection of abstraction can generally be reduced if they are easy to use in practice. Take an example from my own studies; I utilized the simplest Latent Dirichlet Allocation (LDA) model to extract ten topics from free-text comments on a cancer research institution's Facebook page between January 2009 and June 2014. What surprised me is that, the paper with less than 2 impact factors has attracted wide attention and was selected as one of the 15 candidate best papers (among 32,958 papers) in the cancer informatics section of the 2018 IMIA (International Medical Informatics Association) Yearbook. In this article, I only explain in detail how to use the LDA model via specific data. That is to say, the purpose of publications is to indicate how the data would be analyzed. And now, we have a better way—Data as a Service (DaaS) that goes beyond building algorithms to fill data lakes. Note that DaaS is also known as data fabric; data fabric isn't defined rigorously by the industry but means an integration platform that connects the data with its services.

Data as a Service (DaaS) provides a similar but reusable approach to the combination of data and analytics and makes it available across geographies, involving how to collect it, mine it, visualize it, humanize it and, most importantly, act on it. The first step in building out economic analytics, for example, requires gathering both qualitative and quantitative data. But under specific institutions and policies instead of an ideal society, collecting data to fit a single model alone is not enough. To glean data insights, economists need to roll the data back in and close the loop. DaaS not only offers such a data looping

[6] *See:* PiKetty, T. translated by Goldhammer A. Capital in the Twenty-First Century. Cambridge, MA: the Belknap Press of Harvard University Press; Reprint edition (August 14, 2017), 2017: 816 pages.

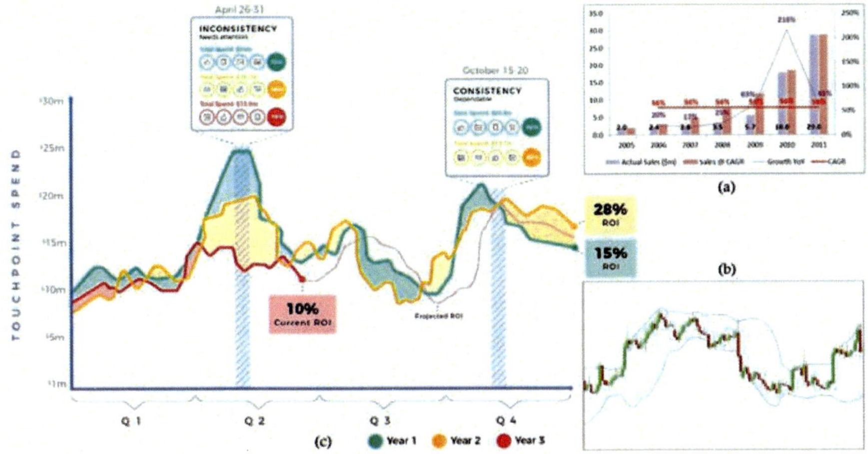

Figure 13.3 Three types of chart to visualize year-over-year (YoY) vs. convergence or divergence bands. Note: Author modified from a post at Footnote 7, and got the approval of permission from Barlow B and Greene J

process but breaks down borders, silos, and vendor walls for data collection. Let's compare three types of charts[7] on data visualization to understand the data loop process. Figure 13.3a as a commonly used chart draws how returns and profit margins, such as year-over-year (YoY) growth and compound annual growth rate (CAGR), have progressed over the years, spikes in profits included. While being valuable, figure 13.3a lacks depth and detail specifically around spending. Some track sales alone but then continue to allocate spending evenly across the board without taking seasonality, consumer preference trends, and other factors into consideration. Allocating spending evenly in this way could cost opportunities by continuing to put money toward channels and strategies that don't give a healthy enough return.

Figure 13.3b, commonly seen in Wall Street, is used to break down and forecast macro market momentum as well as micro-events through Bollinger bands. Bollinger bands are a trading band based standard analysis tool, used as trading pattern recognition. Inspired by Bollinger bands, figure 13.3c offers a way to analyze the efficacy of spend to ROI (Return on Investment) health across each spending segment. When plotting our convergence or divergence bands, two metrics (i.e., two touchpoints spending on the Y-axis with different time on the X-axis) can be paired side by side. We then examine which year had the best return on investment based on how the amount spent synced across a pattern of

[7] This borrow the example posted on the webpage of Rocketsource.Co, and and got its authors' approval of permission. ***See:*** Barlow B, Greene J. Data as a Service: The what, why, how and when. Rocketsource.Co. 2018. Retrieved from https://www.rocketsource.co/blog/data-as-a-service.

divergence or convergence. With the three lines of year plotted on the top of each other, we can quickly see where they start to come together (i.e., converge) and where they drift apart (i.e., diverge). Convergence signals consistency; from 1 year to the next, the spend in that season didn't need to shift due to the consistency. Areas of divergence show inconsistency in spend and are opportunities to look deeper into what caused the change and what effects it had on ROI. From this, we can use additional models to uncover where to slow down or accelerate spending. Such a chart doesn't have to be limited to hard financial metrics. Like figure 13.3c, we are iterating on data visualization insider the data looping processing, through many strategies that go into which metrics to plot and how to compare them, then to gather more insights than before.

Apart from deeper unbiased insights, lower costs, and faster paths to innovative ventures, reusable DaaS also offers data democratization that makes economic science much closer to practical applications. Data democratization means that everybody has use data at any time to make decisions with no barriers to access or understanding. Anyone can not only have personalized experience but also add a time to internal structure to increase collaboration.

Someone might ask how long does DaaS work for economic science. Based on my knowledge, there may exist two weathercocks. One is the Gartner Hype Cycle; and the other comes from the *Nature* (journal). We found that DaaS sits in the "peak of inflated expectations" of Gartner Hype Cycle for Data Management since 2017 to now (note that 2019 report merged DaaS into data fabric).[8] That means about 5–10 years from the plateau of productivity where it's estimated that high-growth adoption will kick in.

Second, Nature announced in early 2019 that will soon appoint an economics editor to ensure economists and natural scientists working together.[9] This initiative pulls closer to scientific research once again, following the World Health Organization (WHO) to consider a chief-economist post.

> *Economists have been pointing out for some years that we don't fully understand why the results of research and innovation — which have ushered in the digital age along*

[8] Since Gartner didn't permit any graphics from its branded research, interested readers can find the Hype Cycle for Data Management reports as primary sources. ***See:*** (1) Hype Cycle for Data Management, 2017. Gartner.com. July 26, 2017. Retrieved from https://www.gartner.com/en/documents/3769157/hype-cycle-for-data-management-2017; (2) Hype Cycle for Data Management, 2018. Gartner.com. July 25, 2018. Retrieved from https://www.gartner.com/en/documents/3884077/hype-cycle-for-data-management-2018; (3) Hype Cycle for Data Management, 2019. Gartner.com. July 31, 2019. Retrieved from https://www.gartner.com/en/documents/3955768/hype-cycle-for-data-management-2019.

[9] ***See:*** Economists and scientists: solve big societal problems by working together. February 26, 2020. Retrieved from https://www.nature.com/articles/d41586-020-00532-4.

> ***with other transformations — are not benefiting everyone in society, as seen for example in wage stagnation and widening inequality. But relatively few natural scientists or engineers have taken up this question.***

Think about data science, partial contributions of its current prosperity are due to the introduction to big data from *Nature* that spends the entire volume on September 4, 2008.[10] I think the "future" is not far away. I wonder if I need to change the title of this section as "For the Next Decade of Economic Science."

[10]The is a whole journal. ***See:*** Nature, Volume 455 Issue 7209, 4 September 2008. Retrieved from https://www.nature.com/nature/volumes/455/issues/7209.

List of Tables and Illustrations

List of Case Study Examples

List of Explaining Data Sources

Data Source 2.1 *Literature Data* Literature data is any single part of a scientific paper (e.g., titles, abstracts, backgrounds, methods, results, conclusions, or the whole article) collected by searching of several specific databases (e.g., DBLP: Computer Science Bibliography can be accessed at https://dblp.uni-trier.de, MEDLINE/PubMed, ScienceDirect), library catalog services (e.g., Harvard's HOLLIS), or online facilities (e.g., Google Scholar) during a given period of time. Quantitative approaches to archives and texts developed by digital humanists have offered one such expansion. These approaches often treat literature as a text mine. Each paper included in the analysis can be classified as either a paper describing article type (e.g., a description, or a review of the literature) or other information: year of publication, journal title, specialty area, the field of study, and characteristics given by authors to data reuse.

Data Source 3.1 *Market Data* Market data is price and trade-related data for a financial instrument (e.g., equities, fixed-income products, derivatives, and currencies) reported by a trading venue such as a stock exchange.

Data Source 3.2 *Day-by-Day Data* Day-by-Day data is recorded in the computer systems according to human routine rhythm, such as stock trading on each closing price in a week as $\{C_1, C_2, \ldots, C_7\}$. Table II.4 shows the day-by-day behavior data, namely a collection of behaviors which include each person's daily behavior. Trade-related data for a financial instrument reported by a trading venue such as a stock exchange. Market data allows traders and investors to know the latest price and see historical trends for instruments such as equities, fixed-income products, derivatives, and currencies.

Data Source 3.3 *Social Media Data* Social media data refers to all of the raw insights and information collected from individuals (e.g. prospects and customers) social media activity. Social media data tracks how individuals engage with a specific content or channels like Facebook, Twitter, Instagram, and LinkedIn.

Data Source 4.1 *Event-Related Potential Data* An event-related potential (ERP) is the measured brain response that is the direct result of a specific sensory, cognitive, or motor event. More formally, it is any stereotyped electrophysiological response to a stimulus. ERPs are measured by means of electroencephalography (EEG). ERP data's format depends on the manufacturers of EEG, such as Brain Products .eeg files and Neuroscan .cnt files.

Data Source 4.2 *Functional Magnetic Resonance Imaging Data* Functional magnetic resonance imaging (fMRI) measures brain activity by detecting changes associated with blood flow. This technique relies on the fact that cerebral blood flow and neuronal activation are coupled. When an area of the brain is in use, blood flow to that region also increases. The primary form of fMRI uses the blood-oxygen-level dependent (BOLD) contrast, discovered by Seiji Ogawa in 1990. DICOM as a format for neuroimaging with fMRI.

Data Source 4.3 *Eye Tracking Data* Eye tracking is the process of measuring either the point of gaze (where one is looking) or the motion of an eye relative to the head. The format of data generated by different eye trackers, the devices for measuring eye positions and eye movement, is various but mostly can be exported as the text.

Data Source 4.4 *Positron Emission Tomography Data* A brain positron emission tomography (PET) scan is typically an outpatient procedure of imaging test that allows doctors to see how the brain is functioning. The scan captures images of the activity of the brain after radioactive "tracers" have been absorbed into the bloodstream. These tracers are "attached" to compounds like glucose (sugar). Glucose is the principal fuel of the brain. Active areas of the brain will be utilizing glucose at a higher rate than inactive areas. PETs' data format depends on their manufacturers.

List of Definitions

Definition 2.1 *Production* Production is a process of combining various material inputs and immaterial inputs in order to make something for consumption (output).

Definition 2.2 *Producer* A producer is any organization (e.g., corporations, industry bodies, government specifies) who plan, coordinates, and performs production.

Definition 2.3 *Economic Resource* An economic resource is the (input) factors of production used in producing goods or providing services.

Definition 2.4 *Investment* Investment is what happens when a producer either acquires a fixed asset or spends resources (money, effort, raw materials) to improve it.

Definition 2.5 *Asset* An asset is an economic resource that is expected to provide a benefit over a period of time.

Definition 2.6 *Fixed* Asset A fixed asset is an asset that results from using up resources in the process of its production.

Definition 2.7 *Rate of Profit* The profit rate is the relative profitability of a whole investment project, but sometimes expresses the share of profits that may be relevant with income from capital.

Definition 2.8 *Interest Rate* An interest rate is the amount of interest due per period, as a proportion of the amount lent, deposited, or borrowed (called the principal sum), which varies widely depending on the identity of the lender or borrower.

Definition 2.9 *Rate of Return on Capital* The rate of return on capital measures the yield on capital over the course of a year regardless of its legal form (e.g., profits, interest, rents, dividends, royalties, capital gains, etc.), expressed as a percentage of the value of capital invested.

Definition 2.10 *Confidence* Given an association rule that is an implication expression of the form $X \rightarrow Y$, where X and Y are disjoint itemsets, i.e., $X \cap Y = \emptyset$. Confidence determines how frequently items in Y appear in patterns that contain X.

Definition 2.11 *Mind Map* A mind map is a hierarchical diagram to be an image-like aid. It is often created around a single concept, drawn as an image in the center of a blank page, to which associated representations of keywords, ideas, tasks or other items in a radial way. Mind maps are considered to be a type of spider diagram.

Definition 3.1 *EBITDA* EBITDA is a metric that excludes expenses associated with debt by adding back interest expense and taxes to earnings.

Definition 3.2 *Intellectual Capital* Intellectual capital is the intangible value of a business, covering its people (human capital), the value relating to its relationships (relational capital), and everything that is left when the employees go home (structural capital), of which intellectual property is but one component. It is the sum of everything everybody in a company knows that gives it a competitive edge.

Definition 3.3 *Meaning Capital* Meaning capital involves several forms of capital (e.g., spiritual, psychological, and experiential capital), which more directly connect to an individual or a collective sense of purpose and what emotions are.

Definition 3.4 *Cross-Border Capital Flows* Cross-border capital flows, at the global level, can be calculated as the sum of capital inflows or the sum of capital outflows, excluding financial derivatives but including flows related to international banking transactions.

Definition 3.5 *Data Asset* A Data asset is the data asset as data exceeding a certain scale that is owned and controlled by a specific agent, collected from the agent's past transactions involved in information processes, and capable of bringing future economic benefits to the agent.

Definition 3.6 *Data Production* Data Production is a process of combining various analyzable data inputs in order to make something for consumption (output). Something for consumption (output) here means data product and the others.

Definition 3.7 *Data Producer* A data producer is any agent (e.g., corporations, industry bodies, government specifies, individuals) who has the ability to pursue adding value to row data or other primary data products.

Definition 3.8 *Data Capital* Data capital, for a specific agent, can be calculated as the sum of add-valued data asset inflows or outflows.

Definition 3.9 *Digital Currency* The digital currency is the blanket term used to describe all digital or electronic, and intangible money, which includes both virtual currency and cryptocurrency. It can be regulated or unregulated and is used online.

Definition 3.10 *Virtual Currency* Virtual currencies are a type of digital currency, typically controlled by its creators and used and accepted among the members of specific virtual community.

Definition 3.11 *Cryptocurrency* "Crypto" in cryptocurrency refers to the fact that many encryption algorithms and cryptographic techniques are used to ensure security across the network. This level of security also makes crypto-currencies hard to counterfeit.

Definition 3.12 *Spatio-Temporal Visualization* The spatio-temporal visualization incorporates the legacy from the conventional cartography to track the kind of changes occurring over time, including: existential changes (e.g., appearance and disappearance), changes of spatial properties (e.g., location, sharp or/and size, orientation, altitude, height, gradient, and volume), and changes of thematic properties expressed through values of attributes (e.g., qualitative changes and changes of ordinal or numeric characteristics: increase and decrease). Of which, timeline is a graphical way of displaying a list of events in chronological order. Some timelines display as a straight (e.g., horizontal or vertical) line, while others can be a curve (e.g., a spire).

Definition 3.13 *Data credibility* Data credibility is the extent to which the good faith of a provider of data or source of data can be relied upon to ensure that the data really represents is what the data is supposed to represent, and that there is no intent to misrepresent what the data is supposed to represent.

Definition 4.1 *Atomic Data* Atomic data is an atomic state of information that cannot be broken down into smaller pieces or sources, such as unit sales (e.g., for a particular model of the washer), gross revenue (that cannot be calculated from anything even if being broken down by transaction; note that net revenue is based on multiple inputs and is not considered atomic data), and base salary (as a hard number that is not typically calculated from the others).

Definition 4.2 *Data Item* A data item describes an atomic data concerning a specific property at a certain time point. Values of a data item are not necessarily atomic. The complexity of a value (represented by symbols like numbers, texts, images, sounds, videos or others) depends on the complexity of the property and time component.

Definition 4.3 *Data Object* A data object is a unique location in a computer memory (i.e., a specific memory address) that contains the value of *something*. This *something* can be accessed by giving a unique name that is known as the identifier of the variable that contains the data object. Data objects are often composed of a collection of limited data items, which are identified by object (o), property (p) and time (t), while the value (v) is a function of o, p, and t, $v = \boldsymbol{F}(o,p,t)$.

Definition 4.4 *Dataset* A dataset (or a data set) is a collection of data objects. Generally, data objects in a specific dataset are limited. Even though building a dataset that contains an infinite stream, there exist limited data objects at a certain (arbitrary) time point.

Definition 4.5 *Data* Data as a general concept refers to the fact that some existing information or knowledge is digitally coded in some form suitable for better usage or processing. The data can be represented as an atomic data, a data item, a data object, or a dataset arbitrarily. The size of the data is measured in bytes (i.e., *B*). Think of *NULL* data is with an empty value, of which the size is zero *B*.

Definition 4.6 *Metadata* Metadata is the data that provides information about other data. There exist multiple distinct types of metadata, including descriptive metadata, structural metadata, administrative metadata, reference metadata, and statistical metadata.

Definition 4.7 *Data Tool* A data tool involves algorithms, data models, programs, and software, which can run in a computer system. Data tools belong to a specific data object designed for some purposes (e.g., an analytical purpose). The data tool itself is also data, which is usually used to process data and can also be processed by other data tools.

Definition 4.8 *Data Resource* As data organizations connected by logics and data areas assembled by data volumes reach a certain scale, the massive distinct datasets become a data resource.

Definition 4.9 *Data Nature* All data in cyberspace constitutes data nature, but the computer systems (or other IT physical facilities) as carriers do not belong to the part of data nature.

Definition 4.10 *Data Experiment* Data experiment is a scientific method to discovering how things (e.g., specific natural phenomena, or any rules in a designated dataset and/or data tool) in the universe work though the direct intervention and control of datasets and/or data tools. Three major types of data experiments include taking datasets, data tools, or simulated natural phenomena (e.g., natural disasters, urban or business operations) as the experimental object, respectively. Data experiment is characteristic that the experimental objects and tools of such a research approach are both composed of data: data verifies data through data.

Definition 4.11 *Repeatability* The measurement can be obtained with stated precision by the same team using the *same* measurement procedure, the same measuring system, under the same operating conditions, in the same location on multiple trials. For data experiments, this means that a researcher can reliably repeat her own computation.

Definition 4.12 *Replicability* The measurement can be obtained with stated precision by a different team using the *same* measurement procedure, the same measuring system, under the same operating conditions, in the same or a different location on multiple trials. For data experiments, this means that an independent group can obtain the same result using the author's own artifacts.

Definition 4.13 *Reproducibility* The measurement can be obtained with stated precision by a different team, a *different* measuring system, in a different location on multiple trials. For data experiments, this means that an independent group can obtain the same result using artifacts which they develop completely independently.

Definition 4.14 *Raw Data* Raw data (sometimes called source data or atomic data) is data that has not been processed for use. Raw data that has undergone processing is sometimes referred to as cooked data.

Definition 4.15 *Data Products* Data products (i.e., cooked data), such as models, intermediate results, and annotated training corpora, are the outcomes from data preparation, processing, and analysis (e.g., statistical analysis, data mining, and machine or deep learning). Data products also include visualizations and dashboards created by the artistic manual work of data scientists to assist in the interpretation of the analysis in an actionable way.

Definition 4.16 *Stream Graph* A stream graph, as a variation of a stacked area graph, displays the changes in data (between the maximum and minimum) over time of different categories through the use of flowing, organic shapes that somewhat resemble a river-like stream.

Definition 5.1 *Data Sovereign* Data sovereign is a data producer with the ability to defend data resources against any attack.

Definition 5.2 *Data Migration* Data migration is a project by means of which data will be moved or copied from one environment to another, and removed or decommissioned in the source.

Definition 5.3 *Data Capital Migration* Data capital migration involves multiple data sovereigns, which makes data capital flowing in one or more directions (and there may be numerous migrations taking place simultaneously). So, its representatives will be moved or copied from one environment to another, and removed or decommissioned in the source.

Definition 6.1 *Feature Engineering* Feature engineering is the process of using domain knowledge to extract features from raw data via data technologies. These features can be used to improve the performance of algorithms.

Definition 6.2 *Labeled Data* Labeled data is a group of samples that have been tagged with one or more labels. Labeling typically takes a set of unlabeled data and augments each piece of that unlabeled data with meaningful tags that are informative.

Definition 6.3 *Wordle* A wordle is a visual depiction of the words contained in a piece of free text. The more frequent a particular word was within the source text, the bigger it's displayed in the wordle—font and color variation, as well as giving weight to particular words, which are positioned vertically or horizontally. It can help filter out both low-frequency and low-quality information to help a viewer understand content by just glancing it.

Definition 6.4 *Barrier to Entry* Barriers to entry, an economic term, describe the existence of high startup costs or other obstacles that prevent new competitors from quickly entering an industry or area of business. Barriers to entry may be natural (high startup costs to drill a new oil well), created by governments (licensing fees or patents stand in the way), or by other companies (monopolists can buy or compete away startups).

Definition 12.1 *Data Jigsaw* Data jigsaw is a behavior that is to splice multiple data fragments or numerous times to restore the data to be integral.

Definition 12.2 *Algorithmic Bias* Algorithmic bias describes repeatable errors caused by algorithms and models that create unfair outcomes. Bias can emerge due to many factors, including but not limited to the design of the algorithm or the unintended or unanticipated use or decisions relating to the way data is collected, selected, or used to train the algorithm. Algorithmic bias can have impacts ranging from inadvertent privacy violations to reinforcing social prejudices of race, gender, sexuality, and ethnicity.

Index

© Springer Nature Switzerland AG 2021

C. Tang, *Data Capital*, https://doi.org/10.1007/978-3-030-60192-8